An Introduction to
Probability and Statistics
Using BASIC

STATISTICS: Textbooks and Monographs

A SERIES EDITED BY

D. B. OWEN, Coordinating Editor
Department of Statistics
Southern Methodist University
Dallas, Texas

PAUL D. MINTON
Virginia Commonwealth University
Richmond, Virginia

JOHN W. PRATT
Harvard University
Boston, Massachusetts

OTHER VOLUMES IN PREPARATION

An Introduction to Probability and Statistics Using BASIC

RICHARD A. GROENEVELD

Department of Statistics
Iowa State University
Ames, Iowa

MARCEL DEKKER, INC.   New York and Basel

Library of Congress Cataloging in Publication Data

Groeneveld, Richard A   [Date]
   An introduction to probability and statistics using
BASIC.

   (Statistics, textbooks and monographs ; v. 26)
   Bibliography: p.
   Includes index.
   1. Probabilities--Data processing. 2. Mathematical
statistics--Data processing. 3. Basic (Computer program
language) I. Title.
QA273.G765   519'.028'5424   78-31493
ISBN 0-8247-6543-5

MARCEL DEKKER, INC.
270 Madison Avenue, New York, New York 10016

Current printing (last digit):
10 9 8 7 6 5 4 3 2 1

PRINTED IN THE UNITED STATES OF AMERICA

To the Memory of my Father

CONTENTS

v

This text has been written for an introductory course in probability
and statistics. A central theme throughout is the introduction of
theoretical ideas in probability and statistics by means of examples.
Generally it is difficult to obtain "real-world" data quickly and
inexpensively enough to motivate and illustrate such ideas. Here the
strengths of an interactive computing language (BASIC) are exploited
to illustrate probabilistic and statistical ideas. However, no great
background in computing is required; all that is needed can be learned
within a short period of time. Appendix A gives a description of
the BASIC language adequate for the use of the reader. The aim is
to present the basic ideas in probability and statistics generally
included in a first course. It is the author's belief that interactive computing can provide a bridge between statistical theory and
practice, yielding better insight into both.

The author sees several ways in which the strength of computing
can be used in an introductory course. The ability of the computer
to simulate rapidly a large number of repetitions of a random experiment and to summarize quickly and present succinctly the results
of these experiments is useful in illustrating a wide range of
probabilistic and statistical ideas. The computer, additionally, can
be utilized to carry out routine but tedious computations, allowing
concentration upon ideas and not upon calculations. Furthermore,
extensive use of computing provides a valuable introduction to modern

methods of data analysis. A magnetic tape of the computer programs
in the text may be obtained at the cost of reproduction and mailing
from the author at Iowa State University.

A word here about the audience to whom this text is addressed
and the background assumed is appropriate. The author sees this
text as addressed to students who have had a two-quarter introductory
sequence in the differential and integral calculus of functions of
a single variable (or the equivalent). The author believes that
many students who are introduced to the calculus are not given an
opportunity to take a course demonstrating the power of this mathematics at a sufficiently early stage. Those with interests outside
the physical sciences often become impatient with applications in
these areas and do not pursue additional work in mathematics. The
text includes those topics described by the Committee on the Undergraduate Program in Mathematics (CUPM) Report of 1965 for Mathematics
2P. However, much more of the text is devoted to statistics than
is suggested there. On the one hand, it is intended that this text
contain sufficient mathematical rigor that a reader may appreciate
probability theory. A coequal aim is to provide an adequate background in statistics to serve as a basis for additional study of
statistical methods in a chosen field of application or for additional study of statistical theory. In the larger sense, the text represents an attempt to blend mathematical theory with statistical
practice. The author believes that such a background will become
increasingly important for the understanding of a complex technologi-
cal society.

I am indebted to the Biometrika Trustees for permission to in-
clude Tables B.I, B.III, and B.V, which are abridgments and adapta-
tions of tables published in *Biometrika Tables for Statisticians*.
I am also grateful to the Literary Executor of the late Sir Ronald
A. Fisher, F.R.S., to Dr. Frank Yates, F.R.S., and to the Longman
Group Ltd., London, for permission to use Table III from their
book *Statistical Tables for Biological, Agricultural, and Medical
Research* (6th Ed., 1974), reproduced as Table B.II.

The original draft of this text was written while the author was on leave of absence at Dartmouth College, Hanover, New Hampshire. I would like to thank Dartmouth for making available access to the Kiewit Computation Center and for providing the facilities and environment for writing this text. I would like to thank Helen Hanchett, Nancy French, and Joy Marshall for typing the original draft and Marlene Sposito for typing the revisions and the final draft.

Richard A. Groeneveld

LIST OF BASIC PROGRAMS

1

In this text we shall pursue an understanding of the basic ideas in
the disciplines of probability and statistics. It is clear from the
most casual reading of the morning's newspaper that crucial social
decisions are based upon the results of the sampling of public opinion
and statistical estimates. For example, estimates of the rate of
unemployment and inflation, obtained by statistical means, are cited
as a basis for the fiscal and monetary policies of the government.
Sampling of public opinion on questions of public policy is often
used as the basis for governmental action. At the very least, such
samplings are carefully considered in questions of public policy.

Evidence for the importance of statistics in a modern techno-
logical society confronts us constantly. Due to the importance of
statistics in understanding a modern society one is naturally led to
the consideration of the mathematical basis of statistics. We shall
be interested in questions of how to describe, mathematically, in-
stances of "chance." Additionally, we shall study some of the basic
concepts of statistics with the aim of understanding in what sense
statistical methods may be considered "valid." There is no universal
agreement on these questions, but here we shall adopt the long-run
frequency interpretation of probability, which will be defined more
explicitly in the succeeding chapters. Let us first consider some
basic definitions and the role the computer will play in this text.

1.1   PROBABILITY AND STATISTICS

It is important to distinguish between the disciplines of probability
and statistics. *Probability* is a mathematical discipline which aims
at describing the outcomes of chance experiments. In probability
theory one defines probability models, which are mathematical struc-
tures. We believe that these models describe the essentially important characteristics of certain chance situations. One defines
the basic elements of a model, and axioms are stated concerning relationships between these elements. The implications of the model,
often called *theorems*, are deduced from the basic definitions and
axioms. The reasoning here is essentially *deductive*. One hopes,
of course, that a probability model used gives a reasonably good
description of the chance experiment, so that the mathematical
properties, often summarized as theorems, will have "real-world"
application.

The discipline of statistics is essentially a body of knowledge
for reasoning inductively, i.e., from the part to the whole. A much
quoted definition of statistics is given in the text *The Nature of
Statistics* by Wallis and Roberts: "Statistics is a body of methods
for making wise decisions in the face of uncertainty." The types
of decisions with which we are concerned are to be based on available
numerical values, which give valuable but incomplete information
needed to make the decision. We are used to making decisions of this
kind every day. We know from previous experiences how long the
telephone conversations of the other party on a two-party line have
lasted. If the other party is on the line when we wish to make a
call, we use the data from previous similar situations to help decide
how long to wait before attempting to place the call again. In such
a case, common sense and patience may suffice, but in other more
complicated situations formal rules of *inference* are required.
Statistics, using the mathematics of probability, is the study of
formal rules of inference, with the intention of providing procedures of inference which are scientifically justifiable and useful.
The relationship between probability and statistics may not at the

moment be completely clear. We shall pursue the relationship and distinction between the two disciplines in more detail in the following chapters to improve our understanding as we proceed.

1.2   THE USE OF BASIC PROGRAMS

As mentioned above, there has been an effort to capitalize on the strengths of the time-sharing system in the introduction and explanation of concepts and in the carrying out of routine but tedious computations. A student will be expected to have a working knowledge of the elementary commands in BASIC. Experience has shown that it takes students a very small amount of time to acquire this competence. Appendix A has been included to give a brief description of the BASIC language. All the essentials of the grammar and content of BASIC required for the text are included in Appendix A and this section.

Each of the chapters here is made up of several numbered subsections. In most sections there are references to particular programs written in BASIC. The first time a program is referred to in a section, a listing of the program and a sample run will appear near the end of the section in which it is first referenced. The program is only listed once, so that later references to the program may refer to a listing appearing in a previous section. Additional runs of a program introduced in an earlier section will often appear. Some exercises call for running a program made available with this text. Other exercises require writing simple programs to show an understanding of a particular idea or computation.

The BASIC language provides a statement which allows the selection of a psuedorandom number from the interval $0 < x < 1$. The function is RND which is used in conjunction with the LET statement. RND does not require an argument. We illustrate the use of this command in several examples.

Suppose a trial is made of a chance event which has two outcomes: A and B. For example, in matching two coins, let A represent getting two heads and B any other outcome. Assume the

true relative frequencies of A and B are 1/4 and 3/4. The following
program will simulate 100 such trials and compute the relative frequencies observed.

```
10   FOR I = 1 TO 100
20   LET X = RND
30   IF X > 0.25 THEN 60
40   LET A = A + 1
50   GO TO 70
60   LET B = B + 1
70   NEXT I
80   PRINT "FREQ A = "; A/100, "FREQ B = "; B/100
90   END
```

The following command will select a random number from the
interval (A, B):

```
10   LET X = A + (B - A)*RND
```

The following command will select a random *integer* from K to L for
K < L.

```
20   LET X = K + INT(RND*(L - K + 1))
```

The command RND will cause the same sequence of random numbers to be
selected each time a program is run. This is useful for debugging
programs and for checking homework assignments. However, any true
simulation requires a different set of random numbers for each run.
This will occur if the command RANDOMIZE appears at the beginning of
a program, i.e.,

```
1 RANDOMIZE
```

is the first statement. Substantial use of the simulation capacity
of the BASIC language will be made in the succeeding chapters.

ELEMENTARY PROBABILITY

2.1   RANDOM EXPERIMENTS

We are all aware that the world in which we live is an unpredictable
one. The outcomes of many of our actions cannot be predicted in
advance. A couple has a child; they do not know the child's sex
until birth. An individual makes a visit to a drive-in bank; the
number of cars in line, including his own, upon arrival at the bank
cannot be predicted. A battery is purchased for an automobile; the
number of years that it will function until failure is unknown.
Nevertheless, actions like those described above are taken daily.
Such actions are more formally called *random experiments*, i.e.,
those for which the outcome is not predictable in advance through
knowledge of the conditions under which the experiment is made.

It is the desire to have some kind of reasonable knowledge
about the outcomes of such trials which has led to the development
of the mathematical discipline of probability. We shall concentrate
initially on the empirical basis for probability models. A *probability model* is a mathematical (or axiomatic) description of a
situation involving random experiments. In the model the salient
features of the real-world situation are abstracted and defined
mathematically. Such a model, of course, only can be an approximation of the real situation. There is, for example, probably no
real-world coin which shows heads or tails with exactly equal

likelihood. Nevertheless, the assumption of the existence of such
a theoretical coin permits a mathematical description which has had
many useful real-world applications. The development of formal
probability models, which aim at describing the outcomes of different types of random experiments, has its basis in man's desire to
better understand his uncertain world and to predict the outcome of
his actions.

The first step in coming to an acceptable description of a
random experiment is to concede that one should not concentrate on
the outcome of a particular trial of a random experiment. Rather,
one should first consider the set of possible outcomes or results
associated with such an experiment. We will denote such sets of
outcomes by capital letters. For the examples given in the first
paragraph we might define the following outcome sets for the three
experiments described:

A = {male, female}
B = {1, 2, 3, ..., 10}
C = {0, 1, 2, ..., 8}

The outcome set B indicates that a maximum of 10 cars may be in line
at the drive-in bank, perhaps due to physical limitations of the
parking lane. The integer n in outcome set C means that the battery
has lasted for a least n years, but not $n + 1$. Hence $n = 0$ implies
failure during the first year of operation. Operation for 9 or more
years is not considered possible.

Of course it is clear that such outcome sets alone are insuffi-
cient to provide an idea of the relative likelihood of the individual
outcomes. We make our first use of the computer here to simulate
outcomes of our first experiment, having a child. The program BIRTH
simulates the birth of N children with the true proportion of male
births assumed to be p. The values of N and p are read from the
DATA statement in line 370. The simulation is repeated 20 times and
the output prints the proportion of female and male births in each
of the 20 simulations. In addition, the graph of the relative
frequency of male births is plotted. The axes will occur in their

traditional orientation if the output is turned 90 degrees in the counterclockwise direction. The vertical axis represents relative frequency with X indicating 0.5, and the horizontal axis represents the trial numbers for the 20 simulations.

In the two runs displayed we have chosen $N = 40$ and $N = 400$ with $p = 0.5$. It is immediately clear that, although it is rare that a particular sample of N births has exactly one-half of the births of each sex, the relative frequency of male births does tend to cluster near 0.5 in both simulations. Additionally, it is apparent from comparison of the two graphs that these relative frequencies cluster more closely about 0.5 for samples of 400 births than for samples of 40 births.

In our second program, entitled CARS, we simulate N visits to the drive-in bank described above. In this case we make the rather unrealistic assumption that the number of cars in line has an equal chance of being any integer from 1 to 10. The program simulates $N = 1000$ visits and the frequencies of 1, 2, ..., 10 cars in line are printed, along with the sum of these relative frequencies and the average number of cars in line in the 1000 visits. The relative frequencies cluster about the theoretical value of 0.1 for each of the possible number of cars in line. In addition, the sum of the frequencies is one, as a little thought convinces us that it must be. The properties of relative frequencies which we observe will lead us to the mathematical definition of a probability model. In particular, if the relative frequencies of the outcomes of a random experiment are defined to be f_i, for i = 1, 2, ..., k (k possible outcomes), we must have $0 \leq f_i \leq 1$ and $\Sigma_{i=1}^{k} f_i = 1$.

For example on September 30, 1976, the results of a Gallup poll on the first Carter-Ford debate were published. In response to the question, "In your opinion, which man did a better job in the debate?" the following was reported:

Ford	*Carter*	*No opinion*
32%	25%	43%

In relative frequency terms, we have $f_1 = 0.32$, $f_2 = 0.25$, and $f_3 = 0.43$. Clearly these relative frequencies satisfy the properties stated in the last paragraph.

Exercises 2.1

1. Assume that a batter is a "300 hitter," and that he comes to bat 400 times in each year. Use BIRTH to simulate the batter's average in the 20 seasons, relabeling the output "HIT" and "NO HIT." Alter the program to calculate his (her) 20-year average. What were his (her) best and worst season averages?

2. The actual frequency of male birth in the United States is about 0.513. Use BIRTH to simulate 1000 births in each of 20 years. Use the program altered to calculate the frequency of male birth for the 20,000 births. Compare this value with p = 0.513.

3. Assume that there is space for 16 cars in line at the drive-in bank described by CARS. Alter CARS to simulate 3000 visits to the bank, printing out the frequency of 1 to 16 cars in line. (You will need a dimension statement.)

4. The statement LET Y = 1 + INT(6*RND) will simulate the roll of a die. Using this, write a program to simulate the sum attained on 2500 tosses of two dice. Compute the relative frequencies of the sums 2 through 12. Which outcome is most frequent and with what frequency was it observed? What do you believe to be the theoretical frequency of this outcome?

2.2   DEFINITION OF A PROBABILITY MODEL IN THE DISCRETE CASE

The stability of the relative frequency of the occurrence of a particular event, when a particular random experiment is repeated a large number of times, serves as motivation for the formal definition of a probability model. The properties of relative frequencies which we have observed previously will be reflected in the definitions which we make.

BIRTH

```
10 DIM M(20),F(20)
20 FOR I = 1 TO 20
30 LET M(I)=0
40 LET F(I)=0
50 NEXT I
60 PRINT "FEMALE","MALE"
70 READ N,P
80 FOR I = 1 TO 20
90 FOR J = 1 TO N
100 LET Z = RND
110 IF Z <=P THEN 140
120 LET F(I)=F(I)+1
130 GO TO 150
140 LET M(I)= M(I)+1
150 NEXT J
160 LET M(I)= M(I)/N
170 LET F(I)= F(I)/N
180 PRINT F(I),M(I)
190 NEXT I
200 PRINT
210 FOR J = 1 TO 29
220 PRINT "+";
230 NEXT J
240 PRINT "X";
250 FOR J = 1 TO 30
260 PRINT "+";
270 NEXT J
280 PRINT "FREQ"
290 FOR I = 1 TO 20
300 LET T = INT(60*M(I))-1
310 PRINT "+";
320 FOR J = 1 TO T
330 PRINT " ";
340 NEXT J
350 PRINT "*"
360 NEXT I
370 DATA 40,.5
380 END
```

READY

```
370 DATA 400,.5
RUN
FEMALE        MALE
0.425         0.575
0.525         0.475
0.675         0.325
0.475         0.525
0.55          0.45
0.55          0.45
0.525         0.475
0.6           0.4
0.425         0.575
0.4           0.6
0.575         0.425
0.45          0.55
0.475         0.525
0.475         0.525
0.4           0.6
0.55          0.45
0.5           0.5
0.45          0.55
0.575         0.425
0.375         0.625
```

```
+++++++++++++++++++++++++++++++++X++++++++++++++++++++++++++++++++++++FREQ
+                                         *
+                                  *
+                        *
+                                     *
+                              *
+                              *
+                           *
+                   *
+                                 *
+                                    *
+                        *
+                              *
+                              *
+                                 *
+                     *
+                          *
+                          *
+                 *
+                                 *

*
```

```
370 DATA 400,.5
RUN
FEMALE              MALE
 0.515              0.485
 0.4825             0.5175
 0.5525             0.4475
 0.4775             0.5225
 0.5175             0.4825
 0.525              0.475
 0.52               0.48
 0.53               0.47
 0.5                0.5
 0.47               0.53
 0.48               0.52
 0.505              0.495
 0.5275             0.4725
 0.5325             0.4675
 0.485              0.515
 0.52               0.48
 0.445              0.555
 0.4825             0.5175
 0.495              0.505
 0.4825             0.5175
```

```
+++++++++++++++++++++++++++++++++X+++++++++++++++++++++++++++++++++FREQ
+                                *
+                                 *
+                             *
+                                *
+                            *
+                            *
+                            *
+                            *
+                             *
+                              *
+                              *
+                            *
+                            *
+                              *
+                            *
+                               *
+                              *
+                             *
+
*
```

CARS

```
10 FOR I = 1 TO 10
20 LET F(I)=0
30 NEXT I
40 READ N
50 FOR I = 1 TO N
60 LET T = 1 + INT(10*RND)
70 LET F(T)=F(T)+1
80 NEXT I
90 PRINT "CARS","REL FREQ"
100 LET A = 0
110 LET B =0
120 FOR I = 1 TO 10
130 LET F(I)=F(I)/N
140 LET B = B+I*F(I)
150 PRINT I,F(I)
160 LET A = A+F(I)
170 NEXT I
180 PRINT
190 PRINT "TOTAL FREQ=";A
200 PRINT "AVERAGE=";B
210 DATA 1000
220 END
*
 RUN
CARS              REL FREQ
  1               0.102
  2               0.103
  3               0.095
  4               0.093
  5               0.103
  6               0.106
  7               0.102
  8               0.106
  9               0.107
 10               0.083

TOTAL FREQ= 1
AVERAGE= 5.471

*
```

We think of an idealized experiment with either a finite or a countable number of outcomes. Countable means that the outcomes can be put into a one-to-one correspondence with the positive integers. The set of all possible outcomes, called the *outcome* (or *sample space*), will be denoted S. The points in S correspond to the outcomes of a real-life experiment and will be denoted as e_i, where the subscript i = 1, 2, ..., n if there are a finite number of outcomes and i = 1, 2, ... in the case of a countable outcome space. Such an outcome space is called *discrete*. Corresponding to the long-run relative frequency in a random experiment will be a number $P(e_i)$ called the *probability associated with the point* e_i. In order that these probabilities reflect the properties of relative frequencies we require that

$$0 \le P(e_i) \le 1 \qquad\qquad (2.2.1)$$

and

$$\sum_{\text{all } i} P(e_i) = 1 \qquad\qquad (2.2.2)$$

Thus a formal probability model is asserted by defining S and the probabilities of the points of S so that (2.2.1) and (2.2.2) are satisfied.

Suppose that we consider a theoretical battery purchase in the previous section with life length described as before so that

$$S = \{0, 1, \ldots, 8\}$$

One possible assignment of values to the points of S is $P(e_i) = 1/9$ for i = 1, 2, ..., 9, where $e_1 = 0$, $e_2 = 1$, ..., $e_9 = 8$. Certainly the requirements (2.2.1) and (2.2.2) are met. This corresponds to the assumption of equal likelihood of failure during each of the first 9 years of life. However, this is not the only possibility by any means. We may define

$$P(e_i) = \frac{(1 - q)q^{i-1}}{1 - q^9} \qquad i = 1, 2, \ldots, 9$$

for any q satisfying 0 < q < 1. As

$$\sum_{i=1}^{9} P(e_i) = \frac{(1 - q)(1 + q + q^2 + \cdots + q^8)}{1 - q^9}$$

$$= \frac{1 - q^9}{1 - q^9} = 1$$

and the $P(e_i)$ are clearly nonnegative. Again (2.2.1) and (2.2.2) are satisfied. It is now clear that, because a different probability assignment occurs for every q, there are an infinity of ways in which the sample space S may be assigned probabilities. This latter assignment has the important property that

$$\frac{P(e_{i+1})}{P(e_i)} = q \qquad i = 1, 2, \ldots, 8$$

so that the probabilities decline geometrically.

The program BATTERY simulates N battery lives assuming this latter model. The values of N and q are determined by the DATA statement at line 270. The output from one run assumes 1000 batteries have lives described by this model with q = 0.5. The relative frequency of the 1000 battery life lengths and the probabilities predicted by the model are printed out. As the output demonstrates, both the frequencies and probabilities sum to 1.

If we change the outcome space of the previous example to

S = {0, 1, 2, 3, ...}

we now consider that batteries may have an arbitrarily long life. Although, of course, this is only theoretically possible, the resulting probability model may nevertheless provide a useful approximation to reality. Suppose we continue to assume that the probabilities decline geometrically, that is,

$$\frac{P(e_{i+1})}{P(e_i)} = q \qquad \text{for } 0 < q < 1$$

Then $P(e_2) = qP(e_1)$, $P(e_3) = qP(e_2) = q^2 P(e_1)$, $P(e_4) = qP(e_3) = q^3 P(e_1)$, and in general,

$$P(e_i) = q^{i-1}P(e_1) \qquad i = 1, 2, \ldots$$

As $\Sigma_{i=1}^{\infty} P(e_i) = P(e_1) \Sigma_{i=0}^{\infty} q^i = P(e_1)/(1 - q)$ (where the expression for the sum of a geometric series has been used), we must have

$$\frac{P(e_1)}{1 - q} = 1$$

Hence $P(e_1) = 1 - q$ and $P(e_i) = (1 - q)q^{i-1}$, $i = 1, 2, \ldots$. As a function of i (instead of e_i), we have $P(i) = (1 - q)q^i$ for $i = 0$, 1, 2, $\ldots$. This is referred to as the geometric distribution on the set S.

Slight changes in the program BATTERY, yield the program GEOM which simulates N observations from this population. Of course the computer cannot provide an infinite number of probabilities, so that the output is terminated when the cumulative probability exceeds 0.999. One run of GEOM is shown with N = 1000 and q = 1/2. A comparison of the frequencies observed and the probabilities computed under the assumption of the geometric model suggests how such a theoretical model may be useful in describing actual random experiments.

Problems 2.2

1. Assume $0 < p < 1$ and let $q = 1 - p$. If $P(e_1) = p^3$, $P(e_2) = 3p^2q$, $P(e_3) = 3pq^2$, and $P(e_4) = q^3$, prove that this represents a valid probability model for $S = \{e_1, e_2, e_3, e_4\}$ for any such value of p.

2. Assume that the probability a batter gets i hits in 5 "at bats" is inversely proportional to i. (Assume the batter gets at least 1 hit.) Find $P(e_i)$, i = 1, 2, 3, 4, 5, where e_i represents getting i hits.

3. Let the probability an animal is successful for the first time on the ith trial of a learning experiment be proportional to the

trial number, that is, $P(e_i) = ip_1$ for i = 1, 2, ..., n. (We assume success on or before trial n.) Find $P(e_i)$, i = 1, 2, ..., n.

4. Prove that $P(i) = 1/i(i + 1)$ is a valid probability model for the positive integers.

5. Let $P(e_i) = k/(i - 1)!$ for i = 1, 2, ... Find the constant k.

6. Find at least two probability models to describe the outcome on the toss of a die, that is, S = {1, 2, 3, 4, 5, 6}.

7. The National Center for Health Statistics published the following information on the cause of death in the United States for 1972.

Diseases of heart and blood vessels	1,036,560
Cancer	346,930
Accidents	113,670
Pneumonia and influenza	61,160
Diabetes	39,070
All other	364,610
Total	1,962,000

Describe a sample space and an appropriate probability assignment for the causes of death in the United States.

Exercises 2.2

1. Assume that a battery lasts at most 5 years. Using the model $P(e_i) = (1 - q)q^{i-1}/(1 - q^k)$, where e_i represents a life length of i - 1 but less than i years for i = 1, 2, ..., k. Simulate the life lengths of 1000 batteries with q = 0.5 using the appropriate value of k. Only slight modifications of the program BATTERY are required.

2. Simulate the life length of 1000 batteries using the geometric distribution with q = 1/3.

3. Consider, as in Problem 2.2.3, an animal which is successful for the first time on trial i of an experiment with probability

BATTERY

```
10 FOR I = 1 TO 9
20 LET F(I)=0
30 NEXT I
40 READ N, P
50 FOR J = 1 TO N
60 LET Z = RND
70 LET C = 0
80 FOR I = 0 TO 8
90 LET T = (1-P)*P†I/(1-P†9)
100 LET C = C+T
110 IF Z > C THEN 130
120 GO TO 140
130 NEXT I
140 LET I = I +1
150 LET F(I)=F(I)+1
160 NEXT J
170 LET C = 0
180 PRINT "LIFE LENGTH","REL FREQ","PROBABILITY","CUMULATIVE PROB"
190 FOR I = 1 TO 9
200 LET T = (1-P)*P†(I-1)/(1-P†9)
210 LET C = T +C
220 PRINT I-1, F(I)/N, T, C
230 LET A = F(I)/N +A
240 NEXT I
250 PRINT
260 PRINT "TOTAL FREQ=";A, "TOTAL PROB=";C
270 DATA 1000,.5
280 END
```

```
*RUN
```

LIFE LENGTH	REL FREQ	PROBABILITY	CUMULATIVE PROB
0	0.497	0.500978	0.500978
1	0.264	0.250489	0.751468
2	0.134	0.125245	0.876712
3	0.056	6.26223 E-2	0.939335
4	0.022	3.13112 E-2	0.970646
5	0.016	1.56556 E-2	0.986301
6	0.007	7.82779 E-3	0.994129
7	0.003	3.91389 E-3	0.998043
8	0.001	1.95695 E-3	1.

TOTAL FREQ= 1   TOTAL PROB= 1.

```
*
```

GEOM

```
5 DIM F(51)
10  FOR I = 1 TO 50
20 LET F(I)=0
30 NEXT I
40 READ N, P
50 FOR J = 1 TO N
60 LET Z = RND
70 LET C = 0
80 FOR I = 0 TO 8
90 LET T = (1-P)*P↑I
100 LET C = C+T
110 IF Z > C THEN 130
120 GO TO 140
130 NEXT I
140 LET I = I +1
150 LET F(I)=F(I)+1
160 NEXT J
170 LET C = 0
180 PRINT "LIFE LENGTH", "REL FREQ", "PROBABILITY", "CUMULATIVE PROB"
190 FOR I = 1 TO 50
200 LET T = (1-P)*P↑(I-1)
210 LET C = T +C
220 PRINT I-1, F(I)/N, T, C
230 LET A = F(I)/N +A
235 IF C > 0.999 THEN 260
240 NEXT I
250 PRINT
260 PRINT "TOTAL FREQ=";A, "TOTAL PROB=";C
270 DATA 1000,.5
280 END
```

*RUN

LIFE LENGTH	REL FREQ	PROBABILITY	CUMULATIVE PROB
0	0.496	0.5	0.5
1	0.263	0.25	0.75
2	0.135	0.125	0.875
3	0.055	0.0625	0.9375
4	0.022	0.03125	0.96875
5	0.018	0.015625	0.984375
6	0.007	7.8125 E-3	0.992187
7	0.002	3.90625 E-3	0.996094
8	0.002	1.95312 E-3	0.998047
9	0	9.76562 E-4	0.999023
TOTAL FREQ= 1.		TOTAL PROB= 0.999023	

*

proportional to the trial number. Assume the first success
occurs on or before trial 10. Write a program to simulate the
trial numbers of the first success for 1000 animals. Print out
the frequencies of first success on trial numbers 1-10 and the
corresponding probabilities.

2.3   PROBABILITIES OF EVENTS IN THE DISCRETE CASE

Often we are interested in the probability associated with a subset
of the sample space S. We make the following formal definition:

DEFINITION 2.3.1. An event A is a subset of the sample S.

It is quite natural to assign to the event A the sum of the probabilities of the points in A. Formally, we state

DEFINITION 2.3.2. The probability of an event A is $P(A) = \sum_{e_i \in A} P(e_i)$,
where $e_i \in A$ means e_i is an element of A.

Suppose we consider, as in our waiting line example, selecting
a random integer from 1 to 10, where, as before, we assume equal
probability of selection of each such integer. A driver might be
interested in one of the following events:

    A: {1}   (He is alone.)
    B: {1, 2, 3, 4, 5}   (Five or fewer cars in line.)
    C: {8, 9, 10}   (Eight or more cars in line.)

It is clear that the probabilities of these events are given by
$P(A) = 0.1$, $P(B) = 0.5$, and $P(C) = 0.3$ using the definitions above.
Referring to the output of the program CARS of Sec. 2.1, we see
that the output yields relative frequencies of 0.102, 0.496, and
0.296, respectively, for the corresponding sets of outcomes in the

simulation for N = 1000. Thus the probability of an event can be
thought of, as in the case of the assignment of probabilities to the
points of S, as an abstraction of the idea of relative frequency in
a large number of trials.

In the case in which we consider the geometric assignment of
probabilities to the set

$$S = \{0, 1, 2, \ldots\}$$

that is

$$P(i) = (1 - q)q^i \qquad i = 0, 1, 2, \ldots$$

[where P(i) stands for the probability of the integer i], we may be
interested in events such as

A = 0, 1, 2, 3   (A battery lasts less than 4 years.)

B = 4, 5, ...   (A battery lasts at least 4 years.)

Here we see that

$$P(A) = (1 - q)(1 + q + q^2 + q^3) = \frac{(1 - q)(1 - q^4)}{(1 - q)}$$
$$= 1 - q^4$$

For B, we have

$$P(B) = (1 - q)q^4 \left(\sum_{i=0}^{\infty} q^i \right) = \frac{(1 - q)q^4}{1 - q}$$
$$= q^4$$

Of course, the fact that P(A) + P(B) = 1 is evident here, so that
the latter calculation would be unnecessary. We will next consider
the relationships between the probabilities of various events
(subsets) of S.

Problems 2.3

1. What is the probability of an even outcome in the toss of a fair
    die? An odd outcome? An outcome less than 5?

2. Suppose for an unbalanced die, $P(e_1) = P(e_2) = P(e_3) = P(e_4) = 0.15$ and $P(e_5) = P(e_6) = 0.20$, where e_i represents the outcome i. Answer the questions in Problem 2.3.1.

3. In Problem 2.2.2 find the probability of more than two hits by the batter in the 5 "at bats."

4. Let $p = 1/2$ in Problem 2.2.1. Let $A = \{e_1\}$, $B = \{e_1, e_4\}$, and $C = \{e_2, e_3, e_4\}$. Find $P(A)$, $P(B)$, and $P(C)$.

5. Let $P(i) = 1/i(i + 1)$ represent the probability integer i is selected from the positive integers. Find the probability the integer selected is between 4 and 6 inclusive.

6. A card is selected at random from a standard bridge deck. What is the probability of a deuce? A heart? A red card?

7. In Problem 2.2.3 find the probability the first success occurs on or before trial k. For $n = 10$, what is the smallest value of k for which this probability is at least 0.5?

8. Let the geometric distribution given by $P(i) = (1 - q)q^i$ for $i = 0, 1, 2, \ldots$ describe life lengths of batteries as before. What is the probability the life length is an odd number? If $p = 1/2$, what is the value of this probability? If $0 < q < 1$, show that the probability of an odd number is less than 0.5.

2.4 PROBABILITIES OF COMPOSITE EVENTS IN A DISCRETE SAMPLE SPACE S

We will assume that the reader has had an introduction to the algebra of subsets of a given set. The references Kemeny et al. (1974) and Scheid (1962) contain good introductions. We will, however, review here briefly the basic definitions and some important theorems concerning the events or subsets of S.

DEFINITION 2.4.1. If A and B are events in S, then A is called *a subset of* B if for any point $e \in A$ we have $e \in B$. We write $A \subset B$.

DEFINITION 2.4.2. Two events A and B in S are equivalent or equal
if $A \subset B$ and $B \subset A$. We write $A = B$

DEFINITION 2.4.3. If A and B are subsets of S, then the *union* of A
and B is the set containing those points in A or B including those
points common to A and B. We write the union as $A + B$.

DEFINITION 2.4.4. If A and B are two events in S, then the *intersection* of A and B is the event containing those points in both A
and B. We write the intersection as AB.

DEFINITION 2.4.5. If A is an event of S, then the *complement* of A
is the set of those points of S not in A. We write the complement
as $\bar{A}$.

DEFINITION 2.4.6. The *empty set* is the subset of S without any
points. We write the empty set as $\emptyset$. From Definition 2.4.1,
$\emptyset \subset A$ for any A since the conditions of Definition 2.4.1 are met
vacuously.

We are interested in finding formulas for the probabilities of $A + B$,
AB, $\bar{A}$, and $\emptyset$. In words, $A + B$ is the event that A or B occurs, AB
is the event that A and B occur, and $\bar{A}$ is the event that A does not
occur.

   There are numerous theorems concerning the algebra of subsets
which can be proved using the preceding definitions. We state a
number of these without proof in the following theorem.

THEOREM 2.4.1. For any events A, B, and C in S,

(a) A + B = B + A	(j) A + $\emptyset$ = A
(b) AB = BA	(k) A$\emptyset$ = $\emptyset$
(c) (A + B) + C = A + (B + C)	(l) A + $\bar{A}$ = S
(d) (AB)C = A(BC)	(m) A$\bar{A}$ = $\emptyset$
(e) (A + B)C = AC + BC	(n) AS = A
(f) A(B + C) = AB + AC	(o) A + S = S
(g) A + BC = (A + B)(A + C)	(p) AA = A
(h) $\overline{(A + B)}$ = $\bar{A}\bar{B}$	(q) A + A = A
(i) $\overline{AB}$ = $\bar{A}$ + $\bar{B}$	

These theorems can be proved by direct arguments from the above definitions or demonstrated convincingly using Venn diagrams. We refer the reader to references on finite mathematics in the bibliography.

Let us consider a list of 100,000 registered voters who are partitioned according to three dichotomous classifications: male or female, married or unmarried, and Democrat or Republican. We will use the following notation: A represents the set of males, B the married individuals, and C the set of Democrats. There are eight categories into which individuals may be classified. These are displayed in Fig. 2.4.1a.

A

A	
AB$\bar{C}$ 10,000	$\bar{A}$B$\bar{C}$ 4,000
ABC 30,000	$\bar{A}$BC 36,000
A$\bar{B}$C 6,000	$\bar{A}\bar{B}$C 3,000
A$\bar{B}\bar{C}$ 2,000	$\bar{A}\bar{B}\bar{C}$ 9,000

B

C

A	
AB 40,000	$\bar{A}$B 40,000
A$\bar{B}$ 8,000	$\bar{A}\bar{B}$ 12,000

B

(a)   (b)

Fig. 2.4.1

The event $AB\overline{C}$, for example, contains married male Republicans, ABC contains married male Democrats, $\overline{A}BC$ contains married female Democrats, and so forth. In Fig. 2.4.1b, the 100,000 individuals have been classified into four categories: married males, married females, unmarried males, and unmarried females, that is, AB, $\overline{A}B$, $A\overline{B}$, and $\overline{A}\overline{B}$. The numbers in both diagrams of course represent the number of individuals in each category. We shall use this example to illustrate some of the elementary theorems of probability, where we assume each person has an associated probability of 1/100,000.

We shall first require a definition.

DEFINITION 2.4.7. Events A and B are called *disjoint* if AB = $\emptyset$, i.e., there are no points common to A and B.

In our example the set of females $\overline{A}$ and the set of married males AB are disjoint. The events A + B (married or male) and $\overline{A}\overline{B}$ (unmarried females) are disjoint. This is, of course, clear from the diagrams in Fig. 2.4.1 but can be formally shown by using Theorem 2.4.1. For example, $\overline{A}(AB) = (\overline{A}A)B = \emptyset B = \emptyset$ using appropriate parts of the theorem. Similarly,

$$\overline{A}\overline{B}(A + B) = (\overline{A}\overline{B})A + (\overline{A}\overline{B})B = (\overline{B}\overline{A})A + \overline{A}(\overline{B}B)$$
$$= \overline{B}(\overline{A}A) + \overline{A}(\overline{B}B) = \overline{B}\emptyset + \overline{A}\emptyset$$
$$= \emptyset + \emptyset = \emptyset$$

We now turn to some basic theorems of elementary probability.

THEOREM 2.4.2 If A and B are disjoint, then P(A + B) = P(A) + P(B).

Proof: The statement of the theorem is equivalent to

$$\sum_{e_i \in A+B} P(e_i) = \sum_{e_i \in A} P(e_i) + \sum_{e_i \in B} P(e_i)$$

Because AB = $\emptyset$, the points in A + B are separated into those in A and those in B, where there are no points common to A and B. Hence both sides represent the sum of the same probabilities and must therefore be equal.

COROLLARY 1. $P(A) + P(\bar{A}) = 1$.

Proof: From Theorem 2.4.1(l) and (m), $A + \bar{A} = S$ and $A\bar{A} = \emptyset$. Hence the current theorem implies that $P(A) + P(\bar{A}) = P(A + \bar{A}) = P(S) = 1$.

COROLLARY 2. If A_1, A_2, ..., A_k are pairwise disjoint (that is, $A_i A_j = \emptyset$ for $i \neq j$), then $P(\Sigma_{i=1}^{k} A_i) = \Sigma_{i=1}^{k} P(A_i)$.

Proof: This follows from a straightforward mathematical induction from the theorem. [Scheid (1962) contains a good discussion of mathematical induction.]

We have seen that in our example (and in general) $\bar{A}$ and AB are disjoint. Hence $P(\bar{A} + AB) = P(\bar{A}) + P(AB) = 0.52 + 0.40 = 0.92$. In words, 92% of the individuals on our list are female or married males. We have seen that A + B and $\overline{AB}$ are disjoint. However Theorem 2.4.1(h) implies that these events are complements. Hence $P(A + B) + P(\overline{AB}) = 1$ by Corollary 1. Thus, $P(A + B) = 1 - P(\overline{AB})$, or $P(A + B) = 1 - 0.12 = 0.88$. In words, 88% of the individuals are male or married.

COROLLARY 3. $P(\emptyset) = 0$.

Proof: By Theorem 2.4.1(j) and (k), we have $S + \emptyset = S$ and $S\emptyset = \emptyset$. Thus from the current theorem, $P(S) + P(\emptyset) = P(S)$ or $1 + P(\emptyset) = 1$. Canceling 1 from each side yields $P(\emptyset) = 0$.

THEOREM 2.4.3. For any events A and B,

$$P(A + B) = P(A) + P(B) - P(AB)$$

From Fig. 2.4.1b it is clear that A = AB + A$\overline{B}$, B = AB + $\overline{A}$B, and A + B = AB + A$\overline{B}$ + $\overline{A}$B, where each right-hand side is a union of disjoint sets. Hence we can write

$$P(A + B) = P(AB) + P(A\overline{B}) + P(\overline{A}B)$$

However $P(A) = P(AB) + P(A\overline{B})$ and $P(B) = P(\overline{A}B) + P(AB)$. Thus

$$P(A + B) = P(AB) + P(A\overline{B}) + P(\overline{A}B) + P(AB) - P(AB)$$
$$= P(A) + P(B) - P(AB)$$

using the expressions for P(A) and P(B).

COROLLARY 1. For any three events A, B, and C,

$$P(A + B + C) = P(A) + P(B) + P(C) - P(AB) - P(AC)$$
$$- P(BC) + P(ABC)$$

Proof: Using A + B and C as two events, the theorem yields

$$P((A + B) + C) = P(A + B) + P(C) - P((A + B)C)$$
$$= P(A + B) + P(C) - P(AC + BC)$$
$$= P(A) + P(B) - P(AB) + P(C) - [P(AC)$$
$$+ P(BC) - P(ACBC)]$$
$$= P(A) + P(B) + P(C) - P(AB) - P(AC)$$
$$- P(BC) + P(ABC)$$

In our example $P(A + B + C)$ represents the probability of a male or a married person or a Democrat. This probability is

$$P(A) + P(B) + P(C) - P(AB) - P(AC) - P(BC) + P(ABC)$$

which from Fig. 2.4.1a is given by

$$0.48 + 0.80 + 0.75 - 0.40 - 0.36 - 0.66 + 0.30 = 0.91$$

COROLLARY 2. If A_1, A_2, ..., A_k are any k events in S, then

$$P\left(\sum_{i=1}^{k} A_i\right) = \sum_{i=1}^{k} P(A_i) - \sum_{i \neq j} P(A_i A_j) + \sum_{\substack{i \neq j, j \neq m, \\ i \neq m}} P(A_i A_j A_m)$$

$$- \cdots + (-1)^{k-1} P(A_1 A_2 \cdots A_k)$$

This corollary is proved by a straightforward mathematical induction from the main theorem. We leave the proof as a problem, but note that Corollary 1 is a special case of Corollary 2 with k = 3, A_1 = A, A_2 = B, and A_3 = C.

The program VOTERS simulates the selection of N individuals from the population of 100,000 voters described above. In the output of the program the proportion of sampled voters in the eight categories is printed. The categories are identified by a symbolic binary string of length 3. The integer 1 in such a sequence indicates that a letter is present in the product representing the event and a 0 indicates that its complement is present. For example, 101 corresponds to $A\bar{B}C$, 000 to $\bar{A}\bar{B}\bar{C}$, and so forth. The frequencies of these outcomes can be compared with the probabilities of the corresponding events in Fig. 2.4.1. The probability of any composite event such as A + B can be compared with the corresponding frequency which results in a sample of size N. The probability of A + B, for example, was found to be 0.88. Its frequency in the output of VOTERS using N = 1000 is 0.897. The frequency of the outcomes male, married, and Democrat are printed and may be compared with the corresponding probabilities of 0.48, 0.80, and 0.75, respectively. The remaining output concerns conditional outcomes which we discuss in the next section.

Problems 2.4

1. In the example given in the section find the probability of the following events: AC, A + C, $\overline{A + C}$, $\bar{A}\bar{C}$.

VOTERS

```
10 FOR I = 1 TO 8
20 LET F(I)=0
30 NEXT I
40 FOR K = 1 TO 7
50 READ N(K)
60 NEXT K
70 READ N1
80 FOR J = 1 TO N1
90 LET C = 0
100 LET V = 1 + INT(10+5*RND)
110 FOR K = 1 TO 7
120 LET C = C +1
130 IF V > N(K) THEN 150
140 GO TO 170
150 NEXT K
160 LET C = 8
170 LET F(C)=F(C)+1
180 NEXT J
190 PRINT "FREQUENCIES IN";" SAMPLE OF SIZE"; N1
200 PRINT
210 PRINT "111","110","101","011"
220 PRINT F(1)/N1, F(2)/N1, F(3)/N1, F(4)/N1
230 PRINT
240 PRINT "100","010","001","000"
250 PRINT F(5)/N1, F(6)/N1, F(7)/N1, F(8)/N1
260 PRINT
270 PRINT "FREQUENCY OF","MALE","MARRIED","DEMOCRAT"
280 LET A = F(1)+F(2)+F(3)+F(5)
290 LET B = F(1)+F(2)+F(4)+F(6)
300 LET C = F(1)+F(3)+F(4)+F(7)
310 PRINT "", A/N1, B/N1, C/N1
320 PRINT "CONDITIONAL ";"FREQUENCIES"
330 LET A3= F(1)+F(2)
340 LET A1= F(1)+F(3)
350 LET A2= F(1)+F(4)
360 PRINT
370 PRINT "MARRIED/MALE","MARRIED/FEMALE"
380 PRINT A3/A, (B-A3)/(N1-A)
390 PRINT "MARRIED/DEM","MARRIED/REP"
400 PRINT A2/C, (B-A2)/(N1-C)
410 PRINT "MALE/DEM","MALE/REP"
420 PRINT A1/C, (A-A1)/(N1-C)
430 DATA 30000, 40000, 46000, 82000, 84000, 88000, 91000
440 DATA 1000
450 END
*
```

FREQUENCIES IN SAMPLE OF SIZE 1000

111	110	101	011
0.3	0.093	0.067	0.365

100	010	001	000
0.025	0.047	0.023	0.08

FREQUENCY OF   MALE   MARRIED   DEMOCRAT
                 0.485   0.805   0.755
CONDITIONAL FREQUENCIES

MARRIED/MALE   MARRIED/FEMALE
 0.810309   0.8
MARRIED/DEM   MARRIED/REP
 0.880795   0.571429
MALE/DEM   MALE/REP
 0.486093   0.481633

*

2. Assume that S represents the set of students at a college. Let
    A represent the set of students taking at least one mathematics
    course and B represent the set of students taking at least one
    language course. Suppose P(A) = 0.7, P(B) = 0.6, and P(AB) =
    0.5. Find

    (a) P(A+B)   (e) $P(\bar{A}\bar{B})$

    (b) $P(\bar{A})$ (f) $P(A\bar{B})$

    (c) $P(\bar{B})$ (g) $P(\bar{A}B)$

    (d) $P(\bar{A} + \bar{B})$ (h) $P(A\bar{B} + \bar{A}B)$

    Interpret these events in words.

3. Assume we select at random a three-digit number from 001 to 400.
    Call the selected number N. Find the following probabilities:

    (a) $P(N \geq 390)$

    (b) P(N is odd)

    (c) $P(N \text{ is odd or } N \geq 390)$

    (d) P(Left-hand digit of N is odd)

    (e) P(Middle digit of N is odd)

    (f) P((d) or (e) is true)

4. Using Theorem 2.4.1, prove that for any events A, B, and C in S,

    $$\overline{A + B + C} = \bar{A}\bar{B}\bar{C} \qquad \text{and} \qquad \overline{ABC} = \bar{A} + \bar{B} + \bar{C}$$

    Interpret the events $\bar{A}\bar{B}\bar{C}$ and $\overline{ABC}$ in words for the example given
    in Fig. 2.4.1a.

5. Prove that for A_1, A_2, ..., A_k mutually disjoint events, that
    is, $A_i A_j = \emptyset$ for $i \neq j$,

    $$P\left(\sum_{i=1}^{k} A_i\right) = \sum_{i=1}^{k} P(A_i)$$

    Use mathematical induction.

6. Use Theorem 2.4.3 to prove that $P(A + B) \leq P(A) + P(B)$. By
    mathematical induction, prove that for any events A_1, A_2, ...,
    A_k, we have

$$P(\sum_{i=1}^{k} A_i) \leq \sum_{i=1}^{k} P(A_i)$$

7. Use mathematical induction to prove that for any events A_1, A_2, ..., A_k, we have

$$P(\sum_{i=1}^{k} A_i) = \sum_{i=1}^{k} P(A_i) - \sum_{i<j} P(A_i A_j) + \cdots$$
$$+ (-1)^{k-1} P(A_1 A_2 \cdots A_k)$$

8. Assume three events A, B, and C are given such that $P(A) = 0.5$, $P(B) = 0.6$, $P(C) = 0.4$, $P(AB) = 0.3$, $P(AC) = 0.1$, $P(BC) = 0.2$, and $P(ABC) = 0.05$. Find

(a) $P(AB\bar{C})$ (e) $P(\bar{A}B\bar{C})$

(b) $P(A\bar{B}C)$ (f) $P(A\bar{B}\bar{C})$

(c) $P(\bar{A}BC)$ (g) $P(\bar{A}\bar{B}\bar{C})$

(d) $P(\bar{A}\bar{B}C)$ (h) $P(A + B + C)$

Draw a diagram similar to Fig. 2.4.1a labeling the eight basic events with the appropriate probabilities.

Exercises 2.4

1. Use Nl = 2000 in the program VOTERS (line 440) to sample this number of voters from the population of 100,000 voters. Indicate which of the eight frequencies are added together to give the relative frequency of the male, married, and Democratic individuals selected. Indicate the set equality which describes the events A, B, and C, in terms of the eight basic sets (e.g., $AB = ABC + AB\bar{C}$).

2. In Problem 2.4.8 let A represent the set of homeowners, B the set of Democrats, and C the set of dog owners among 100,000 heads of household. Use the probabilities of Problem 2.4.8 as the true proportions in each of the eight basic events in S. Alter the program VOTERS at line 430 to represent this population. Delete

lines 320-420 and change the PRINT statement in line 270 appro-
priately to sample 1000 individuals from this population. What
frequencies are observed for Democratic home owners, Democratic
dog owners, and dog-owning home owners. Compare these with the
corresponding probabilities.

3. Write a program to simulate the sum on the toss of two dice 2000
    times (as in Exercise 2.1.3). What is the observed frequency of
    an odd sum? What is the observed frequency of 7 or 11? What is
    the observed frequency of a sum less than 6? Assuming the 36
    outcomes (i, j), $i = 1, 2, \ldots, 6$ and $j = 1, 2, \ldots, 6$, have
    equal probability, find the theoretical probabilities correspond-
    ing to these events.

2.5   CONDITIONAL FREQUENCY AND CONDITIONAL PROBABILITY

In the example given in the preceding section we might be interested
in comparing the proportion of married persons among females to the
proportion of married persons among males for the population of
100,000 persons. In sampling from the population we would, of course,
take the proportion of married males among the sampled females as
estimates of the population proportions. If we take a sample of size
N, we would estimate the proportion of males who are married as

$$\text{Frequency (married/male)} = \frac{\text{number of married males}}{\text{number of males}}$$

Dividing the numerator and denominator by N, we can express this
conditional frequency as

$$\text{Frequency (married/male)} = \frac{\text{sample frequency of married males}}{\text{sample frequency of males}}$$

Similarly the conditional frequency of married given a female is
given by the ratio of the frequency of married females to the frequency of females. The program VOTERS prints out these particular
conditional frequencies as well as several others for a sample of
size N = 1000.

We are naturally led by consideration of conditional frequencies to the idea of conditional probability. If we replace sample frequency by probability in the foregoing, we are led to the following definition.

DEFINITION 2.5.1. For any two events A and B for which $P(A) > 0$, we define the conditional probability of B given A by

$$P(B|A) = \frac{P(AB)}{P(A)}$$

The requirement that $P(A) > 0$ ensures that division by zero does not occur. As we have $AB + A\bar{B} = A$ and $P(AB) + P(A\bar{B}) = P(A)$, we see that $P(AB) \le P(A)$ or $0 \le P(AB)/P(A) \le 1$, so that a conditional probability is a true probability. Using the definitions of $A \equiv$ males, $B \equiv$ married individuals, and $C \equiv$ Democrats in our example, with equal probability assignment as before, we find

$$P(B|A) = \frac{P(AB)}{P(A)} = \frac{0.40}{0.48} = 0.833$$

and

$$P(B|\bar{A}) = \frac{P(\bar{A}B)}{P(\bar{A})} = \frac{0.40}{0.52} = 0.769$$

These conditional probabilities are not equal, but also are not very different, so that it is not surprising that the sample estimates of 0.810309 and 0.8, respectively, from the program VOTERS do not differ greatly. However,

$$P(B|C) = \frac{P(BC)}{P(C)} = \frac{0.66}{0.75} = 0.88$$

and

$$P(B|\bar{C}) = \frac{P(B\bar{C})}{P(\bar{C})} = \frac{0.14}{0.25} = 0.56$$

The sample estimates of these conditional probabilities given by the program VOTERS are 0.881 and 0.571. These are reasonable estimates of the corresponding conditional probabilities. It is clear that the conditional probability of an event (being married in this case) can be altered greatly by a conditioning event (Democrat or Republican in this case).

The conditional probabilities of the points of S with respect to a *fixed* event A for which $P(A) > 0$ define a new probability assignment to S. Clearly, if $e_i \in A$, then $P(e_i|A) = P(e_i)/P(A) \leq 1$, and if $e_i \not\in A$, $P(e_i|A) = 0$. Hence,

$$\sum_{e_i \in S} P(e_i|A) = \sum_{e_i \in A} P(e_i|A) = \sum_{e_i \in A} \frac{P(e_i)}{P(A)} = \frac{P(A)}{P(A)} = 1$$

Thus we have

$$0 \leq P(e_i|A) \leq 1 \qquad \text{for } e_i \in S \tag{2.5.1}$$

and

$$\sum_{e_i \in S} P(e_i|A) = 1 \tag{2.5.2}$$

for any event A for which $P(A) > 0$. This can be also thought of as an assignment of probabilities to the event A by itself [as if $e_i \not\in A$, $P(e_i|A) = 0$], consistent with the original assignment of probabilities to S. It is often useful to consider a conditional probability assignment to S with respect to a fixed set A.

Let us recall the battery example in which we assumed the geometric distribution, that is, $P(i) = (1 - q)q^i$ for i = 0, 1, 2, ...
Let p = 1/2, so that $P(i) = (1/2)^{i+1}$ for i = 0, 1, 2, ... Suppose we consider the conditioning event that the life length is less than x years. The probability that the life length is x or more years is

$$\sum_{i=x}^{\infty} (\tfrac{1}{2})^{i+1} = (\tfrac{1}{2})^x$$

Hence the conditional probability of a life of i years given a life length less than x years is

$$f(i|x) = P(i|\text{life length} < x \text{ years}) = \frac{(1/2)^{i+1}}{1 - (1/2)^x}$$

Table 2.5.1   Probabilities for Life Lengths Less Than x Years

Life length	P(i)	P(i|2)	P(i|4)	P(i|8)
0	0.5	0.666667	0.533333	0.501961
1	0.25	0.333333	0.266667	0.250980
2	0.125		0.133333	0.125490
3	0.0625		0.066667	0.062745
4	0.03125			0.031373
5	0.015625			0.015686
6	0.007813			0.007843
7	0.003906			0.003922

for $i = 0, 1, 2, \ldots, x - 1$. Table 2.5.1 indicates how the probabilities of the various life lengths are affected by knowledge of
the fact that the life length is less than x years.
If the maximum life length is very short, the conditional probabilities differ markedly from those of the geometric distribution. If
the maximum life length is fairly long (as in the case $x = 8$), the
conditional probabilities and the probabilities given by the geometric
distribution are quite similar, indicating again that a model with an
infinite simple space may be useful in describing real-world phenomena.

The equality expressed in Definition 2.5.1 is often written in
the form

$$P(AB) = P(A)P(B|A)$$

and referred to as the multiplication law. It also follows that

$$P(ABC) = P(AB)P(C|AB) = P(A)P(B|A)P(C|AB)$$

Similarly, it can be shown by mathematical induction that for any
events $A_1, A_2, \ldots, A_k$,

$$P(A_1A_2A_3\cdots A_k) = P(A_1)P(A_2|A_1)P(A_3|A_1A_2)\cdots P(A_k|A_1A_2\cdots A_{k-1})$$

For example, suppose that an individual is asked to perform a task
three times. Assume that the probability he succeeds increases with

the trial number. For example, let p_1 = 1/4, p_2 = 1/2, and p_3 = 3/4
be the probabilities of success on trials 1, 2, and 3, respectively.
Let S_i represent success on trial i and F_i represent failure on trial
i. The probability of success on the first two trials is given by
$P(S_1 S_2)$ = $P(S_1)P(S_2|S_1)$ = (1/4)(1/2) = 1/8. The probability of at
least two successes in the three trials would be given by $P((S_1 S_2)$ +
$F_1 S_2 S_3$ + $S_1 F_2 S_3)$ = $P(S_1 S_2)$ + $P(F_1 S_2 S_3)$ + $P(S_1 F_2 S_3)$ as these events
are mutually disjoint. In turn we may write this expression as
$P(S_1 S_2)$ + $P(F_1)P(S_2|F_1)P(S_3|F_1 S_2)$ + $P(S_1)P(F_2|S_1)P(S_3|S_1 F_2)$ = 1/8 +
(3/4)(1/2)(3/4) + (1/4)(1/2)(3/4) = 1/2.

As another example of conditional probability let us assume
that the probability that a program runs correctly after n debugging
efforts is given by n/(n + 1). The probability that the program
runs correctly, of course, approaches 1 as the number of debugging
efforts n becomes large. The probability that one trial is required
is, of course, 1/2. The probability that two trials are required
until success in the notation of the previous paragraph is $P(F_1 S_2)$ =
(1/2)(2/3) = 1/3. The probability that three trials are required is
$P(F_1 F_2 S_3)$ = (1/2)(1/3)(3/4) = 1/8. The probability of n required
trials is given by

$$P(F_1 F_2 \cdots F_{n-1} S_n) = (\tfrac{1}{2})(\tfrac{1}{3})\cdots(\tfrac{1}{n})(\frac{n}{n + 1}) = \frac{n}{(n + 1)!}$$

It seems reasonable that $\sum_{n=1}^{\infty}$ n/(n + 1)! = 1, i.e., that the probability of a successful run at some trial is 1. This is, in fact,
the case. Can you show that it is true?

Problems 2.5

1. Suppose that a fair die is tossed twice and the sum recorded.
    If the events (i, j), i = 1, 2, ..., 6 and j = 1, 2, ..., 6
    are equally likely, find

    (a) The probability the sum is 7.
    (b) The probability the sum is 7 given the sum is odd.
    (c) The probability the sum is 7 given the sum is odd but not 11.

2. Two selections are made from a standard bridge deck without replacement. Find the probability of

   (a) Two spades in the two draws.
   (b) Two cards of the same suit.
   (c) Two cards of the same denomination.
   (d) No spades in the two draws.

3. Suppose a basketball player shoots three foul shots with probability of success on trial 1 equal to 1/4, on trial 2 equal to 1/2, and on trial 3 equal to 3/4. Find the probability the player makes 0, 1, 2, or 3 shots in the three attempts.

4. Suppose one of four similar appearing keys opens a lock. The keys are chosen successively until the correct one is found. If a key fails, it is put aside. What is the probability that 1, 2, 3, or 4 trials are required?

   Consider the case of n keys, one of which opens a lock. What is the probability of 1, 2, ..., or n required trials until success?

5. Suppose we have the situation described in Problem 2.4.2. Find

   (a) $P(A|B)$ (c) $P(B|A)$
   (b) $P(A|\bar{B})$ (d) $P(B|\bar{A})$

6. Prove that for any two events A and B in a sample space S, $P(A|B) + P(\bar{A}|B) = 1$, assuming $P(B) > 0$.

7. Given the geometric distribution with

   $$P(i) = (1 - p)p^i \qquad i = 0, 1, 2, \ldots$$

   prove that

   $$P(i \geq (x + y)|i \geq x) = P(i \geq y)$$

   where x and y are nonnegative integers. Interpret this in terms of the length of life of a battery.

8. Consider $S = \{1, 2, 3, \ldots, 20\}$. Let A represent the set of prime numbers in S, B represent the set of odd numbers in S,

and C represent the set of numbers evenly divisible by 3. Assuming each integer is assigned probability 1/20, find

(a) $P(A|B)$ (c) $P(A|C)$

(b) $P(B|A)$ (d) $P(B|C)$

9. Suppose that we make successive trials which result in success or failure and that the probability the first success occurs *on* or *after* the nth trial is 1/n for n = 1, 2, ...

(a) Find an expression for the probability that the first success occurs on trial n.

(b) What is the conditional probability that the first success occurs on trial n given at least n trials are required?

Exercises 2.5

1. Alter the program VOTERS in lines 330 through 420 to print out the conditional frequencies of Democrat/married and Democrat/unmarried, Democrat/male and Republican/male, and Democrat/female and Republican/female. Run the program using N = 1000. What is the sum of the last two frequencies? Why?

2. Expand the program written for Exercise 2.4.3 to print out the conditional frequency of a sum of 7, the conditional frequency of a sum of 7 given an odd outcome, and the conditional frequency of a sum of 7 given an odd outcome not equal to 11. Compare with the probabilities in Problem 2.5.1.

2.6   INDEPENDENT EVENTS

An important idea in probability is that of independence. We shall again use the output of the program VOTERS to introduce this concept. If we consider again the outcomes male, married, or Democrat, we see that the conditional frequency of married individuals is considerably larger among Democrats than among Republicans. On the other hand, the conditional frequency of males among Democrats is almost the same as for Republicans. Thus the conditional frequency of males among Democrats in the sample of 1000 (0.486) is almost the same as the frequency of males among all of the 1000 sampled individuals (0.485). Thus we have

$$\text{Frequency (males)} \doteq \text{frequency(males}|\text{Democrat)}$$
$$\doteq \text{frequency(males}|\text{Republicans)}$$

In short the frequency of males does not seem to depend upon party preference.

The probabilistic counterpart of the approximate equalities given above for the events $A \equiv$ male and $C \equiv$ Republican would be

$$P(A) = P(A|C) = P(A|\bar{C})$$

In fact, if we compute these probabilities using equal probability assignment as before, we find

$$P(A|C) = \frac{P(AC)}{P(C)} = \frac{0.36}{0.75} = 0.48$$

$$P(A|\bar{C}) = \frac{P(A\bar{C})}{P(\bar{C})} = \frac{0.12}{0.25} = 0.48$$

and $P(A) = 0.48$. The three probabilities are equal. On the other hand, we have previously found $P(B|C) = 0.88$, $P(B|\bar{C}) = 0.56$, and $P(B) = 0.80$. The conditional probabilities are not equal in this case. Thus we are led to the following provisional:

DEFINITION 2.6.1'.   In a sample space S, two events A and B are said to be *independent* if $P(A|B) = P(A)$.

Notice that this implies $P(AB)/P(B) = P(A)$ or $P(AB) = P(A)P(B)$, where we have tacitly assumed $P(B) > 0$. Conversely, for any event B for which $P(B) > 0$, $P(AB) = P(A)P(B)$ implies $P(A|B) = P(A)$. We would like our definition of independence to include the case $P(B) = 0$ and to assert that A is independent of B when $P(B) = 0$. The equation $P(A)P(B) = P(AB)$ holds if $P(B) = 0$. The left side is clearly 0, and as $0 \leq P(AB) \leq P(AB) + P(AB) + P(\bar{A}B) = P(B) = 0$, the right side is 0 also. Hence the generally accepted definition of independence is

DEFINITION 2.6.1. In a sample space S, two events A and B are *independent* if $P(AB) = P(A)P(B)$.

If $P(A) > 0$, $P(B) > 0$, and A and B are independent, then $P(AB)/P(A) = P(B|A) = P(B)$ and $P(AB)/P(B) = P(A|B) = P(A)$. The conditional probability of B given A does not depend on A, and the conditional probability of A does not depend on B. Hence the definition treats A and B symmetrically.

To illustrate this idea once more let us consider the following example. Suppose we consider two baseball players, who we number 1 and 2. Let the event labeled A_i mean that batter i gets a hit and B_i mean that there are men on base when he is credited with an "at bat." Suppose that both men are credited with 450 at bats during the year with the following results

If we consider these to be two theoretical outcome spaces and assign equal probability to each at bat, we obtain

$$P(A_1) = P(A_2) = \frac{1}{3} \qquad P(B_1) = P(B_2) = \frac{1}{3}$$

However, $P(A_1|B_1) = P(A_1B_1)/P(B_1) = (1/9)/(1/3) = 1/3$, but $P(A_2|B_2) = P(A_2B_2)/P(B_2) = (2/15)/(1/3) = 2/5$. The events A_1 and B_1 are independent [note $P(A_1)P(B_1) = 1/9 = P(A_1B_1)$], while the events A_2 and B_2 are not. Apparently batter 2 does better with men on base.

The idea of independence for three events A, B, and C is contained in the following:

DEFINITION 2.6.2 In a sample space S, three events A, B, and C are said to be *mutually independent* if

$$P(AB) = P(A)P(B)$$
$$P(AC) = P(A)P(C)$$
$$P(BC) = P(B)P(C)$$

and

$$P(ABC) = P(A)P(B)P(C)$$

It is possible for the first three of these conditions to hold but the last to fail. This would mean that A and B, A and C, and B and C are *pairwise independent* but not mutually independent. For events A_1, A_2, ..., A_r to be mutually independent, we require that the probability of the intersection of *any* combination of these events equal the product of the probabilities of the events in the intersection. The most useful of these equalities is

$$P(A_1 A_2 \cdots A_r) = P(A_1)P(A_2) \cdots P(A_r)$$

Consider a batter who gets a hit with probability 1/3 (i.e., his batting average is 0.333). If he is charged with three at bats the following eight outcomes can occur: HHH, HHO, HOH, OHH, HOO, OHO, OOH, and OOO, where H indicates a hit and O, no hit with the position of the letter indicating the first, second, or third at bat. If the trials are considered independent, we obtain

$$P(HHH) = \left(\frac{1}{3}\right)^3 = \frac{1}{27}$$

$$P(HHO) = P(HOH) = P(OHH) = \left(\frac{2}{3}\right)\left(\frac{1}{3}\right)^2 = \frac{2}{27}$$

$$P(OOH) = P(OHO) = P(HOO) = \left(\frac{2}{3}\right)^2\left(\frac{1}{3}\right) = \frac{4}{27}$$

$$P(OOO) = \left(\frac{2}{3}\right)^3 = \frac{8}{27}$$

Hence the probability of x hits in the three trials is given by

x	P(x)
0	8/27
1	12/27
2	6/27
3	1/27

where $\sum_{x=0}^{3} P(x) = 1$. This is an example of an important distribution in probability, called the *binomial distribution*, which will be described in detail in the next chapter.

Problems 2.6

1. Assume that in two-child families, male and female births occur independently with probability 1/2. What is the probability such a family has

    (a) Two male children?
    (b) Two children of the same sex?
    (c) At least one male child?

2. Assume that an outcome space S describes the blood types of individuals with the types O, A, B, and AB occurring with probabilities 0.55, 0.25, 0.15, and 0.05, respectively. Assuming independence find the following probabilities for two randomly selected individuals.

    (a) The individuals have blood type O.
    (b) The individuals have the same blood type.
    (c) Neither has blood type AB.
    (d) The individuals have different blood types.

3. A machine performs an operation (say multiplication) correctly, with probability 0.9. The machine performs three successive multiplications. Assuming independence, find the probability of

    (a) Three correct multiplications
    (b) Exactly one failure in the three multiplications

(c) No correct multiplications

(d) Exactly one success in the three multiplications

What is the sum of the probabilities in (a) through (d)? Why?

4. In a triangle test one individual is asked to taste three simi-
    lar appearing items and to choose the one with a different taste
    than the other two. One of the three items has a different
    composition from the other two. It may contain a preservative,
    for example, but the other two items do not. Assume that three
    individuals make their choices, which we assume to be independ-
    ent, but the preservative cannot be detected by taste.

    (a) What is the probability an individual correctly identifies
         the item with the preservative by chance?
    (b) What is the probability that at least two of the three
         individuals identify the item with the preservative?
    (c) If five individuals make a selection, what is the proba-
         bility at least one will select correctly?

5. The statement LET T = 1 + INT(10*RND) selects an integer from
    1 to 10 with equal probability. If two integers are selected
    independently, what is the probability that

    (a) The integers will both be odd?
    (b) The integers will both be even?
    (c) Of the two integers, one will be odd, and the other even?
    (d) The two integers will be the same?
    (e) The second integer will be larger than the first?

6. If $P(A) > 0$ and $P(\bar{A}) > 0$, show that A and $\bar{A}$ cannot be indepen-
    dent. Note the distinction between disjoint events and inde-
    pendent events.

7. Prove that if A is independent of B, then $\bar{A}$ is independent of
    $\bar{B}$. Use the fact that $P(\bar{A}\bar{B}) = P(\overline{A + B})$ and the definition of
    independence.

8. Let a sample space S consist of binary sequences as indicated:
    S = {100, 010, 001, 111}. A sequence is selected at random.
    The following events are defined:

    > A ≡ the first digit in the sequence is 1.
    > B ≡ the second digit in the sequence is 1.
    > C ≡ the third digit in the sequence is 1.

    Show that A, B, and C are pairwise independent but not mutually
    independent.

9. (a) Let three independent trials of an experiment be made in
         which the probability of success is 2/3. What is the
         probability of three successes?
    (b) Let five independent trials of an experiment be made in
         which the probability of a success is 4/5. What is the
         probability of five successes?
    (c) Let n independent trials be made with success probability
         (n - 1)/n. What is the limiting value of the probability
         of n successes as n → ∞?

10. Consider tossing a "fair" coin until a head occurs. The outcome
    space S = {1, 2, 3, ...}. Assuming the outcomes of the tosses
    to be independent, find the probability of

    (a) One required toss.
    (b) Two required tosses.
    (c) x required tosses.
    (d) Assume the probability the coin shows heads is $0 < p < 1$
         with $q = 1 - p$, find the probability of x required tosses.
         Show that these probabilities sum to 1.

Exercises 2.6

1. Run the program VOTERS as altered in Exercise 2.5.1. What are
    the frequencies of Democrat, Democrat/male, and Democrat/female?
    Compare these frequencies with the corresponding probabilities
    implied by Fig. 2.4.1a.

2. Write a program which selects pairs of integers randomly from
    the set of integers from 0 to 9. Use a loop so that 1000 pairs
    are selected. Compute the frequency of the following outcomes:

    (a) The first integer of the pair is odd.
    (b) The second integer of the pair is odd.
    (c) Both integers in the pair are odd.

    Do the outcomes described by items (a) and (b) appear to be
    independent?

3. Write a program which simulates the number of trials required to
    obtain a head on the toss of a fair coin. Repeat this simulation
    1000 times and compare the frequency of i required tosses with the
    probabilities in Problem 2.6.10. Calculate the average number of
    tosses required. The program must allow only a finite number of
    tosses, of course. Let the maximum number be 20.

4. Write a program as in Exercise 2.6.3 for a coin which shows heads
    with probability p. Run the program using p = 1/3. What is the
    average number of tosses required. Can you guess the theoretical
    average as a function of p?

3

PROPERTIES OF DISCRETE RANDOM VARIABLES

3.1   DISCRETE RANDOM VARIABLES

It is often the case that we naturally associate a number with each
outcome of a random experiment. This idea was illustrated in our
examples of Sec. 2.1.1. The outcomes of the trip to the drive-in
bank were naturally identified with the number of cars in line at
the bank, and the outcomes of the battery purchase were identified
by the number of full years the battery functioned. It was not as
traditional to assign numerical values to the outcome set {male,
female}, but it is clearly equivalent to code a male birth as 1 and
a female birth as 2 and speak of the frequency of the outcome 1 as
the frequency of a male birth. Such an assignment of a number to
each outcome of a theoretical sample space S is called a *random
variable*. In this chapter we will again consider discrete sample
spaces only. Formally we give the definition of a random variable
as

DEFINITION 3.1.1. A *random variable* X is a function which assigns
to every point e_i, in a sample space S, a numerical value $X(e_i)$.

A random variable is often referred to by the letter X alone, but
it is important to remember that X, contrary to notation often

used in mathematics, is a real-valued function defined on S. Much
of formal probability theory is concerned with the properties of
random variables.

As an example, let us consider the theoretical sample space
associated with the "at bats" in three trials of a batter who gets
a hit 1/3 of the time. Using the notation of Sec. 2.6, the sample
space describing the outcome of the three at bats is

S = {HHH, HHO, HOH, OHH, HOO, OHO, OOO}

The random variable X giving the number of hits in the three trials
has values $X(HHH) = 3$, $X(HHO) = X(HOH) = X(OHH) = 2$, $X(HOO) = X(OHO)$
$= X(OOH) = 1$, and $X(OOO) = 0$. Notice that by defining Y to be the
number of outs made in the three at bats we would obtain $Y(HHH) = 0$,
$Y(HHO) = 1$, etc., and in general, $Y(e_i) = 3 - X(e_i)$ for each $e_i \in S$.
Hence more than one random variable may be defined on the same sample
space S, so that one cannot speak of the *random variable* associated
with S. As a second example, consider the 95 Justices of the Supreme
Court who have completed their service on the bench through 1976 as
a theoretical outcome space. A random variable, say L, assigns to
each Justice his length of service on the court in years. For
example, the first Chief Justice would be assigned the value 5, as
he served for 5 years.

Recalling the assignment of probabilities to the sample space
associated with three at bats of the 0.333 hitter, we have seen that
an assignment of probabilities to S implies an assignment of probabilities to the values of the random variable X defined on S. Under
the assumption of independence of the three at bats, the probabilities
of x hits for x = 0, 1, 2, 3 were found to be 8/27, 12/27, 6/27, and
1/27, respectively. In order to find the probability assigned to any
x we use the rule $P(X = x) = \Sigma \, P(e_i)$ for those e_i for which $X(e_i) = x$.

The function $P(X = x)$, often written as $f(x)$, gives the probability the random variable X achieves the value x. From the properties of the probabilities assigned to the points of S, we see that
$\Sigma_{\text{all } x} f(x) = 1$, as is easily checked in the baseball example.
Formally, we state the following.

DEFINITION 3.1.2. The probability function $f(x)$ of a random variable
X gives $P(X = x)$ for all values x which the random variable X may
attain. Such a probability function satisfies

1. $0 \leq f(x) \leq 1$ for all x.
2. $\Sigma_{\text{all } x} f(x) = 1$.

The probability function for the random variable L is given below:

l	f(l)	l	f(l)	l	f(l)	l	f(l)
1	3/95	9	3/95	18	3/95	28	3/95
2	2/95	10	4/95	19	3/95	29	1/95
3	2/95	11	2/95	20	4/95	30	1/95
4	5/95	12	1/95	21	2/95	31	1/95
5	8/95	13	3/95	22	2/95	32	2/95
6	5/95	14	3/95	23	3/95	33	1/95
7	4/95	15	5/95	26	3/95	34	4/95
8	4/95	16	6/95	27	1/95	37	1/95

We see for example that one Justice, William O. Douglas, served for
37 years, while a proportion of $20/95 = 0.210$ served 5 or fewer
years. Simple addition yields $\Sigma_{\text{all } l} f(l) = 1$
    Let us consider random variables associated with the countably
infinite sample space $S = \{0, 1, 2, \ldots\}$. The geometric distribution, as given in Sec. 2.3, defines such a random variable with
probability function $f(k) = (1 - q)q^k$ for $k = 0, 1, 2, \ldots$. There
is a different probability function defined for each value of q,
$0 < q < 1$, with the class of distributions referred to as the geometric family. We can obtain another family by using the fact that
for any $t > 0$,

$$e^t = \sum_{k=0}^{\infty} \frac{t^k}{k!}$$

so that

$$1 = \sum_{k=0}^{\infty} \frac{e^{-t}t^k}{k!}$$

If t > 0, each term in this last infinite series is positive, and it is possible to define a probability function for a random variable on S by $f(k) = e^{-t}t^k/k!$, k = 0, 1, 2, ..., for any value of t > 0. Again there is a family of distributions, one for each value of t > 0. Finally, we might consider a random variable on S for which $X(e_i) = 1$ if e_i is an even integer and $X(e_i) = 0$ if e_i is an odd integer. This is an example in which S is infinite, but X takes on only two values. If S is assigned geometric probabilities with q = 1/2, then the probability function for this random variable is given by f(1) = 2/3 and f(0) = 1/3.

We consider lastly one additional function which is used to describe a random variable.

DEFINITION 3.1.3. The cumulative distribution function of the random variable X is given by

$$F(x) = \sum_{t \leq x} f(t)$$

where f(x) is the probability function for X.

In words, F(x) is the probability that the random variable X has a value of x or less. For discrete random variables the cumulative distribution function has a graph given by a step function which jumps at x by the amount f(x). For the random variable X giving the number of hits in the three trials above, we have

$$F(x) = 0 \quad \text{for } -\infty < x < 0$$

$$F(x) = \frac{8}{27} \quad \text{for } 0 \leq x < 1$$

$$F(x) = \frac{20}{27} \quad \text{for } 1 \leq x < 2$$

$$F(x) = \frac{26}{27} \quad \text{for } 2 \leq x < 3$$

and

$$F(x) = 1 \quad \text{for } 3 \leq x$$

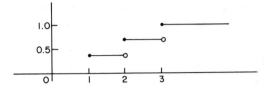

Fig. 3.1.1

The graph of this function is given in Fig. 3.1.1.

It is apparent that the cumulative distribution function of a random variable X satisfies the following properties:

1. $F(x_1) \leq F(x_2)$ for $x_1 < x_2$
2. $\lim_{x \to \infty} F(x) = 1$.
3. $\lim_{x \to -\infty} F(x) = 0$.

The first property states that for $x_1 < x_2$ the probability of X achieving a value of x_1 or less is not greater than the probability of X achieving a value of x_2 or less. The reader is invited to make similar probabilistic interpretations for the other two properties. We turn next to properties of discrete random variables.

Problems 3.1

1. Assume a fair die has one face numbered 1, two faces numbered 2, and three faces numbered 3. Find the probability function f(x) and cumulative distribution function F(x) for the random variable X which gives the number observed on the toss of such a die.

2. Suppose a lot of 10 items has three defectives. Three items are chosen randomly with replacement of the item before another selection. Find the probability function f(x) for the random variable X giving the number of defectives in the three selections.

3. Answer Problem 3.1.2 if the items drawn are not replaced after each selection.

4. Suppose that S consists of all half-open intervals on [0, 1) of form [(k - 1)/1000, k/1000) for k = 1, 2, ..., 1000. Label the kth interval as k and define f(k) = 1/1000 for k = 1, 2, ..., 1000.

   (a) What probability does f(k) assign to the intervals with union [0, 0.230)? [0, 0.500)?
   (b) What probability does f(k) assign to the intervals with union [0, 0.250)? [0.250, 0.500)? [0.75, 1.00)?

5. Assume that a girl has written letters to four boyfriends and has addressed four envelopes with their names. In haste she places the letters into the envelopes in random order. Consider the addressed envelopes to be in positions 1, 2, 3, 4, and let the order of the letters placed into the envelopes be given by a listing of the letters A, B, C, D. For example, ACBD means that boy 2 gets the letter for 3 and boy 3, the letter for 2, while the others receive the correct letters.

   (a) How many listings are possible?
   (b) Let X be a random variable giving the number of boys who receive their letter. Find f(x), the probability function for X.

6. Suppose a fair coin is tossed until a head appears.

   (a) Find the sample space S and a probability function defining the random variable X which gives the number of tosses until the first head.
   (b) Answer part (a) for the random variable Y giving the number of tosses until a 6 occurs on repeated tosses of a fair die.

7. An employee is twice as likely to be absent on Monday and Friday as on the other 3 days of a 5-day work week. Let f(x), x = 1, 2, 3, 4, 5, give the probability of absence on Monday through Friday. Find and graph f(x) and F(x).

8. It is known that $\sum_{k=1}^{\infty} 1/k^2 = \pi^2/6$. Find c so that $f(k) = c/k^2$ is a probability function on $\{1, 2, 3, \ldots\}$.

9. Suppose $f(k)/f(k-1) = 1/k$ for $k = 2, 3, \ldots$, and that $f(k)$ is a probability function on the positive integers.

    (a) Find $f(1)$.

    (b) Find $f(k)$ for $k \geq 1$.

10. For what value or values of x can a probability function satisfy $f(k) = kx^k$ for all positive integers k.

Exercises 3.1

1. Write a program to simulate the outcome of tossing the die described in Problem 3.1.1. Find the frequency of a 1, 2, or a 3 in 1000 repetitions. Compare these frequencies with the probabilities calculated in Problem 3.1.1.

2. Write a program which simulates the situation described in Problem 3.1.5 by programming a method to produce a random ordering of the integers 1, 2, 3, 4. Compare the ordering with the natural ordering 1 2 3 4, and count the number of matches. Run the simulation 1000 times and find the frequency of 1, 2, 3, or 4 matches. Compare these frequencies with the probabilities calculated in Problem 3.1.5.

3.2   THE MOMENTS OF A DISCRETE RANDOM VARIABLE

In this section we wish to consider some important parameters of a discrete random variable with distribution given by the probability function $f(x)$. The parameters serve to describe important character- istics of the random variable X. The first moment of a random vari- able, called the *expected value* of X, written $E(X)$, gives information about an average value of the random variable.

DEFINITION 3.2.1. The expected value of a discrete random variable with probability function f(x) is given by

$$E(X) = \sum_{\text{all } x} xf(x), \text{ if } \sum_{\text{all } x} |x|f(x) \text{ converges}$$

For example the expected number of hits in three at bats by the batter described in Sec. 2.6 would be given by

$$\sum_{x=0}^{3} xf(x) = 0\cdot\frac{8}{27} + 1\cdot\frac{12}{27} + 2\cdot\frac{6}{27} + 3\cdot\frac{1}{27} = 1$$

Similarly, we find $E(L) = 14.7$ in the Supreme Court Justice example. As another example, let a random variable X have probability function $f(x) = 1/n$ for $x = 1, 2, \ldots, n$. This is referred to as the *discrete uniform random variable*. We find

$$E(X) = \sum_{x=1}^{n} \frac{x}{n} = \frac{n(n+1)}{2n} = \frac{n+1}{2}$$

The expected value is often denoted by $\mu = E(X)$.

In order to indicate that this definition of $E(X)$ is not arbitrary, let us consider the random experiment described in the program CARS. The program simulates 1000 visits to a drive-in bank, assuming the number of cars in line is equally likely to be any integer from 1 to 10. The program prints the average number of cars in line in the 1000 trials, which is given by

$$\frac{\text{Total number of cars in line}}{1000} = \frac{1n_1 + 2n_2 + 3n_3 + \cdots + 10n_{10}}{1000}$$

$$= 1f_1 + 2f_2 + 3f_3 + \cdots + 10f_{10}$$

Here n_i is the number of times that i cars are observed in line in the 1000 trials and $f_i = n_i/1000$ is the frequency of i cars in line. The last line states that the average number of cars is a weighted average of the values 1 through 10, where the weights are the frequencies of these numbers. As the theoretical counterpart of frequency is probability, it is reasonable to define $E(X)$ as a weighted

CARS1

```
5 FOR R = 1 TO 10
10 FOR I = 1 TO 10
20 LET F(I)=0
30 NEXT I
40 LET N = 1000
50 FOR I = 1 TO N
60 LET T = 1 + INT(10*RND)
70 LET F(T)=F(T)+1
80 NEXT I
100 LET A = 0
110 LET B =0
120 FOR I = 1 TO 10
130 LET F(I)=F(I)/N
140 LET B = B+I*F(I)
160 LET A = A+F(I)
170 NEXT I
200 PRINT "AVERAGE=";B
210 NEXT R
220 END
*

 RUN
AVERAGE=  5.471
AVERAGE=  5.487
AVERAGE=  5.543
AVERAGE=  5.563
AVERAGE=  5.517
AVERAGE=  5.542
AVERAGE=  5.407
AVERAGE=  5.475
AVERAGE=  5.509
AVERAGE=  5.575

 *
```

average of the values that the random variable X achieves, where the weights are the probabilities assigned to these values by f(x). This is the definition of E(X) given in the preceding paragraph. Ten runs of the simulation CARS gave the following averages: 5.471, 5.487, 5.543, 5.563, 5.517, 5.542, 5.407, 5.475, 5.509, and 5.575. These approximations to the theoretical expectation given in the paragraph above for n = 10, that is (10 + 1)/2 = 5.5. The output of CARS1 shows these approximations.

Let us consider the geometric distribution defined by the probability function

$$f(x) = pq^x \quad x = 0, 1, 2, \ldots$$

for $0 < q < 1$ and $p = 1 - q$. In this case we compute

$$E(X) = \sum_{x=0}^{\infty} xpq^x = pq \sum_{x=1}^{\infty} xq^{x-1}$$

$$= pq \frac{d}{dq} \left(\frac{1}{1 - q} \right) = \frac{pq}{p^2} = \frac{q}{p}$$

We have used the fact that

$$\frac{d}{dq} \left(\frac{1}{1 - q} \right) = \frac{d}{dq} (1 + q + q^2 + \cdots) = 1 + 2q + 3q^2 + \cdots$$

By a slight alteration in the program GEOM we can compute the average life length of 1000 batteries assuming the geometric distribution. The averages obtained in 10 runs using q = 0.5 were 0.948, 0.949, 0.978, 1.027, 0.98, 1.027, 0.956, 0.99, 1.0008, and 1.047. The theoretical expectation is, of course, 1 in this case. We shall later return to the statistical question of how good such an approximation of μ by a calculated average may be.

It is quite natural to define the expected value of a function of a random variable X, say g(X), by

DEFINITION 3.2.2. The expected value of a function g(X) of discrete random variable with probability function f(x) is given by

$$E(g(X)) = \sum_{\text{all } x} g(x)f(x)$$

if the latter converges absolutely.

There are two sets of functions of a random variable which have particularly important expected values. These are X^k, $k = 1, 2, 3$, $\ldots$, and $(X - \mu)^k$, $k = 1, 2, \ldots$. The parameters

$$m_k = E(X^k) = \sum_{\text{all } x} x^k f(x)$$

and

$$\mu_k = E(X - \mu)^k = \sum_{\text{all } x} (x - \mu)^k f(x)$$

are called the kth *moments* and kth *central moments*, respectively, of the random variable X if they exist. It is clear that $m_1 = \mu = E(X)$ and

$$\mu_1 = E(X - \mu) = \sum_{\text{all } x} xf(x) - \sum_{\text{all } x} \mu f(x) = \mu - \mu = 0$$

when E(X) exists. While we shall not continue to repeat this last proviso, it is not an empty one in the case of a random variable with an infinite number of values. For example, if $f(x) = 1/x(x + 1)$ for $x = 1, 2, \ldots$, it can be shown that $\sum_{x=1}^{\infty} f(x) = 1$, but in attempting to compute E(X), we obtain $\sum_{x=1}^{\infty} 1/(x + 1)$ which diverges. On the other hand, the expectation of a real-valued function of a random variable with a finite number of values will always exist.

The importance of the moments of a random variable X lies in the information which these parameters convey about the distribution of the random variable. The expectation E(X) clearly contains information about an average value of a random variable. The second central moment $\mu_2 = E(X - \mu)^2$ conveys information about the variability of a random variable about μ.

DEFINITION 3.2.3. The *variance* of a random variable X is μ_2 = $E(X - \mu)^2$, the second central moment of X, often denoted as σ^2 or Var(X).

Suppose we assume the error X in prediction of the size of a crowd (actual-predicted) is equally likely to achieve any of the values -k, -k + 1, ..., -1, 0, 1, 2, ..., k. It is apparent that E(X) = 0, whatever the value of k. On the other hand, it seems clear that the variation of this random variable will be larger for larger values of k, the assumed maximum absolute error. This is reflected in the following computation of the variance:

$$\sigma^2 = E(X - \mu)^2 = E(X^2) = \frac{1}{2k + 1} \sum_{j=1}^{k} 2j^2 = \frac{k(k + 1)}{3}$$

which clearly increases with k. It may appear that the choice of $(X - \mu)^2$ is quite arbitrary and that $|X - \mu|$ or some other increasing function of $|X - \mu|$ would be equally appropriate. However, the variance has proven to be tractable mathematically and to be useful in applications.

Let us consider three random variables X_1, X_2. X_3 defined on the integers -k, -k + 1, ..., -1, 0, 1, 2, ..., k. Let these be defined by the probability functions

$$f_1(x) = \frac{1}{2k + 1}$$

$$f_2(x) = \frac{|x|}{k(k + 1)}$$

$$f_3(x) = \frac{(k + 1) - |x|}{(k + 1)^2}$$

on the specified set of integers. Graphs of these probability functions are given below in the case k = 2.

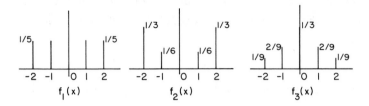

$f_1(x)$ $f_2(x)$ $f_3(x)$

The variance of X_1 has been found to be $k(k + 1)/3$ previously. It would appear that the variance of X_2 would exceed the variance of X_1, while the variance of X_3 would be less, for the same value of k. Calculations yield

$$\sigma^2_{X_2} = E(X_2^2) = 2 \sum_{x=1}^{k} \frac{x^3}{k(k + 1)} = \frac{k(k + 1)}{2}$$

$$\sigma^2_{X_3} = E(X_3^2) = 2 \sum_{x=1}^{k} \frac{(k + 1 - x)x^2}{(k + 1)^2} = \frac{k(k + 2)}{6}$$

It is easy to check that for $k \geq 1$,

$$\frac{k(k + 2)}{6} < \frac{k(k + 1)}{3} < \frac{k(k + 1)}{2}$$

or

$$\sigma^2_{X_3} < \sigma^2_{X_1} < \sigma^2_{X_2}$$

Thus the variance reflects the relative spread of the values of a random variable with respect to μ. Greater probability assigned to large deviations from μ will tend to increase the variance. While other moments of a random variable X can be important, the expectation and variance of X are the most important in applications, so we shall not pursue the properties of the higher-order moments here.

We now discuss some properties of the expectation operator which are implied in Definition 3.2.2. We state these properties in the form of a theorem, leaving most of the proofs as exercises.

THEOREM 3.2.1. If X is a discrete random variable defined on a sample space S, then if $E(g(X))$ and $E((h(X))$ exist, it follows that

1. $E(kg(X)) = kE(g(X))$ for any constant k.
2. $E(g(X) + h(X)) = E(g(X)) + E(h(X))$.

COROLLARY 1. For random variables X_1, X_2, ..., X_k defined on a sample space S, for which $E(X_i)$ exist, i = 1, 2, ..., k,

$$E\left(\sum_{i=1}^{k} X_i\right) = \sum_{i=1}^{k} E(X_i)$$

Proof: This follows by induction using part 2 of Theorem 3.2.1.

COROLLARY 2. For any constant k, $E(k) = k$.

Proof:

$$E(k) = \sum_{\text{all } x} kf(x) = k \sum_{\text{all } x} f(x) = k$$

COROLLARY 3. For any constant k, $Var(k) = 0$.

Proof:

$$Var(k) = E(X - k)^2 = E(k - k)^2 = E(0)^2 = 0$$

COROLLARY 4. $Var(kX) = k^2 Var(X)$.

COROLLARY 5. $Var(X) = E(X^2) - \mu^2$, for $\mu = E(X)$.

The last expression is sometimes helpful in computing the variance of a random variable. For example, if $f(x) = 1/n$ for $x = 1, 2, ...,$ n, we have seen that $E(X) = (n + 1)/2$, and we find

$$E(X^2) = \sum_{x=1}^{n} \frac{x^2}{n} = \frac{(n+1)(2n+1)}{6}$$

Hence

$$Var(X) = \frac{(n+1)(2n+1)}{6} - \left(\frac{n+1}{2}\right)^2$$

$$= \frac{n^2 - 1}{12}$$

To conclude this section we consider an important expectation associated with a discrete random variable X.

DEFINITION 3.2.4. The *moment generating function* (mgf) of a random variable X is defined by $E(e^{Xt}) = m_X(t)$ when this expectation exists.

Note that the mgf is a function of the real variable t. To see that the name is appropriate, consider derivatives of $m_X(t)$ at t = 0.

$m'(t) = E(Xe^{Xt})$ and hence $m'(0) = E(X)$.

$m''(t) = E(X^2 e^{Xt})$ and hence $m''(0) = E(X^2)$.

In general, $m^{(k)}(0) = E(X^k)$, that is, the kth derivative of the mgf at t = 0 is the kth moment of X. To obtain these derivatives, we have assumed that the operations expectation and differentiation could be interchanged and that the required derivatives actually exist. This will be true for the cases in which the mgf is used in this text.

Consider the geometric random variable with $f(k) = pq^k$, k = 0, 1, 2, ..., where p = 1 - q.

$$m_X(t) = \sum_{k=0}^{\infty} p(qe^t)^k = \frac{p}{1 - qe^t} \qquad |qe^t| < 1$$

$$m'_X(0) = \frac{pqe^0}{(1 - qe^0)^2} = \frac{q}{p} = E(X)$$

agreeing with the previous result. Moreover, we find

$$m_X''(0) = \frac{q(1 + q)}{p^2} = E(X^2)$$

Thus the variance of the geometric random variable is given by

$$E(X^2) - (E(X))_r^2 = \frac{q(1 + q)}{p^2} - \frac{q^2}{p^2} = \frac{q}{p^2}$$

We note here two properties of the moment generating function in the following theorem.

THEOREM 3.2.2.

1. $M_{(a+X)}(t) = e^{at}M_X(t)$ for any constant a.
2. $M_{(kX)}(t) = M_X(kt)$ for any constant k.

We prove part 2 and leave part 1 as an exercise.

Proof of part 2:

$$M_{(kX)}(t) = E(e^{(kt)X}) = M_X(kt)$$

We will make use of these properties of the mgf in Chap. 5.

Problems 3.2

1. Consider the fair die with one face numbered 1, two faces numbered 2, and three faces numbered 3. Let X be a random variable representing the outcome of a toss of such a die. Find E(X) and Var(X).

2. Find E(X) for Problems 3.1.2 and 3.1.3.

3. Find the expected number of boys who receive their own letter in Problem 3.1.5.

4. Suppose $f(x) = kx$ for $x = 1, 2, \ldots, N$ is a probability function for a random variable X.

    (a) Find k, E(X).
    (b) Find Var(X).

5. Suppose that visits are made to the drive-in bank with the distribution of the number of cars in line given by $f(x) = 1/10$, $x = 1, 2, \ldots, 10$.

    (a) What is the variance of the number of cars in line?
    (b) What is the expected total number of cars in line in two visits and in k visits?

6. Let a fair die be tossed until a head appears. What is the expected number of tosses?

7. Let $f(x) = 1/(e - 1)x!$ for $x = 1, 2, \ldots$ be the probability function for a random variable X. Find E(X).

8. Let $f(x) = 6/(x\pi)^2$ for $x = 1, 2, \ldots$ be the probability function for a random variable X. What can be said about E(X) and Var(X)?

9. Prove that if X is a discrete random variable defined on a sample space S, then if $E(g(X))$ and $E(h(X))$ exist,

    (a) $E(kg(X)) = kE(g(X))$ for a constant k.
    (b) $E(g(X) + h(X)) = E(g(X)) + E(h(X))$.

10. Using Theorem 3.2.1 prove that

    (a) $Var(X) = E(X^2) - E^2(X)$.
    (b) $Var(kX) = k^2 Var(X)$ for a constant k.

11. Assume that Var(X) exists for a random variable X.

    (a) Prove that $Var(-X) = Var(X)$.
    (b) Prove that $Var(X) \geq 0$.

12. Show by example that in general, $E(g(X)) \neq g(E(X))$ for a function g.

13. Assume that $E(X)$ and $E(Y)$ exist. By considering $E(X + tY)^2$ show that $E^2(XY) \leq E(X^2)E(Y^2)$, assuming these expectations exist.

14. The *covariance* of two random variables X and Y is defined as $Cov(X, Y) = E((X - \mu_X)(Y - \mu_Y))$.

    (a) Prove $Cov(X, Y) = E(XY) - E(X)E(Y)$.

    (b) Prove that if $Cov(X, Y) = 0$, then $Var(X + Y) = Var(X) + Var(Y)$.

15. A fair coin is tossed until head appears. Let X represent the number of tosses required. Find $m_X(t)$ and use it to find $E(X)$.

16. (a) Find the moment generating function of the discrete uniform distribution, that is $f(x) = 1/n$, $k = 1, 2, \ldots, n$.

    (b) Use the result in part (a) to find the expected value and variance of the number showing on the toss of a fair die.

17. Prove part 1 of Theorem 3.2.2.

18. If X is a random variable with a moment generating function $m_X(t)$, show that $m(0) = 1$.

19. Suppose a random variable X has $m_X(t) = e^{-a(1-e^t)}$, $a > 0$. Find $E(X)$ and $Var(X)$.

Exercises 3.2

1. Write a program to simulate the selection of an observation from the distribution described by the probability function

    $$f(x) = \frac{1}{19} \quad \text{for } x = 1, 2, \ldots, 19$$

    Find the average of 1000 selections and compare this with the theoretical value.

2. Write a program to simulate the tossing of a coin until a head occurs. Find the average number of tosses required in 1000 such trials. Compare this average with the expectation found in Problem 3.2.5.

3.3   THE BINOMIAL RANDOM VARIABLE

Let us consider the arrivals of flights of Omega Airlines at O'Hara
Airport. An arrival is considered to be on time if the plane arrives
at its designated gate within 10 min. of its scheduled time. Suppose
that experience has shown that three of four flights arrive on time.
On a given weekday 12 Omega flights are scheduled to arrive at O'Hara.
What can be said about the likelihood of 0, 1, 2, ..., 12 of the
flights arriving on time. We use the program FLIGHTS to simulate the
flight arrivals on 1000 weekdays. The program prints out the frequency
of 0, 1, 2, ..., N on-time flights with the theoretical population
of on-time flights equal to p for the 1000 days. The program has
been run for N = 12 and p = 0.75. Although it is perhaps not surprising, the frequency of on-time arrivals is largest for nine
arrivals (0.263 in the sample run). Additionally, the frequency of
six or more on-time arrivals is quite large (0.985 in the example).
Apparently the number of days with less than six on-time arrivals is
small under the assumptions made here.

The situation described above is a simulation of a probability
model called the *binomial trials model*. We say that n trials are
binomial trials if

1.   Each trial results either in a success or a failure.
2.   The probability of success at any trial is the constant p,
     and hence of failure is 1 - p.
3.   The trials are independent.

The sample space for this theoretical experiment is all sequences of
S's and F's of length n. For example, in the case of n = 3, the
sequences are FFF, SFF, FSF, FFS, SSF, SFS, FSS, and SSS. The binomial random variable B which counts the number of successes in
the three trials has values

$$B(FFF) = 0, \qquad B(SFF) = 1, \qquad ..., \qquad B(SSS) = 3$$

Due to the independence of the trials, the probability of any outcome is a product of p's and q's in which the number of p's is equal

FLIGHTS

```
10 DIM F(50)
20 FOR I = 1 TO 50
30 LET F(I)=0
40 NEXT I
50 READ N, P
60 FOR J = 1 TO 1000
70 LET X = 0
80 FOR I = 1 TO N
90 LET T = RND
100 IF T > P THEN 120
110 LET X = X +1
120 NEXT I
130 LET Y = X +1
140 LET F(Y)=F(Y)+1
150 NEXT J
160 PRINT "NUM ON TIME", "FREQUENCY"
170 FOR I = 1 TO N+1
180 LET A= A+(I-1)*F(I)/1000
190 PRINT I-1, F(I)/1000
200 NEXT I
210 PRINT
220 PRINT "AVERAGE=  "; A
230 DATA 12, .75
240 END
*
 RUN
```

NUM ON TIME	FREQUENCY
0	0
1	0
2	0
3	0.001
4	0.003
5	0.011
6	0.038
7	0.097
8	0.189
9	0.263
10	0.238
11	0.126
12	0.034

AVERAGE= 9.03

*

to the number of S's and the number of q's is equal to the number of
F's in the particular sequence. Hence for n = 3,

$$P(FFF) = q^3$$
$$P(SFF) = P(FSF) = P(FFS) = pq^2$$
$$P(SSF) = P(SFS) = P(SSF) = p^2q$$
$$P(SSS) = p^3$$

Thus in this case the probability function for B is given by $f(0;3,p) = q^3$, $f(1;3,p) = 3pq^2$, $f(2;3,p) = 3p^2q$, and $f(3;3,p) = p^3$. Here $f(x;n,p)$
represents the probability of x successes in n trials with success
probability p at each trial.

In the general case of n trials, we seek the probability function
for the random variable B which counts the number of successes in n
independent trials. We consider all sequences of length n of S's
and F's with exactly x successes. For example,

$$\underbrace{SSSS\cdots S}_{x}\underbrace{FF\cdots F}_{n-x}$$

is such a sequence in which the successes come first. Each such
sequence has probability $p^x q^{n-x}$, and every sequence with x S's and
n - x F's will have the same probability. To find the probability
of x successes we need only count the number of ways in which x S's
can be placed in the n locations $(\underbrace{123\cdots n}{})$, and multiply this number
times $p^x q^{n-x}$. The answer to this question is given by the binomial
coefficient $\binom{n}{x} = n!/x!(n - x)!$, where $0! = 1$ [see Kemeny et al.
(1974)]. Hence the probability function for the random variable B
is given by

$$f(x;n,p) = \binom{n}{x}p^x q^{n-x} \qquad x = 0, 1, 2, \ldots, n$$

If n = 3 and p = 1/3 as in our baseball player example of Sec. 2.6,
we obtain

$$f(0;\ 3,\ \tfrac{1}{3}) = \binom{3}{0}(\tfrac{1}{3})^0(\tfrac{2}{3})^3 = \tfrac{8}{27}$$

$$f(1;\ 3,\ \tfrac{1}{3}) = \binom{3}{1}(\tfrac{1}{3})^1(\tfrac{2}{3})^2 = \tfrac{12}{27}$$

$$f(2;\ 3,\ \tfrac{1}{3}) = \binom{3}{2}(\tfrac{1}{3})^2(\tfrac{2}{3}) = \tfrac{6}{27}$$

$$f(3;\ 3,\ \tfrac{1}{3}) = \binom{3}{3}(\tfrac{1}{3})^3(\tfrac{2}{3})^0 = \tfrac{1}{27}$$

Let us consider some properties of the binomial distribution. Note that the sum of the binomial probabilities is 1 as

$$\sum_{x=0}^{n} f(x;n,p) = \sum_{x=0}^{n} \binom{n}{x} p^x q^{n-x} = (p + q)^n = 1$$

by the binomial theorem. From the definition of $f(x;n,p)$, we obtain

$$f(x;n,p) = \binom{n}{x} p^x (1 - p)^{n-x} = \binom{n}{n-x}(1 - p)^{n-x} p^x$$

$$= f(n - x; n, 1 - p)$$

This expresses the equality of the probability of x successes in n independent trials with success probability p, with the probability of n - x failures in n independent trials with failure probability 1 - p. This allows the calculation of binomial probabilities using values of p on the interval $[0, 0.5]$ only. We find the formula for $E(X)$ when X has the binomial distribution [we write $X \sim B(n,p)$] in the following theorem.

THEOREM 3.3.1. If $X \sim B(n,p)$, then $E(X) = np$.

Proof:

$$E(X) = \sum_{x=0}^{n} x f(x;n,p) = \sum_{x=1}^{n} \frac{xn!}{x!(n - x)!} p^x q^{n-x}$$

$$= np \sum_{x=1}^{n} \frac{(n - 1)!}{(x - 1)!(n - x)!} p^{x-1} q^{n-x}$$

$$= np \sum_{y=0}^{n-1} \binom{n-1}{y} p^y q^{n-1-y} \qquad \text{where } y = x - 1$$

$$= np \sum_{y=0}^{n-1} f(y; n - 1, p) = np$$

The latter summation is 1 because it is the sum of all binomial probabilities in the case of n - 1 trials with success probability p.

The variance of $X \sim B(n,p)$ can be obtained from the definition of Var(X) as E(X) has been obtained here, but the calculation is tedious. Another method which is useful here is based on the idea of indicator random variables.

DEFINITION 3.3.1. An *indicator random variable* I takes on only the values 1 and 0. The indicator variable is 1 when the event "indicated" occurs and 0 otherwise. We define n indicator random variables as follows for $k = 1, 2, \ldots, n$.

$$I_k = \begin{cases} 1 & \text{if the kth binomial trial is a success.} \\ 0 & \text{if the kth binomial trial is a failure.} \end{cases}$$

Here $P(I_k = 1) = p$ and $P(I_k = 0) = q$ for all k. Thus if $X \sim B(n,p)$, we see that

$$X = \sum_{k=1}^{n} I_k$$

Hence

$$E(X^2) = E(\sum_{k=1}^{n} I_k)^2 = \sum_{k=1}^{n} E(I_k^2) + \sum_{k \neq j} E(I_k I_j).$$

Clearly I_k^2 and $I_k I_j$ are also indicator variables with $E(I_k^2) = P(I_k^2 = 1) = P(I_k = 1) = p$. Also, $E(I_k I_j) = P(I_k I_j = 1) = P(I_k = 1)P(I_j = 1) = p^2$, where the independence of the trials has been used. Hence, $E(X^2) = np + n(n - 1)p^2$, and thus

$$\text{Var}(X) = E(X^2) - (np)^2 = np + n(n - 1)p^2 - (np)^2$$
$$= np - np^2 = np(1 - p) = npq$$

The program BINOM calculates the values of $f(x;n,p)$ and the values of the cumulative distribution function $F(x;n,p) = \sum_{t=0}^{x} f(t;n,p)$ for values of n and p requested by the program. It is necessary only to run the program and supply the requested values

```
BINOM

10 PRINT "BINOMIAL";" DISTRIBUTION"
20 PRINT
30 PRINT "WHAT IS THE ";"VALUE OF N"
40 INPUT N
50 PRINT
60 PRINT "WHAT IS THE";" VALUE OF P"
70 INPUT P
80 PRINT
90 PRINT " X SUCCESSES","PROB OF X","CUM PROB OF X"
100 LET Q = 1-P
110 LET Q1= Q↑N
120 LET C = Q1
130 FOR I = 1 TO N+1
140 IF Q1 < 10E-6 THEN 180
150 PRINT I-1,Q1,C
160 IF ABS(C-1)<10E-10 THEN 220
170 IF I > N THEN 210
180 LET Q1=(N-I+1)/I*Q1
190 LET Q1 = Q1*P/Q
200 LET C = C + Q1
210 NEXT I
220 END
*
 RUN
BINOMIAL DISTRIBUTION

WHAT IS THE VALUE OF N
? 12

WHAT IS THE VALUE OF P
? .75
```

X SUCCESSES	PROB OF X	CUM PROB OF X
2	3.54052 E-5	3.76105 E-5
3	3.54052 E-4	3.91662 E-4
4	2.38985 E-3	2.78151 E-3
5	1.14713 E-2	1.42528 E-2
6	4.01495 E-2	5.44022 E-2
7	0.103241	0.157644
8	0.193578	0.351221
9	0.258104	0.609325
10	0.232293	0.841618
11	0.126705	0.968324
12	3.16764 E-2	1

```
*
```

of n and p. The values of p should be chosen from the interval
[0, 0.5] using the relationship f(n - x; n, 1 - p) = f(x;n,p)
for cases for which n > 25 and p > 0.5. Table 3.3.1 compares the
simulated frequency of on-time arrivals in 1000 days with the probability function for X ~ B(12, 0.75) (accurate to three decimal
places). The frequencies approximate the probabilities quite well.
The program FLIGHTS also computes the average number of flights
which are on time in the 1000. days. Ten runs of the program yielded
the values 9.03, 8.914, 9.023, 9.03, 8.99, 8.995, 8.976, 8.989,
9.009, and 8.944, which are approximations to the expected value
np = 12(3/4) = 9.

The binomial probability model has proven to be satisfactory
in a wide variety of situations. One example is the triangle test
which is used to determine whether individuals can detect which of

Table 3.3.1   Simulated Frequency and Probability Function

x	Frequency	f(x; 12, 0.75)
0	0	0.000
1	0	0.000
2	0	0.000
3	0.001	0.000
4	0.003	0.002
5	0.011	0.011
6	0.038	0.040
7	0.097	0.103
8	0.189	0.194
9	0.263	0.258
10	0.238	0.232
11	0.126	0.127
12	0.034	0.032

three similar appearing items has a distinctive feature. For example,
a cheese manufacturer may wish to know if a preservative can be tasted.
Each of n individuals is given three pieces of cheese, only one of
which has the preservative. Each individual is asked to indicate
which of the three pieces of cheese tastes unlike the other two. If
we suppose the trials are independent, then the number of persons out
of n who correctly identify the item with the distinctive feature is
binomially distributed, that is $X \sim B(n,p)$. Here p is the probability an individual can correctly identify the distinctive item.
If the preservative actually cannot be detected, then this probability is 1/3. Hence, in this case, $X \sim B(n,1/3)$.

The program BINOM has been run to yield the binomial distribution for the triangle test situation assuming p = 1/3 with n = 18
and n = 32, respectively. As the reader will note, for n = 18, no
probabilities are printed out for $x \geq 16$. The program is written
so that any value of $f(x;n,p)$ less than 10^{-6} will not be printed
in the output. This prevents the listing of large numbers of very
small probabilities in the case of large n. In order to suggest
how the binomial distribution might be used to decide whether the
preservative can be tasted, consider the case n = 32. If the choices
are random, then we expect n/3 = 10.67 correct decisions. Suppose
in fact the number of successes is 16. If p = 1/3, the probability
of 16 or more correct decisions would be $1 - P(X \leq 15) = 0.047653$.
One is presented with the choice of accepting p = 1/3 and concluding
that a rather unusual event has occurred or deciding that p > 1/3.
A reasonable decision might be for the latter. This discussion is
an example of statistical reasoning, which will be discussed in
detail in the later chapters.

The two runs of BINOM for the triangle test give examples of a
property of binomial probabilities. In the case n = 18, there is a
single maximum probability at x = 6, while for n = 32, the binomial
distribution has two values for which a maximum is attained, namely
x = 10 and x = 11. It can be shown that if p + np is not an integer,
then the binomial probabilities increase to a single maximum at

```
 RUN
BINOMIAL DISTRIBUTION

WHAT IS THE VALUE OF N
? 18

WHAT IS THE VALUE OF P
? .333333

 X  SUCCESSES     PROB OF X          CUM PROB OF X
 0               6.76646  E- 4       6.76646  E- 4
 1               6.0898  F- 3        6.76645  E- 3
 2               2.58816  E- 2       3.26481  F- 2
 3               6.90175  F- 2       0.101666
 4               0.129408            0.231073
 5               0.18117             0.412244
 6               0.196268            0.608512
 7               0.168229            0.776741
 8               0.115657            0.892398
 9               0.064254            0.956652
10               2.89143  E- 2       0.985567
11               1.05143  E- 2       0.996081
12               3.06666  E- 3       0.999147
13               7.07689  E- 4       0.999855
14               1.26373  E- 4       0.999981
15               1.68497  E- 5       0.999998

     *
```

```
    RUN
BINOMIAL DISTRIBUTION

WHAT IS THE VALUE OF N
? 32

WHAT IS THE VALUE OF P
?   .33333333

    X SUCCESSES      PROB OF X           CUM PROB OF X
    0                2.31782 E-6       • 2.31782 E-6
    1                3.70851 E-5         3.94029 E-5
    2                2.8741 E-4          3.26813 E-4
    3                1.43705 E-3         1.76386 F-3
    4                5.2093 F-3          6.97316 E-3
    5                0.014586            2.15592 E-2
    6                3.28186 E-2         5.43778 E-2
    7                6.09488 E-2         0.115327
    8                9.52325 E-2         0.210559
    9                0.126977            0.337536
    10               0.146023            0.483559
    11               0.146023            0.629582
    12               0.12777             0.757353
    13               9.82848 E-2         0.855637
    14               6.66933 F-2         0.922331
    15               0.040016            0.962347
    16               2.12585 E-2         0.983605
    17               0.010004            0.993609
    18               4.16833 F-3         0.997777
    19               1.5357 E-3          0.999313
    20               4.99103 E-4         0.999812
    21               1.42601 E-4         0.999955
    22               3.56502 E-5         0.999991
    23               7.75004 E-6         0.999998
    24               1.45313 E-6         1.
    25               2.32501 E-7         1.

    *
```

```
RUN
BINOMIAL DISTRIBUTION

WHAT IS THE VALUE OF N
? 10

WHAT IS THE VALUE OF P
? .5

 X SUCCESSES       PROB OF X        CUM PROB OF X
 0                 9.76562 E-4      9.76562 E-4
 1                 9.76562 E-3      1.07422 E-2
 2                 4.39453 E-2      5.46875 E-2
 3                 0.117187         0.171875
 4                 0.205078         0.376953
 5                 0.246094         0.623047
 6                 0.205078         0.828125
 7                 0.117187         0.945312
 8                 4.39453 E-2      0.989258
 9                 9.76562 E-3      0.999023
10                 9.76562 E-4      1

*
```

```
  RUN
BINOMIAL  DISTRIBUTION

WHAT IS THE VALUE OF N
?  20

WHAT IS THE VALUE OF P
?  .5

  X  SUCCESSES      PROB OF X        CUM PROB OF X
  0               9.53674  E-7      9.53674  E-7
  1               1.90735  E-5      2.00272  E-5
  2               1.81198  E-4      2.01225  E-4
  3               1.08719  E-3      1.28841  E-3
  4               4.62055  E-3      5.90897  E-3
  5               1.47858  E-2      2.06947  E-2
  6               3.69644  E-2      5.76591  E-2
  7               7.39288  E-2      0.131588
  8               0.120134         0.251722
  9               0.160179         0.411901
 10               0.176197         0.588099
 11               0.160179         0.748278
 12               0.120134         0.868412
 13               7.39288  E-2      0.942341
 14               3.69644  E-2      0.979305
 15               1.47858  E-2      0.994091
 16               4.62055  E-3      0.998712
 17               1.08719  E-3      0.999799
 18               1.81198  E-4      0.99998
 19               1.90735  E-5      0.999999
 20               9.53674  E-7      1

  *
```

$[p + np] = x_0$, where $[x]$ represents the *greatest integer in* x, and then decrease. If $p + np = x_0$ is an integer, then the binomial probabilities increase to two adjacent maximum values at $x_0 - 1$ and x_0 and then decrease.

We describe here several additional examples of real-world situations for which the binomial distribution has proved to be a successful model. In certain circumstances it is possible to employ the model even though one or more of the assumptions are violated.

1. Hybrid Crosses in Genetics

    If we assume that there are two types of genes, G (dominant) and g (recessive), controlling a certain trait, then a hybrid (Gg) carries one of each. In a hybrid cross the offspring are assumed to acquire one gene at random from each parent. Hence the geno-types GG, Gg, and gg among the offspring of such crosses are taken to have probabilities 1/4, 1/2, and 1/4, respectively. As both GG and Gg types display the dominant trait, we expect that the number of offspring X displaying the dominant trait among n such crosses will have a $B(n, 3/4)$ distribution.

2. True-False Test

    A true-false test with n questions can be considered as an example of n binomial trials. The probability of getting a question correct would be taken as the parameter p. The assump-tion of independence of the trials and the equality of the probability of a correct answer for each question might be suspect. However, if a person knows nothing whatever about the questions and arbitrarily answers them true or false, the binomial model with $X \sim B(n, 1/2)$ would be appropriate. We have run BINOM for $p = 0.5$ and $n = 10$ and $n = 20$. In the case of a 10-question test with 70% representing passing, there is a not insignificant probability of 0.172 of such a respondent passing. If the number of questions is increased to 20, the probability of 70% or more correct answers is a more tolerable 0.058. Making the test longer would further reduce the probability of passing

"by guessing." Note, however, that the expected number of
correct answers by guessing is n/2 or 50%.

3. Roulette Wheels

A standard roulette wheel has 38 slots. Thirty-six of these
are numbered randomly with the integers 1, 2, ..., 36, while the
remaining two are numbered 0 and 00. Eighteen of the numbered
slots are colored red and 18 are colored black. If a ball ends
up on a number, then those betting on that number and its color
win the amount they have bet. The other bettors lose to the
house. If the ball ends up in 0 or 00, the house wins. Suppose
we make n \$1 bets on a color. The number of wins in the n trials
has a binomial distribution with p = 18/38 = 9/19. Note that in
n trials the expected winnings are (9/19)n, but the expected
losings are (10/19)n. Hence the average advantage to the house
is n/19.

4. Defectives in a Lot

In order to determine the quality of a large lot of N items a
sample of size n is often taken. The items selected are either
classified as defectives or good. Let X denote the number of
defectives in the lot. A decision to accept the lot or to reject
it may be based on the quality of the *sample*. (An alternative
to rejecting the lot might be a complete evaluation of the
quality of all items in the lot.) The random variable X is often
considered to be $B(n,p)$ with p representing the proportion defective in the lot.

A decision rule might be to accept the lot if $X \leq d$ for d a
fixed constant. Note that we can find the probabilities of accepting a lot as a function of p for any such plan. For example, if
n = 25 and d = 3, we find $P(B(25,p) \leq 3)$. The values of certain of
these probabilities are tabulated in Table 3.3.2. We see that using
this decision rule the probability of accepting a lot with 5% defectives is 0.9659. The probability of accepting a lot with 25% or

Table 3.3.2   Probability as a function of p

p	P(B(25,p) < 3
0.05	0.9659
0.10	0.7636
0.15	0.4711
0.20	0.2340
0.25	0.0962
0.30	0.0332
0.35	0.0097
0.40	0.0024
0.45	0.0005
0.50	0.0001
0.55 → 1.0	0.0000

more defectives is less than 0.0962. The properties of decision
rules is an important aspect of the subject of statistics which we
study in detail in the second half of this text.

In this example the assumptions of the binomial model do not
hold exactly. If one item is removed from a lot, it alters the
probability of obtaining a defective for the next selected item.
Hence the trials are not truly independent. However if N is very
large in relation to n, this dependence will be slight and thus have
little effect upon the distribution of X. We consider this situation in more detail in the next section.

Problems 3.3

1. Assume 10% of the income tax forms received by the IRS are
    audited. Let X represent the number of forms out of the next
    100 processed which will be audited. Assuming independence,
    find E(X) and Var(X).

2. Which is more probable, three heads in six tosses of a fair coin or two heads in four tosses of a fair coin.

3. Assume a fair coin is tossed $2n$ times. Find $f(n;2n,1/2)$, the probability of exactly n heads. Find $f(n + 1; 2n + 2, 1/2)$ and show that this probability is less than $f(n;2n,1/2)$.

4. A basketball player makes 80% of his foul shots. Assuming independence, find the probability of x successes in his next four shots for $x = 0, 1, 2, 3, 4$.

5. Show that $f(n - x; n, 1/2) = f(x;n,1/2)$ and interpret this equality in terms of tossing a coin.

6. Suppose animals undergoing a particular experimental procedure survive with probability 0.8. What is the smallest number of animals with which to start in order to ensure a probability of at least 0.9 that at least five animals survive.

7. Show that the variance of $X \sim B(n,p)$ is greatest for fixed n when $p = 1/2$.

8. Use the fact that $X \sim B(n,p)$ can be written as a sum of indicator variables to prove that $E(X) = np$.

9. If $X \sim B(n,p)$, show that $Z = (X - np)/\sqrt{np(1 - p)}$ satisfies $E(Z) = 0$ and $Var(Z) = 1$.

10. Prove that $f(X;n,p)/f(x - 1; n, p) = (n - x + 1)p/x(1 - p)$. Using this fact show that

    (a) If $np + p = 1$, then $f(0;n,p) = f(1;n,p) > f(2;n,p) > \cdots > f(n;n,p)$.

    (b) If $np + p = k$ an integer greater than 1, then $f(0;n,p) < \cdots < f(k - 1; n, p) = f(k;n,p) > \cdots > f(n;n,p)$.

    (c) If $np + p$ is not an integer, show that for $k_0 = [np + p]$, $f(0;n,p) < \cdots < f(k_0; n, p) > f(k_0 + 1; n, p) > \cdots > f(n;n,p)$.

11. (a) Show that the moment generating function of the binomial
          random variable is $(pe^t + q)^n$.

     (b) Verify that $E(X) = np$ and $VAR(X) = npq$ for $X \sim B(n,p)$
          using the result in part (a).

     Exercises 3.3

1. Run the program FLIGHTS to simulate the number of daily on-time
    arrivals of an airline over a 1000-day period at an airport.
    Assume an on-time arrival rate of 0.80 and 15 flights a day
    arrive at the airport.

2. Run BINOM with the values of $n = 15$ and $p = 0.8$. Compare the
    frequencies observed in the previous exercise with the theoret-
    ical probabilities. Compare the average number of daily arrivals
    in the simulation with the expected number of on-time arrivals.

3. [Answer part (a) before computations.]
    (a) If $n = 16$ and $p = 3/8$ in the binomial distribution, for
         what value or values of x will $f(x;16,3/8)$ be greatest.
         If $n = 29$ and $p = 1/2$, for what value or values of x will
         $f(x;29,1/2)$ be maximum.
    (b) Verify your answers in part (a) by running BINOM with the
         appropriate values of n and p.

4. An assembly line is shut down for repairs if a sample of size
    20 contains two or more defectives. Assume that the sample can
    be treated as 20 independent trials. If $g(p)$ is the probability
    of two or more defective items, when the assembly line is actu-
    ally producing $100p\%$ defectives, find $g(0.05)$, and $g(p)$ for
    $p = k/10$, $k = 1, 2, \ldots, 10$. Plot $g(p)$ as a function of p.

3.4   THE HYPERGEOMETRIC DISTRIBUTION

Let us consider a finite population of N persons of whom R are
Republicans and N - R are Democrats. Suppose that a sample of size
n is taken from this population in such a way that each individual

has equal probability of selection, but once an individual is chosen
he is not considered again. This is called *random sampling without
replacement*. We define the random variable H to be the number of
Republicans in the sample of n. If we consider that there are $\binom{N}{n}$
combinations of size n of the N persons, it is clear that each such
combination has equal probability of selection. In order to compute
the probability of x Republicans, we must find the number of different combinations having exactly x Republicans. This number is
$\binom{R}{x}\binom{N-R}{n-x}$ because the first factor counts the number of combinations
of x Republicans and the second factor counts the number of combinations of n - x Democrats. Hence the probability of exactly x
Republicans is given by

$$\binom{R}{x}\binom{N-R}{n-x} \bigg/ \binom{N}{n}$$

This is an example of the hypergeometric random variable. The probability function for the hypergeometric random variable is defined in
the following.

DEFINITION 3.4.1. From a finite population of size N of which R
elements have a property and N - R do not, we select a random sample
of size n without replacement. The probability function for the
number of elements *in the sample* with a property is given by

$$h(x;N,R,n) = \binom{R}{x}\binom{N-R}{n-x} \bigg/ \binom{N}{n}$$

where

$$\max(0,\ n - N + R) \le x \le \min(R,n)$$

The program HYPER generates the probabilities for the hypergeometric distribution with values of N, R, and n provided in a DATA
statement. An example run illustrates the sampling of 10 individuals
from a finite population of 50 Republicans and 50 Democrats. The
probabilities h(x;100,50,10) are printed out as well as the values
of the cumulative distribution function

$$H(x) = \sum_{t=0}^{x} h(t;100,50,10) \qquad \text{for } x = 0, 1, 2, \ldots, 10$$

It seems intuitively clear that, if the population size N is large in relation to the sample size n, then sampling without replacement should not be very different from sampling with replacement. However, the number of Republicans obtained in a sample of size n, if we sample *with* replacement from a population with a proportion of R/N Republicans, will just be an example of n binomial trials with probability of success R/N at each trial. In fact it can be shown that if $N \to \infty$ and $R \to \infty$ in such a way that R/N = p, then for fixed n,

$$\lim_{\substack{N \to \infty \\ R/N = p}} h(x;N,R,n) = f(x;n,R/N)$$

where $f(x;n,R/N)$ is the probability function for the binomial distribution. The proof of this is sketched in Problem 3.4.4.

To illustrate this idea we compare the hypergeometric probabilities for N = 100, 200, 1000 with R = 50, 100, 500, respectively, with the binomial probabilities in the case of 10 trials with p = 0.5.

x	h(x;100,50,10)	h(x;200,100,10)	h(x;1000,500,10)	f(x;10,0.5)
0	0.00059	0.00077	0.00093	0.00098
1	0.00724	0.00847	0.00953	0.00977
2	0.03799	0.04103	0.04337	0.04395
3	0.11310	0.11529	0.11683	0.11719
4	0.21141	0.20820	0.20570	0.20508
5	0.24933	0.25247	0.24733	0.24609

The probabilities for x > 5 need not be compared as both distributions are symmetric about x = 5. While it may be observed that the hypergeometric probabilities are approaching the binomial probabilities, the approximation is not very satisfactory until the ratio N/n, the population size to sample size, is quite large. In our example the ratio of N/n = 100 provides reasonable accuracy.

HYPER

```
10  READ N, R, N1
20  DIM F(100)
30  PRINT "HYPERGEOMETRIC"; " DISTRIBUTION"
40  PRINT
50  PRINT "X", "PROB OF X", "CUM PROB OF X"
60  LET C = 1
70  FOR I = 0 TO (N1-1)
80  LET C = C*(N-I)/(N1-I)
90  NEXT I
100 LET C = 1/C
110 LET T = 1
120 IF N1<N-R THEN 150
30  LET H = 1
140 GO TO 200
150 LET H = 1
160 FOR J = 0 TO (N1-1)
170 LET H = H*(N-R-J)/(N1-J)
180 NEXT J
190 GO TO 320
200 FOR J = 0 TO N1-N+R-1
210 LET H = H*(R-J)/(N1-N+R-J)
220 NEXT J
230 LET F(N1-N+R)=H*C
240 FOR K=N1-N+R TO R
250 IF K>N1 THEN 440
260 LET F(K+1)=F(K)*(R-K)*(N1-K)/((K+1)*(N-R-N1+K+1))
270 LET L = L+F(K)
280 PRINT K, F(K), L
290 IF ABS(L-1)<10E-8 THEN 440
300 NEXT K
310 GO TO 440
320 LET F(0)=H*C
330 FOR K=0 TO R-1
340 IF K>N1 THEN 370
350 LET F(K+1)=F(K)*(R-K)*(N1-K)/((K+1)*(N-R-N1+K+1))
360 NEXT K
370 FOR K= 0 TO R
380 IF K >N1 THEN 440
3 90 LET L= L+F(K)
400 PRINT K, F(K), L
410 IF ABS(L-1)<10E-8 THEN 440
420 NEXT K
430 DATA 10, 8, 8
440 END
*
```

```
430  DATA 100, 50, 10
RUN
HYPERGEOMETRIC DISTRIBUTION

X                    PROB OF X           CUM PROB OF X
  0                   5.9342 E- 4         5.9342 E- 4
  1                   7.23683 E- 3        7.83024 E- 3
  2                   3.79933 E- 2        4.58236 E- 2
  3                   0.113096            0.15892
  4                   0.211413            0.370333
  5                   0.259334            0.629667
  6                   0.211413            0.84108
  7                   0.113096            0.954176
  8                   3.79933 E- 2        0.99217
  9                   7.23683 E- 3        0.999407
 10                   5.9342 E- 4         1

*
```

Let us consider the expected value and variance of the hypergeometric random variable. It is again helpful to use the idea of indicator random variables. We define n indicator random variables as follows:

$$I_k = \begin{cases} 1 & \text{if a Republican is chosen on the kth trial} \\ 0 & \text{if a Democrat is chosen on the kth trial} \end{cases}$$

for k = 1, 2, ..., n. We write the hypergeometric random variable as

$$H = I_1 + I_2 + \cdots + I_n$$

Consider $P(I_k = 1)$. As there are n elements selected without replacement, the number of distinct sequences of n elements chosen from N is given by $N(N - 1)\cdots(N - n + 1)$, which is denoted as $(N)_n$. This is because the first element can be chosen in N ways, the second in N - 1 ways once the first is chosen, and so forth until the nth choice. We count the number of such sequences with a Republican in the kth place by considering that there are R ways that the kth place may be filled and then $(N - 1)_{n-1}$ ways of filling the remaining n - 1 places. Hence $P(I_k = 1) = R(N - 1)_{n-1}/(N)_n = R/N$. This probability is not dependent on k. Hence

$$E(H) = \sum_{k=1}^{n} E(I_k) = \sum_{k=1}^{n} P(I_k = 1) = \frac{nR}{N}$$

Although the algebra is more extensive, it is straightforward to show, using indicator variables, that

$$Var(H) = n \frac{R}{N} \left(1 - \frac{R}{N}\right) \frac{N - n}{N - 1}$$

Again we see the close correspondence with binomial distribution $X \sim B(n,p)$, for if we write p = R/N, we obtain

$$E(H) = np \quad \text{and} \quad Var(H) = np(1 - p) \frac{N - n}{N - 1}$$

The expectation matches the expectation of the binomial distribution, and if N is very much larger than n, we have $\text{Var}(H) \doteq np(1 - p)$ the binomial variance. The factor $(N - n)/(N - 1)$ is referred to as the finite population correction factor.

That binomial probabilities can be used as good approximations to the corresponding hypergeometric probabilities is of much more than theoretical interest. We often sample from a large population, taking a small fraction of it in a sample of size n. Common examples are polls such as the Gallup Poll, sampling plans used to decide to accept or reject a large lot of items, and catches of fish. In each case we may be interested in the distribution of the number of elements in the sample which have a certain characteristic. For the preceding examples the characteristics could be having voted at the last presidential election, being a defective item, and belonging to the bass family, respectively. N, the population size, is almost never actually known, so hypergeometric probabilities cannot be used. However the sample size n is known, and using an approximation for p, probabilistic statements can be made about X, the number of elements in the sample possessing the characteristic in question.

As another example of the hypergeometric distribution let us consider a college class of size 500 of whom 150 are women. We take a random sample of size 20. If we sample without replacement, the expected number of women in the sample is $E(H) = 20(150/500) = 6$ and the $\text{Var}(H) = 20(0.3)(0.7)(480/499) = 4.040$. If we sample with replacement, the expected number of women is given by $np = 20(0.3) = 6$ and variance $np(1 - p) = 20(0.3)(0.7) = 4.2$, using the expectation and variance of the binomial distribution. The programs HYPER and BINOM have been run to obtain the hypergeometric distribution with $N = 500$, $R = 150$, and $n = 20$ and the binomial distribution $B(20, 0.3)$. Again these should be compared to get an idea of the accuracy of the binomial approximation to the hypergeometric probabilities.

```
430 DATA 500,150,20

RUN
HYPERGEOMETRIC DISTRIBUTION
```

X	PROB OF X	CUM PROB OF X
0	6.74398 E-4	6.74398 E-4
1	6.11237 E-3	6.78677 E-3
2	2.60604 E-2	3.28472 E-2
3	6.94945 E-2	0.102342
4	0.12999	0.232332
5	0.181288	0.413619
6	0.195586	0.609205
7	0.167147	0.776352
8	0.114914	0.891266
9	0.06418	0.955446
10	2.92774 E-2	0.984724
11	1.09273 E-2	0.995651
12	3.33091 E-3	0.998982
13	8.24698 E-4	0.999806
14	1.6422 E-4	0.999971
15	2.58945 E-5	0.999997
16	3.15729 E-6	1.
17	2.8688 F-7	1.

*

```
RUN
BINOMIAL DISTRIBUTION

WHAT IS THE VALUE OF N
? 20

WHAT IS THE VALUE OF P
?   .3

X SUCCESSES      PROB OF X         CUM PROB OF X
0                7.97923 E-4       7.97923 E-4
1                6.83934 E-3       7.63726 E-3
2                2.78459 E-2       3.54831 F-2
3                7.16037 E-2       0.107087
4                0.130421          0.237508
5                0.178863          0.416371
6                0.191639          0.60801
7                0.164262          0.772272
8                0.114397          0.886669
9                6.53696 E-2       0.952038
10               3.08171 E-2       0.982855
11               1.20067 E-2       0.994862
12               3.85928 E-3       0.998721
13               1.01783 E-3       0.999739
14               2.18107 E-4       0.999957
15               3.73898 E-5       0.999995
16               5.00756 E-6       1.
17               5.04964 E-7       1.

    *
```

Problems 3.4

1. Assume that a group consists of 10 persons of whom seven are men and three are women. A committee of three persons is chosen without replacement. Find the probability of x women on the committee for x = 0, 1, 2, 3.

2. The number of registered voters in a city is 100,000. Of these 32% are Republicans, 56% are Democrats, and 12% are independents. If a random sample of 50 is taken without replacement from the list of registered voters, find the expected number $E(X)$ of Republicans and $Var(X)$.

3. (a) If $n \leq N/2$ for even N, show that $h(x;N,N/2,n) = h(n - x; N, N/2, n)$ for x = 0, 1, 2, ..., n.
   (b) If $X \sim h(x;N,N/2,n)$, find $E(X)$ and $Var(X)$.

4. If N and R are very much larger than n, show that
   (a) $h(x;N,R,n) = \binom{n}{x}\dfrac{R}{N}\dfrac{(R - 1)}{N - 1}\cdots\dfrac{(R - x + 1)}{(N - x + 1)}\dfrac{(N - R)}{(N - x)}\cdots$
   $$\frac{(N - R - n + x + 1)}{(N - n + 1)}$$
   (b) If $R \to \infty$ and $N \to \infty$, so that $R/N = p$, show that
   $$\lim_{\substack{N\to\infty \\ R\to\infty}} h(x;N,R,n) = \binom{n}{x}p^X(1 - p)^{n-x}$$
   for x = 0, 1, 2, ..., n.

5. (a) If $X \sim B(10,1/2)$, find σ_X.
   (b) If Y has the hypergeometric distribution, find σ_Y/σ_X for X in part (a) when n = 10 and (1) R = N/2 = 10, (2) R = N/2 = 20, and (3) R = N/2 = 100.

6. (a) Prove directly that
   $$\sum_{j=0}^{R} \binom{R}{j}\binom{N-R}{n-j} = \binom{N}{n}$$
   by comparing the coefficients of the polynomials

$$(1 + t)^R (1 + t)^{N-R} = (1 + t)^N$$

Assume $n \leq R$ and $n \leq N - R$.

(b) What is the importance of this equality?

Exercises 3.4

1. Suppose that a college class of 400 has 300 men and 100 women. In a sample taken without replacement of 20 persons from this class, find the distribution of the number of women.

2. Compare the probabilities in Exercise 3.4.1 with the appropriate binomial probabilities.

3.5   THE POISSON DISTRIBUTION

Another probability distribution of importance related to the binomial distribution is the Poisson distribution which assigns probabilities to the set $S = 0, 1, 2, 3, \ldots$. We consider the binomial trials model in the case that the number of trials $n \to \infty$ while the probability of success $p \to 0$ in such a way that $np = \mu$ is held constant. The binomial probability function is written

$$f(x;n,p) = \binom{n}{x} p^x (1 - p)^{n-x} \qquad x = 0, 1, 2, \ldots, n$$

$$= \frac{n(n - 1) \cdots (n - x + 1)}{x!} \left(\frac{\mu}{n}\right)^x \left(1 - \frac{\mu}{n}\right)^n \left(1 - \frac{\mu}{n}\right)^{-x}$$

$$= \frac{\mu^x}{x!} \left(1 - \frac{\mu}{n}\right)^{-\left(\frac{n}{\mu}\right)(-\mu)} \left(\frac{n}{n} \frac{n - 1}{n} \cdots \frac{n - x + 1}{n}\right) \left(1 - \frac{\mu}{n}\right)^{-x}$$

For fixed x, if $n \to \infty$, the last two terms have limit equal to 1, while the second term has limit $e^{-\mu}$ as $\lim_{z \to 0} (1 + z)^{1/z} = e$. This fact has been used with $z = -\mu/n$. Hence

$$\lim_{\substack{n \to \infty \\ P \to 0}} f(x;n,p) = \frac{e^{-\mu} \mu^x}{x!} \qquad x = 0, 1, 2, \ldots$$

We have already seen in Sec. 3.1 that $\sum_{k=0}^{\infty} e^{-\mu} \mu^x / x! = 1$ for any fixed $\mu > 0$. Hence we can make the following definition:

DEFINITION 3.5.1. The probability function for the Poisson random variable X with parameter $\mu > 0$ is given by $f(x;\mu) = e^{-\mu}\mu^x/x!$, $x = 0, 1, 2, \ldots$. If X has the Poisson distribution with parameter μ, we write $X \sim P(\mu)$.

The expectation and variance of $X \sim P(\mu)$ can be obtained by direct computations using the definitions of expectation and variance.

$$E(X) = \sum_{x=0}^{\infty} x \frac{e^{-\mu}\mu^x}{x!} = \sum_{x=1}^{\infty} \frac{e^{-\mu}\mu^{x-1}}{(x-1)!}$$

and letting $y = x - 1$, we have

$$E(X) = \mu \sum_{y=0}^{\infty} \frac{e^{-\mu}\mu^y}{y!} = \mu$$

This result is very plausible because the expectations of each of the binomial distributions having the Poisson as the limiting distribution were $np = \mu$. To find the variance we find

$$E(X(X-1)) = \sum_{x=2}^{\infty} x(x-1) \frac{e^{-\mu}\mu^x}{x!} = \mu^2 \sum_{x=2}^{\infty} \frac{e^{-\mu}\mu^{(x-2)}}{(x-2)!}$$

and letting $y = x - 2$, we have

$$E(X(X-1)) = \mu^2 \sum_{y=0}^{\infty} \frac{e^{-\mu}\mu^{-y}}{y!} = \mu^2$$

As $Var(X) = E(X(X-1)) + E(X) - E^2(X)$, we find the variance of $X \sim P(\mu)$ as $Var(X) = \mu^2 + \mu - \mu^2 = \mu$. This result is also plausible as the variances of the binomial distributions with the Poisson limit are $np(1-p)$ which has limit $\mu = np$ as $p \to 0$ and $n \to \infty$ with $np = \mu$.

The Poisson distribution should be a good approximation to the binomial distribution when the number of trials n is large and p is small. There are many situations in which this is the case. Consider sending a message which consists of a binary string of n zeros and

ones. We assume that the message received may be altered by noise
in the transmission system, so that one or more of the numbers
received is altered from a 0 to a 1 or a 1 to a 0. We assume that
the frequency of such a transposition is given by p. The program
MESSAGE simulates the sending of 1000 messages of length n. The
proportion of these messages with 0, 1, 2, ... errors is calculated.
The program has been run for n = 100 for two values of p (0.02 and
0.025). The distribution of the number of errors should be well
approximated by the Poisson distribution with $\mu = 2$ and $\mu = 2.5$. It
is only necessary to run the program and supply the values of n and
p as requested.

The program POISSON calculates the probabilities for the Poisson
distribution with expectation μ. Again it is only necessary to supply
the requested value of the expectation μ. The program has been run
for values of $\mu = 2$ and 2.5 to permit comparison with the simulations
of the previous paragraph. It can be shown that the Poisson probabilities increase to a maximum value of $[\mu]$ if μ is not an integer
and then decrease. If μ is a positive integer the Poisson probabilities increase to adjacent maximum values at $x = \mu - 1$ and $x = \mu$ and
then decrease. These properties are illustrated by the two runs of
POISSON for $\mu = 2$ and $\mu = 2.5$ given below.

The program BIPO has been written to compare the binomial probabilities with their corresponding Poisson approximations. It is
only necessary to run the program and answer the questions requiring
the input of p and n. The program prints the binomial probabilities
and the corresponding Poisson probabilities using $\mu = np$. The example
run uses the parameters n = 80 and p = 0.025. As can be seen from
the output, the agreement of the binomial and Poisson probabilities
is excellent.

The Poisson distribution is widely applied as a probability model
to describe the number of "events" of a certain type occurring in a
time period of length t, when the events occur "randomly" during the
time interval. Examples are arrivals of phone calls at a switchboard
during a fixed period of time, arrivals of individuals at a queue for

MESSAGE

```
10 DIM F(101)
20 PRINT "WHAT IS THE ";"ERROR FREQUENCY"
30 INPUT P
40 PRINT "WHAT IS THE ";"MESSAGE LENGTH"
50 INPUT N
60 PRINT "BE PATIENT"
70 FOR I = 1 TO 1000
8 0 LET C = 1
90 FOR J = 1 TO N
100 LET Z = RND
110 IF Z > P THEN 130
120 LET C = C + 1
130 NEXT J
140 LET F(C)=F(C)+1
150 NEXT I
160 LET F1 =0
170 PRINT "ERRORS", "FREQUENCY", "CUM FREQ"
180 FOR I = 1 TO 100
190 LET F = F(I)/1000
2 00 IF F = 0 THEN 230
1🅐 LET F1=F1+F
220 PRINT I-1, F, F1
230 NEXT I
240 END
*
```

```
 RUN
WHAT IS THE ERROR FREQUENCY
? .02
WHAT IS THE MESSAGE LENGTH
? 100
BE PATIENT
```

ERRORS	FREQUENCY	CUM FREQ
0	0.143	0.143
1	0.269	0.412
2	0.267	0.679
3	0.185	0.864
4	0.094	0.958
5	0.031	0.989
6	0.006	0.995
7	0.004	0.999
9	0.001	1

*

```
 RUN
WHAT IS THE ERROR FREQUENCY
? .025
WHAT IS THE MESSAGE LENGTH
? 100

BE PATIENT
ERRORS               FREQUENCY            CUM FREQ
   0                  0.094                0.094
   1                  0.191                0.285
   2                  0.266                0.551
   3                  0.213                0.764
   4                  0.142                0.906
   5                  0.063                0.969
   6                  0.022                0.991
   7                  0.007                0.998
   8                  0.001                0.999
   9                  0.001                1

 *
```

POISSON

```
10 PRINT "POISSON ";"DISTRIBUTION"
20 PRINT "WHAT IS THE ";"VALUE OF MU"
30 INPUT U
40 PRINT
50 IF U > 1 THEN 80
60 LET T = 8
70 GO TO 90
80 LET T=INT(3*U+U)
9 0 LET V = EXP(-U)
1 00 PRINT "X","PROB OF X","CUM PROB OF X"
110 FOR I = 0 TO T
120 LET C = V + C
130 PRINT I,V,C
140 LET V = V*U/(I+1)
1 50 NEXT I
160
170 END
*
```

```
 RUN
POISSON DISTRIBUTION
WHAT IS THE VALUE OF MU
?   2
```

X	PROB OF X	CUM PROB OF X
0	0.135335	0.135335
1	0.270671	0.406006
2	0.270671	0.676676
3	0.180447	0.857123
4	9.02235 E-2	0.947347
5	3.60894 E-2	0.983436
6	1.20298 E-2	0.995466
7	3.43709 E-3	0.998903
8	8.59272 E-4	0.999763

```
*RUN
POISSON DISTRIBUTION
WHAT IS THE VALUE OF MU
? 2.5
```

X	PROB OF X	CUM PROB OF X
0	0.082085	0.082085
1	0.205212	0.287297
2	0.256516	0.543813
3	0.213763	0.757576
4	0.133602	0.891178
5	6.68009 E-2	0.957979
6	2.78337 E-2	0.985813
7	9.94062 E-3	0.995753
8	3.10644 E-3	0.99886
9	8.62901 E-4	0.999723
10	2.15725 E-4	0.999938

*

```
BIPO

10 REM THIS PROGRAM COMPARES THE PROBABILITIES OF THE
20 REM BINOMIAL DISTRIBUTION WITH THOSE OF THE POISSON DISTRIBUTION
30 REM TYPE "RUN" TO EXECUTE THE PROGRAM
40 PRINT "WHAT IS THE";" VALUE OF P?"
50 INPUT P
60 PRINT "WHAT IS THE"; " VALUE OF N?"
70 INPUT N
80 PRINT "HOW MANY";" POISSON";" PROBABILITIES DO"; " YOU WANT?"
90 INPUT C
100 LET Q = 1-P
110 LET Q1=Q^N
120 LET M=N*P
130 LET Q2=EXP(-M)
140 PRINT "  X  ","BINOMIAL PROB","POISSON PROB"
150 FOR I = 1 TO C+1
160 PRINT I-1,Q1,Q2
170 IF I > N THEN 210
180 LET Q1=(N-I+1)/I*Q1
190 LET Q1=Q1*(P/Q)
200 GO TO 220
210 LET Q1=0
220 LET Q2=Q2*M/I
230 NEXT I
240 END
*
 RUN
WHAT IS THE VALUE OF P?
 ?.025
WHAT IS THE VALUE OF N?
 ?80
HOW MANY POISSON PROBABILITIES DO YOU WANT?
 ?10
  X            BINOMIAL PROB   POISSON PROB
 0             0.13194         0.13534
 1             0.27064         0.27067
 2             0.27411         0.27067
 3             0.18274         0.18045
 4             9.01990E-2      9.02235E-2
 5             3.51545E-2      3.60894E-2
 6             1.12675E-2      1.20298E-2
 7             3.05418E-3      3.43709E-3
 8             7.14601E-4      8.59272E-4
 9             1.46585E-4      1.90949E-4
 10            2.66859E-5      3.81898E-5


*
```

a bus in a fixed time interval, the number of suicides in a city during a week, and other similar situations. The following axioms are sufficient to guarantee that the number of events in a time period have a Poisson distribution. We write $P_k(t)$ to mean the probability of k events in a time period of length t.

1. The number of arrivals of events in two nonoverlapping time periods are independent.

2. The probability of the arrival of one event in a short period of time is proportional to the length of the time interval. We write $P_1(h) = \lambda h + o(h)$, where $o(h)$ is a function with the property $\lim_{h \to 0} o(h)/h = 0$.

3. The arrival of more than one event in a short time interval is essentially zero: $P_k(h) = o(h)$ for $k \geq 2$.

These properties are sufficient to guarantee that $P_k(t) = e^{-\lambda t}(\lambda t)^k/k!$, $k = 0, 1, 2, \ldots$. The parameter $\mu = \lambda t$ is the average number of events in a time period of length t. Hence if calls arrive at the rate of 2 per minute, the probability of no calls in 2 min. is e^{-4}. The probability of one call in a 2-min period would be $4e^{-4}$.

Problems 3.5

1. Assume that a typist makes a keyboard error once every 1000 times she strikes the keyboard. If a page consists of 25 lines of 80 characters each, find the expected number of errors per page. Find the probability of 0, 1, 2, 3, or 4 errors per page.

2. Assume that customers arrive in a queue at the rate of one every 5 min.
    (a) What is the probability of zero arrivals in a 5-min period?
    (b) What is the probability of more than four arrivals in a 5-min period?

3. A good outfielder has a fielding percentage of 0.99. What is the probability he handles his next five chances without an error?

4. Suppose that the probability a calculator makes a numerical error
    is 10^{-10}. What distribution can be used to describe the number
    of errors in 10^{10} calculations? What is the probability of more
    than five errors in 10^{10} calculations?

5. Find the moment generating function for the random variable
    $X \sim P(\mu)$.

6. Use the result in Problem 3.5.5 to verify that $E(X) = \text{Var}(X) = \mu$
    for the Poisson distribution.

7. (a) Using the axioms 1 through 3 concerning $P_k(t)$ show that
         $(P_0(t + h) - P_0(t))/h = (-\lambda + o(h)/h)P_0(t)$.
    (b) Using the assumption that $P_0(0) = 1$, prove that $P_0(t) = e^{-\lambda t}$.

8. If $X \sim P(\mu)$, show that $f(x;\mu)/f(x - 1; \mu) = \mu/x$ for $x \geq 1$. Let
    $[\mu] = n$.
    (a) If $n = 0$, show that $f(0;\mu) > f(1;\mu) > \cdots > f(x;\mu) > \cdots$.
    (b) If $\mu = 1$, show that $f(0;\mu) = f(1;\mu) > f(2;\mu) > \cdots >$
         $f(x;\mu) > \cdots$.
    (c) If μ is an integer greater than 1, show that $f(0;\mu) <$
         $f(1;\mu) < \cdots < f(\mu - 1; \mu) = f(\mu;\mu) > f(\mu + 1; \mu) > \cdots$.
    (d) If $[\mu] = n > 1$ but $[\mu] \neq \mu$, then show that $f(0;\mu) < f(1;\mu)$
         $< \cdots < f(n;\mu) > f(n + 1; \mu) > \cdots$.

Exercises 3.5

1. Using an error proportion of $p = 0.01$ run the program MESSAGE to
    simulate the frequency of errors in 1000 messages of 100 charac-
    ters each.

2. Run the program POISSON with the appropriate value of μ to compare
    the frequences observed in Exercise 3.5.1 with the theoretical
    probabilities given by the POISSON model.

3. Use the program BIPO to compare the binomial and Poisson proba-
    bilities in the case that $n = 100$ and $p = 0.01$, $n = 20$ and $p = 0.05$, $n = 10$ and $p = 0.1$, and $n = 5$ and $p = 0.2$. For which
    values of n and p is the Poisson approximation best? worst?

3.6   THE STANDARD DEVIATION AND THE CHEBYSHEV INEQUALITY

We expand upon the meaning of the variance of a random variable as
a measure of the spread or variability of a random variable. Notice
that as $Var(X) = E(X - \mu)^2$, the units of $Var(X)$ will be in terms of
squared units of those used to measure the observations of the random
variable. For example, we used the Poisson random variable to represent the distribution of errors in the transmission of binary
strings of length N. Suppose that $E(X) = \mu = 2$, i.e., the expected
number of errors is 2. We have seen before that $Var(X) = \mu$ for the
variance is $2(errors)^2$. In order to obtain a measure of variability
in the same units as those used to measure X, we state the following:

DEFINITION 3.6.1.   The standard deviation of a random variable X is
$\sqrt{Var(X)}$, when the variance exists. The standard deviation is denoted
σ_X or σ alone if it is clear which random variable is involved.

A theorem due to Chebyshev has great theoretical importance in
probability theory and also importance in giving meaning to the concept of the standard deviation of a random variable X.

THEOREM 3.6.1.   If a discrete random variable X has variance σ^2,
then the following inequality holds for any $\varepsilon > 0$.

$$P(|X - \mu| \geq \varepsilon) \leq \frac{\sigma^2}{\varepsilon^2}$$

Before proving the theorem, let us consider its consequences. If we
let $\varepsilon = k\sigma$ for $k > 0$, we obtain the inequality $P(|X - \mu| \geq k\sigma) \leq 1/k^2$.
The following table gives a short list of these bounds.

k	$1/k^2$
1	1
2	1/4
3	1/9
4	1/16
5	1/25

Thus the probability that an observation of a random variable is 3 or more standard deviations from μ cannot exceed 1/9. Hence the probability is at least 8/9 that an observation will be within 3 standard deviations of μ. In general the probability is at least $1 - 1/k^2$ that an observation is within k standard deviations of μ.

Often the Chebyshev bounds are quite conservative. For example, for the Poisson distribution with $\mu = 1$, we find

k	$P(\lvert X - 1 \rvert \geq k)$	$1/k^2$ *(upper bound)*
1	0.6321	1.0000
2	0.0803	0.2500
3	0.0190	0.1111
4	0.0037	0.0625
5	0.0006	0.0400

Hence the probability of obtaining an observation at least 3 standard deviations from $\mu = 1$ is 0.0190, considerably smaller than the upper bound 0.1111. The strength of the Chebyshev bound lies in the fact that it applies to all random variables which have a variance. The bound cannot be improved upon in general, for if X has distribution $P(X = \pm 1) = 1/2$, then $E(X) = 0$, $Var(X) = 1$, and $P(\lvert X \rvert \geq 1) = 1$. Hence the Chebyshev bound holds as an *equality* in this case for $k = 1$.

A proof of Theorem 3.6.1 proceeds as follows. Let X have variance σ^2 and let $\varepsilon > 0$. Then

$$\sigma^2 = E(X - \mu)^2 = \sum_{|x-\mu|<\varepsilon} (x - \mu)^2 f(x) + \sum_{|x-\mu|\geq\varepsilon} (x - \mu)^2 f(x)$$

$$\geq \sum_{|x-\mu|>\varepsilon} (x - \mu)f(x) \geq \varepsilon^2 \sum_{|x-\mu|\geq\varepsilon} f(x)$$

Hence

$$\sigma^2 \geq \varepsilon^2 P(|X - \mu| \geq \varepsilon)$$

using the first and last expression in this string. Thus

$$P(|X - \mu| \geq \varepsilon) \leq \frac{\sigma^2}{\varepsilon^2}$$

As one further example to illustrate the meaning of the Chebyshev inequality, we consider again the random variable with probability function $f(x) = 1/N$ for $x = 1, 2, \ldots, N$. We have seen that $E(X) = (N + 1)/2$ and $Var(X) = (N^2 - 1)/12$. Consider

$$\frac{N + 1}{2} \pm \sqrt{3}\sigma_x = \frac{N + 1}{2} \pm \frac{\sqrt{(N - 1)(N + 1)}}{2}$$

We have

$$\frac{N + 1}{2} + \sqrt{3}\sigma_x > \frac{N + 1}{2} + \frac{N - 1}{2} = N$$

and

$$\frac{N + 1}{2} - \sqrt{3}\sigma_x < \frac{N + 1}{2} - \frac{N - 1}{2} = 1.$$

Hence the open interval $(N + 1)/2 \pm \sqrt{3}\sigma_x$ contains all the integers 1, 2, $\ldots$, N, so that $P(|X - \mu| \geq \sqrt{3}\sigma_x) = 0$, while the Chebyshev bound would yield 1/3 as an upper bound for this probability. The Chebyshev bound is useful in providing a general interpretation for the meaning of the standard deviation σ and for many theoretical purposes, but it is not useful for calculating probabilities for any particular distribution.

Problems 3.6

1. Suppose a fair coin is tossed 400 times. Find upper bounds for
    the following probabilities, where H represents the number of
    heads in the 400 tosses.
    (a) $P(|H - 200| \geq 10)$
    (b) $P(|H - 200| \geq 20)$
    (c) $P(|H - 200| \geq 30)$

2. Let $f = X/n$ represent the proportion of heads in n tosses of a
    fair coin. Use the Chebyshev inequality to show that

    $$\lim_{n \to \infty} P(|f - \tfrac{1}{2}| \geq \varepsilon) = 0 \qquad \text{for any } \varepsilon > 0$$

3. Let X be a Poisson random variable with $\mu = 4$. Compare
    $P(|X - 4| \geq 2k)$ with the Chebyshev bounds for $k = 1, 2, 3$.

4. If X is a random variable with $P(X = +t) = P(X = -t) = 1/2$ for
    $t \neq 0$, show that
    (a) $P(|X - E(X)| \geq \sigma_X) = 1$.
    (b) $P(|X - E(X)| \geq k\sigma_X) = 0$ for $k > 1$.
    (c) What does part (a) imply about the Chebyshev inequality?

CONTINUOUS OUTCOME SPACES

4.1   INTRODUCTION TO CONTINUOUS OUTCOME SPACES

Until now we have considered outcome spaces which contained at most
a countable number of points. In turn, the random variables defined
on these spaces could have at most a countable number of values.
Many random experiments, however, can be considered to have as outcomes, all possible values on an interval of the real number line.
For example, we may measure the length of time that an individual
takes to respond to a stimulus. For instance, the length of time
it takes a child to answer an arithmetic problem. The response time
is generally assumed to be a real number in an interval of possible
times. Similarly, the net weight of a package of breakfast food
labeled 16 oz net weight may be thought of as a value on an interval
centered at 16. The error made in rounding off a number to the
nearest integer (calculated as actual-integer) can be thought of as
any number on the interval (-0.5, 0.5], where we round down if the
observed number is an integer plus 1/2. Appropriate sample spaces
for these experiments can be taken to be, respectively,

$S_1 = \{t\,|\,0 < t \leq 20\}$, t measured in minutes.

$S_2 = \{w\,|\,14 \leq w \leq 18\}$, w measured in ounces.

$S_3 = \{x\,|\,-0.5 < x \leq 0.5\}$.

Measurements such as length, weight, voltage, speed, and many others
can be thought of as attaining values on an interval of the real-
number line.

The outcome spaces for measurements such as those described in
the previous paragraph are called continuous. Formally, we state

DEFINITION 4.1.1. An outcome space S of a random experiment is
called *continuous* if it is composed of an interval or a union of a
countable number of intervals of the real line.

Often the interval is taken to have infinite length, for example,
$(-\infty,\infty)$ or $(0,\infty)$, even though in reality there is a limit to the size
measurements can realistically attain. For example, the length of
time that a light bulb will burn if it is run until failure is often
considered to have an outcome space $S = \{t \mid t > 0\}$, although physically
we know that the bulb cannot function for more than some maximum time
(say 100,000 hr). Nevertheless such outcome spaces are useful in
describing the outcomes of real-world experiments.

The program ANSWER simulates times required by N children to
complete an arithmetic problem. The program requires the average
time of completion M, and the maximum time allowed which is denoted
as T. In the case considered here, we assume an average time $M = 4$
and a maximum time of $T = 20$. The program has been run for $N = 1000$.
ANSWER prints out the relative frequency of times in the intervals
$(k, k + 1]$ for $k = 0, 1, 2, \ldots, 19$. As one can see from the output
these relative frequencies are not equal. An idea of the distribution of these times can be gotten by considering the graph of the
histogram which indicates the relative frequency for each interval.
For example, the height of the bar associated with the first interval
is 0.223, with the second is 0.160, and with the last is 0.005. The
graph of the histogram is shown in Fig. 4.1.1. The average time for
the 1000 trials was 3.89. However the distribution of times has a
long tail to the right. The relative frequency of a time of 4 min

ANSWER

```
10 DIM F(25)
20 FOR I = 1 TO 20
30 LET F(I)=0
40 NEXT I
50 PRINT "HOW MANY";" SIMULATED ";"TIMES"
60 INPUT N
70 PRINT "WHAT IS MAX ";"TIME"
80 INPUT T
90 PRINT "WHAT IS AVERAGE"; " TIME?"
100 INPUT M
110 FOR I = 1 TO N
120 LET Z = RND
130 LET X =-LOG(1-Z)*M
140 FOR J = 1 TO 20
150 IF X > J*T/20 THEN 170
160 GO TO 180
170 NEXT J
180 LET F(J)=F(J)+1
190 NEXT I
200 LET A=0
210 LET B=0
220 PRINT "FREQUENCY"; " COUNTS FOR ";"INTERVALS"
230 PRINT
240 PRINT "INTERVAL NO.","LOWER END","UPPER END","REL FREQ","CUM FREQ"
250 FOR J = 1 TO 20
260 LET A =A+F(J)*(J-1/2)*T/(20*N)
270 LET B = B+F(J)/N
280 PRINT J,(J-1)*T/20,J*T/20,F(J)/N,B
290 NEXT J
300 PRINT
310 PRINT "AVERAGE=";A
320 END
*
```

```
 RUN
HOW MANY SIMULATED TIMES
? 1000
WHAT IS MAX TIME
? 20
WHAT IS AVERAGE TIME?
? 4
FREQUENCY COUNTS FOR INTERVALS
```

INTERVAL NO.	LOWER END	UPPER END	REL FREQ	CUM FREQ
1	0	1	0.223	0.223
2	1	2	0.16	0.383
3	2	3	0.138	0.521
4	3	4	0.111	0.632
5	4	5	0.093	0.725
6	5	6	0.063	0.788
7	6	7	0.046	0.834
8	7	8	0.047	0.881
9	8	9	0.033	0.914
10	9	10	0.022	0.936
11	10	11	0.012	0.948
12	11	12	0.008	0.956
13	12	13	0.008	0.964
14	13	14	0.008	0.972
15	14	15	0.005	0.977
16	15	16	0.007	0.984
17	16	17	0.005	0.989
18	17	18	0.	0.989
19	18	19	0.006	0.995
20	19	20	0.005	1

```
AVERAGE= 3.889
```

*

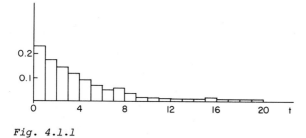

Fig. 4.1.1

or less is 0.632, but a substantial number of children require a long
time to answer the question. Note that the total area of the histogram is 1 as the base of each bar is 1 and the height is the relative
frequency of that interval. Hence the total area is the relative
frequency of all the intervals which is 1.

The program NETWT simulates the net weights of 1000 packages of
a breakfast food. Here the input to the program is the number of
packages N, the true average weight M, and a measure of variation,
the standard deviation. The program prints out the relative frequencies of 17 intervals with the central interval having a midpoint of M.
The program has been run for N = 1000, M = 16, and a standard deviation of 0.5. The intervals all have length 0.25, so that the histogram
of these relative frequencies would have a total area of 0.25. In
order to make the graph comparable with the previous one, we should
have a total area of 1. To accomplish this, each bar of the histogram has a height four times the corresponding relative frequency.
Proceeding in this way the *area* of the bar of the histogram gives
the relative frequency for the interval. Note that the height is no
longer a relative frequency. This histogram appears in Fig. 4.1.2,
which indicates symmetry of the distribution about a central value.
The program also points out the average net weight, which in this
case was 15.9822. This shows close correspondence with the theoretical value, which we would expect for a large number of observations.

The program ERRORS simulates a sample of N rounding errors as
in the first paragraph. The program described has been run for

NETWT

```
10 REM SAMPLE FROM NORMAL
20 DIM F(20)
30 FOR I = 1 TO 18
40 LET F(I)=0
50 NEXT I
60 PRINT "WHAT IS SAMPLE";" SIZE?"
70 INPUT N
80 PRINT "WHAT IS TRUE";" MEAN?"
90 INPUT M
100 PRINT "WHAT IS ";"STD DEV?"
110 INPUT S
120 FOR I = 1 TO N
130 GOSUB 340
140 LET X = M+S*Z
150 FOR J= 1 TO 17
160 IF X > M+(2*J-1)/4*S-4*S THEN 180
170 GO TO 190
180 NEXT J
190 LET F(J)=F(J)+1
200 NEXT I
210 LET A = 0
220 LET B = 0
230 PRINT "FREQUENCY ";"COUNTS FOR ";"INTERVALS"
240 PRINT
250 PRINT "INTERVAL NO.","LOWER END","UPPER END","FREQENCY","CUM FREQ"
260 FOR J= 1 TO 17
270 LET A= A+(F(J)/N)*(M+(J-1)/2*S-4*S)
280 LET B = B+F(J)/N
290 LET L=M+(2*J-3)/4*S-4*S
300 LET R=L+S/2
310 PRINT J,L,R,F(J)/N,B
320 NEXT J
325 PRINT "AVERAGE=";A
330 GO TO 380
340 LET Z1=SQR(-2*LOG(RND))
350 LET Z2=6.2831853*RND
360 LET Z=Z1*COS(Z2)
370 RETURN
380 END
*
```

RUN

WHAT IS SAMPLE SIZE?
? 1000
WHAT IS TRUE MEAN?
? 16
WHAT IS STD DEV?
? .5
FREQUENCY COUNTS FOR INTERVALS

INTERVAL NO.	LOWER END	UPPER END	FREQUENCY	CUM FREQ
1	13.875	14.125	0	0
2	14.125	14.375	0.002	0.002
3	14.375	14.625	0.001	0.003
4	14.625	14.875	0.009	0.012
5	14.875	15.125	0.036	0.048
6	15.125	15.375	0.055	0.103
7	15.375	15.625	0.121	0.224
8	15.625	15.875	0.198	0.422
9	15.875	16.125	0.188	0.61
10	16.125	16.375	0.181	0.791
11	16.375	16.625	0.118	0.909
12	16.625	16.875	0.055	0.964
13	16.875	17.125	0.022	0.986
14	17.125	17.375	0.011	0.997
15	17.375	17.625	0.003	1
16	17.625	17.875	0	1
17	17.875	18.125	0	1

AVERAGE=   15.9822

*

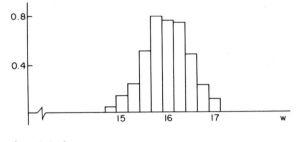

Fig. 4.1.2

N = 1000. ERRORS prints out the relative frequency of observations
in the intervals $(-0.5 + k/10, -0.5 + (k + 1)/10]$ for $k = 0, 1, 2,$
..., 9. As can be seen from the output, in this case the relative
frequency of observations in each of the intervals is close to 0.100
for each interval. In order to have a histogram with total area
equal to 1 we must use a height of ten times the relative frequency
for each interval as the length of each interval is 1/10. Again the
area of a bar of the histogram represents the relative frequency
associated with each interval. A graph of this histogram appears in
Fig. 4.1.3. Note again that the height of a bar of the histogram is
not a relative frequency but a relative frequency multiplied by 10.

It is the properties of the histograms representing relative
frequencies which is mirrored in the formal definition of a probability distribution on an interval of the real line. The first property
is that the relative frequency associated with an interval is given

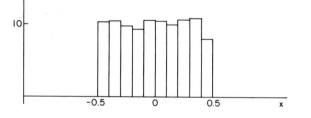

Fig. 4.1.3

ERRORS

```
10 DIM F(25)
20 FOR I = 1 TO 20
30 LET F(I)=0
40 NEXT I
50 PRINT "HOW MANY ";"SIMULATED ";"ERRORS?"
60 INPUT N
70 FOR I = 1 TO N
80 LET Z =RND-.5
90 FOR J= 1 TO 10
100 IF Z>-.5+J/10 THEN 120
110 GO TO 130
120 NEXT J
130 LET F(J)=F(J)+1
140 NEXT I
150 LET A=-.5
160 LET T=1
170 LET B=0
180 PRINT "FREQUENCY"; " COUNTS FOR ";"INTERVALS"
190 PRINT
200 PRINT "INTERVAL NO.","LOWER END","UPPER END","REL FREQ","CUM FREQ"
210 FOR J= 1 TO 10
220 LET A=A+F(J)*(J-1/2)*T/(10*N)
230 LET B = B+F(J)/N
240 PRINT J,-.5+(J-1)*T/10,-.5+J*T/10,F(J)/N,B
250 NEXT J
260 PRINT
270 PRINT "AVERAGE=";A
280 END
*
```

by the area of the histogram above the interval. The second property
is that the total area under the histogram describing relative frequencies is 1 because the total of the relative frequencies is 1.
In the next section we will define continuous probability distributions. The concept of probability will be expressed by area under
a curve. The area under a curve and above an interval will define
the probability associated with the interval. The total area under
the curve, as the total area of the histogram, will be 1.

Exercises 4.1

1. Run the program ANSWER with M = 2, T = 10, and N = 1000, thus
    simulating 1000 times. Plot a histogram by hand for your obser-
    vations so that the area under the histogram is 1.

2. Write a program to simulate N selections of a real number on the
    interval [0, 2] in such a way that intervals of equal length have
    the same probability. Run the program with N = 1000, and plot a
    histogram using intervals of length 0.2 so that the area under
    the histogram is 1.

4.2   CONTINUOUS PROBABILITY MEASURES AND RANDOM VARIABLES

A continuous probability distribution will assign probabilities to a
continuous outcome space S as defined in Sec. 4.1. It is not possible
to proceed as in the case of a discrete outcome space. If an assignment of positive probabilities were made to every point of an interval
$I = \{x \mid a \leq x \leq b\}$ for $b > a$, it can be shown that

$$\sum_{x \, I} P(x) = \infty$$

The interval would be assigned a probability which would be arbitrarily large. Instead we assign probabilities to intervals of the real
number line by means of a nonnegative function $f(x)$. Areas under
$f(x)$ above an interval will represent the probability associated with

the interval. The total area under f(x) must be 1 so that the total probability assigned to the real line is 1. Formally, we state

DEFINITION 4.2.1. A probability model for a continuous outcome space S is defined by a real-valued function f(x) satisfying

1. $f(x) \geq 0$, for all real x.
2. The function f(x) has at most a finite number of discontinuities on any finite interval of the real line.
3. $\int_{-\infty}^{\infty} f(x) \, dx = \int_{S} f(x) \, dx = 1.$

The function f(x) is called a *probability density function* (abbreviated as pdf) and f(x) is *not* a probability.

If $f(x) \geq 0$ is defined only on S satisfying $\int_{S} f(x) \, dx = 1$ and also property 2, the definition of f(x) can be extended to the whole real line by defining $f(x) = 0$ on the complement of S. Hence one can always think of defining f(x) on the whole real line although f(x) may only be positive on a finite interval. Finally, the probability associated with any interval of the real line is *defined* by $\int_{a}^{b} f(x) \, dx$, the area under f(x) above the interval [a, b].

As an example of a class of probability density functions assigning probability to the interval [0, 1], we may consider

$$f(x) = \begin{cases} (k + 1)x^k & 0 \leq x \leq 1 \\ \\ 0 & \text{elsewhere} \end{cases}$$

where we assume that k = 0, 1, 2, ... For each such integer k, a different probability measure is assigned. For k = 0 and k = 1, the graphs of these density functions appear in Fig. 4.2.1. In the case that k = 0, we see that intervals on [0, 1] are assigned a probability equal to the length of the interval as $\int_{a}^{b} 1 \, dx = b - a$. For instance, the probability associated with I = [1/2, 3/4] is 1/4. On the other hand, for k = 1, that is, f(x) = 2x on [0, 1], the probability

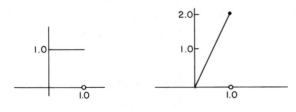

Fig. 4.2.1

associated with the interval I is $\int_{1/2}^{3/4} 2x \, dx = x^2 \Big|_{1/2}^{3/4} = 5/16$.
Another point to note is that for any $x_0 \in [0, 1]$, we have $P(x_0) = \int_{x_0}^{x_0} f(x) \, dx = 0$. The property that points have zero probability
holds for any continuous space S assigned a probability measure by
a probability density function as described in Definition 4.2.1.
Hence for such a probability assignment the intervals (a, b), (a, b],
[a, b), and [a, b] are all assigned the same probability, because
the probability contributed by the endpoints is zero.

As in the discrete case, a random variable is a *function* which
assigns a numerical value to points in S. If S is a continuous outcome space, then the points of S are real values, so that the function
is a real-valued function of a real variable. Perhaps unfortunately,
this function is widely denoted as $X(\omega)$ for $\omega \in S$. Although not
every real-valued function may define a random variable, the excep-
tions are rarely encountered in practice. The most common function
is $X(\omega) = \omega$, the identity function. As before the probabilities
associated with the random variable X are inherited from the probability assignment to S. If $X(\omega) = \omega$, then

$$P(a \le X(\omega) \le b) = P_S(a \le \omega \le b) = \int_a^b f(x) \, dx$$

where $f(x)$ is the probability density function assigning probabilities
to the continuous outcome space S. In this case we say that $f(x)$ is
the probability density function for the random variable X and write
$P(a \le X \le b) = \int_a^b f(x) \, dx$ for any interval [a, b]. For example, the
density function $f(x) = 1$ for [-0.5, +0.5] and $f(x) = 0$ elsewhere

would be the natural pdf for the random variable X representing the
error made in rounding a real number to the nearest integer.

Another important function associated with a probability assignment to a continuous outcome space S is the function

$$F(x) = \int_{-\infty}^{x} f(x) \, dx$$

where $f(x)$ is a pdf. $F(x)$ represents the probability of the event
$\{\omega | \omega \leq x\}$ a subset of S. $F(x) = P(X \leq x)$ and $F(x)$ is called the
cumulative distribution function (cdf) for X as in the discrete case.
Except at the points of discontinuity of $f(x)$, the fundamental
theorem of the calculus yields the relationship

$$F'(x) = f(x)$$

For example, if

$$f(x) = \begin{cases} 1 & \text{in } [0, 1] \\ 0 & \text{elsewhere} \end{cases}$$

the graph of $F(x)$ is given in Fig. 4.2.2. We see that $F(x) = 0$ for
$x < 0$, $F(x) = x$ for $0 \leq x \leq 1$, and $F(x) = 1$ for $x < 1$. We see that
$F'(x) = 0$ for $x < 0$, $F'(x) = 1$ for $0 < x < 1$, and $F'(x) = 0$ for
$x > 1$, while the derivative is not defined at $x = 0$ and $x = 1$.
Hence the density function is given by $F'(x)$ except at the points
of discontinuity of $f(x)$, namely $x = 0$ and $x = 1$.

Let us consider the waiting time T until the first Poisson event
from a time point which we label $t = 0$. The corresponding outcome
space is

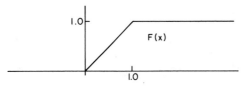

Fig. 4.2.2

$$S = \{t \mid t > 0\}$$

If the rate of occurrence of Poisson events is λ per unit of time, we know that the number of events in a time period of length t has the Poisson distribution with parameter λt. Let $F(t)$ represent the cumulative distribution function for the random variable T. We see that

$$P(T > t) = P(\text{no Poisson events in } [0, t])$$

or

$$1 - F(t) = e^{-\lambda t} \qquad \text{for } t > 0$$

Also $F(t) = 0$ for $t \leq 0$. Hence we can obtain the pdf for T by differentiating $F(t)$ except at $t = 0$. We find

$$f(t) = \begin{cases} \lambda e^{-\lambda t} & \text{for } t > 0 \\ \\ 0 & \text{for } t \leq 0 \end{cases}$$

where the function has arbitrarily been assigned the value of 0 at $t = 0$. That $f(t)$ is a density function can be immediately checked by simple integration. Graphs of the cdf and pdf for T appear in Fig. 4.2.3.

As in the case of a discrete random variable, it is clear that the cumulative distribution function of a continuous random variable satisfies the following properties:

1. $F(x_1) \leq F(x_2)$ for $x_1 < x_2$.

2. $\lim_{x \to \infty} F(x) = 1$.

3. $\lim_{x \to -\infty} F(x) = 0$.

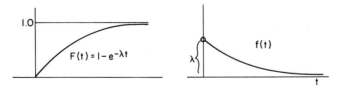

Fig. 4.2.3

These properties are demonstrated by the graphs of the cumulative
distribution functions which appear in Figs. 4.2.2 and 4.2.3. Proofs
of these properties are left to the problems. We also remark that
if two continuous random variables have the same distribution function,
then the corresponding random variables are said to have the same
distribution.

Problems 4.2

1. The probability density function of a random variable X is
    defined as follows:

$$f(x) = \begin{cases} cx^2 & \text{on } [-1,\ 1] \\ 0 & \text{elsewhere} \end{cases}$$

(a) Find the constant c.

(b) Find the cumulative distribution function F(x) of X.

(c) Find $P(-0.5 \le X \le 0.5)$.

2. The cumulative distribution function of the random variable
    X is given by

$$F(x) = \begin{cases} 0 & \text{for } x < 1 \\ 1 - \dfrac{1}{x^3} & \text{for } x \ge 1 \end{cases}$$

(a) Find the pdf for X.

(b) Find $P(X \ge 2)$.

(c) Graph f(x) and F(x).

3. The time required for an individual to complete a task is a
    random variable T (measured in minutes). The pdf for T is
    $f(t) = (1/3)e^{-t/3}$ for $t > 0$ and $f(t) = 0$ elsewhere.

(a) Show that $\int_0^\infty (1/3)e^{-t/3}\, dt = 1$.

(b) Find $P[T \le 3]$.

(c) What is the probability that at least one of two individuals
     chosen at random will complete the task in 3 min or less.

4. Show that the following are probability density functions for
    random variables on the indicated intervals. The functions are
    0 outside these intervals.
    (a) $f(x) = \sin x$ on $(0, \pi/2]$
    (b) $f(x) = 1/(\pi\sqrt{1-x^2})$ on $(-1, 1)$
    (c) $f(x) = 6x(1 - x)$ on $[0, 1]$

5. Find the cumulative distribution functions for the random vari-
    ables of Problem 4.2.4.

6. Assume that a continuous random variable is defined on $(-\infty, \infty)$
    by the pdf $f(x)$. Prove the following:
    (a) $\lim_{x \to \infty} F(x) = 1$, and interpret this statement probabilist-
    ically.
    (b) The requirements on the pdf of a random variable X as defined
    in this section imply that $\lim_{b \to -\infty} P(X \le b) = 0$. Interpret
    this statement probabilistically.

7. The family of exponential distributions is defined by the proba-
    bility density functions:

    $$f(x;\lambda) = \begin{cases} \lambda e^{-\lambda x} & \text{for } x \ge 0 \text{ and fixed } \lambda > 0 \\ \\ 0 & \text{for } x < 0 \end{cases}$$

    (a) Find $P(X > b)$ for a constant b.
    (b) Find $P(X > a + b | X > a)$.
    (c) What family of discrete random variables has the property
    analogous to that in part (b)?

8. Consider two independent selections from the distribution
    described by density function $f(x) = 1$ on $[0, 1]$. Such a selec-
    tion can be thought of as the selection of a point in the square
    $0 \le x \le 1, 0 \le y \le 1$, where the x coordinate represents the
    first selection and the y coordinate the second selection. The

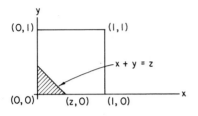

Fig. 4.2.4

area of a subset of this unit square gives the probability of
the subset. We wish to find the pdf for Z = X + Y. The graph
below shows that for $0 < z < 1$, $P[Z \leq z] = z^2/2$ as the area of
the shaded triangle is $z^2/2$.

(a) Find the cdf for Z for $z < 0$, $z > 2$, and $1 < z < 2$.
     [Note: The cdf is not $z^2/2$ on $(1, 2)$.]

(b) Find the pdf for Z.

4.3   MOMENTS OF A CONTINUOUS RANDOM VARIABLE

In the case in which one has a large number of measurements of a
continuous variable, it is customary to classify these observations
in a tabular format called a *frequency table*. The frequency table
is a device for summarizing a large number of observations into a
concise readable format. The interval on which the observations may
fall is divided into nonoverlapping subintervals generally of equal
length. The number of observations in each subinterval and the relative frequency of observations is displayed. The output of the
program ERRORS gives an example of a frequency table. In this example
1000 simulated round-off errors are classified into 10 subintervals
of $(-0.5, 0.5]$. These intervals, $(-0.5 + (k - 1)/10, -0.5 + k/10]$
for $k = 1, 2, \ldots, 10$, each have a length of 0.1 and are nonoverlapping. The corresponding frequency table appears below:

Interval	Boundaries	Class mark	n_i	f_i
1	-0.5, -0.4	-0.45	102	0.102
2	-0.4, -0.3	-0.35	103	0.103
3	-0.3, -0.2	-0.25	95	0.095
4	-0.2, -0.1	-0.15	93	0.093
5	-0.1, 0.0	-0.05	103	0.103
6	0.0, 0.1	0.05	106	0.106
7	0.1, 0.2	0.15	102	0.102
8	0.2, 0.3	0.25	106	0.106
9	0.3, 0.4	0.35	107	0.107
10	0.4, 0.5	0.45	83	0.083

The average value of the 1000 observations is computed from the frequency table as

$$\bar{x} = \sum_{i=1}^{10} x_i f_i$$

The value x_i is the midpoint of the ith subinterval. This value is referred to as the *class mark*. In computing this average all observations in the ith interval are treated as if their true value equaled x_i. If we denote the height of the ith bar of the histogram in Fig. 4.1.3 as h_i and the length as $\Delta \ell_i$, recall that $f_i = h_i \Delta \ell_i$. Hence we can write

$$\bar{x} = \sum_{i=1}^{10} x_i h_i \Delta \ell_i$$

It is this form of the computation of $\bar{x}$ which is mirrored in the definition of the expectation of a continuous random variable X. If we consider observing a very large number of round-off errors, the number of subintervals k can be made large as well. At the same time $\Delta \ell_i = 1/k$ will become small. Hence the expression given above for $\bar{x}$ will be an approximation to the integral of a function with ordinate at x_i of $x_i h(x_i)$. As the probability density function in the theoretical model corresponds to the height of the histogram, it is natural to make the following definition:

DEFINITION 4.5.1.   The expected value of the continuous random variable X with probability density function f(x) is given by

$$E(X) = \int_{-\infty}^{\infty} xf(x) \, dx$$

Technically, we require $\int_{-\infty}^{\infty} |x| f(x) \, dx < \infty$, which will be true in the text except when specifically mentioned.   In those other cases the expectation is said not to exist.

Let us consider the example given in Sec. 4.2 of the family of random variables X_k defined by the probability density functions

$$f_k(x) = \begin{cases} kx^{k-1} & \text{on } [0, 1] \\ \\ 0 & \text{elsewhere} \end{cases}$$

where a different random variable is defined for each positive integer k

$$E(X_1) = \int_0^1 x \cdot 1 \, dx = \frac{1}{2} \qquad E(X_2) = \int_0^1 x \cdot 2x \, dx = \frac{2}{3}$$

and in general,

$$E(X_k) = \int_0^1 kx^k \, dx = \frac{k}{k+1}$$

Before considering another example we introduce a famous integral. The following integral

$$\int_0^{\infty} x^{\alpha-1} e^{-x} \, dx = \Gamma(\alpha)$$

can be shown to converge for $\alpha > 0$ and is called the *gamma function* of the real variable α.   An important property of the gamma function can be demonstrated by letting $u = x^{\alpha-1}$ and $dv = e^{-x} \, dx$ and then integrating by parts.

$$\Gamma(\alpha) = \int_0^{\infty} u \, dv = -x^{\alpha-1} e^{-x} \Big|_0^{\infty} + (\alpha - 1) \int_0^{\infty} x^{\alpha-2} e^{-x} \, dx$$

$$= (\alpha - 1)\Gamma(\alpha - 1) \qquad \text{if } \alpha - 1 > 0$$

It is easy to compute the value of the gamma function for positive integers. First $\Gamma(1) = \int_0^\infty e^{-x} \, dx = -e^{-x} \big|_0^\infty = 1$. Secondly, by repeated integration by parts

$$\Gamma(n) = (n - 1)\Gamma(n - 1) = (n - 1)(n - 2)\Gamma(n - 2)$$
$$= (n - 1)(n - 2)\cdots(2)(1)\Gamma(1) = (n - 1)!$$

Hence the gamma function can be thought of interpolating through the values

$$(1,0!), \ (2,1!), \ (3,2!), \ \ldots, \ (k, \ (k - 1)!), \ \ldots$$

As another example of an expected value computation, consider the random variable T of Sec. 4.2 giving the time until the first Poisson event, where the events have arrival rate λ. The pdf is $f(t) = \lambda e^{-\lambda t}$ for $t > 0$ and $f(t) = 0$ for $t \leq 0$. Hence

$$E(T) = \int_0^\infty \lambda t e^{-\lambda t} dt = \frac{1}{\lambda} \int_0^\infty u e^{-u} du = \frac{\Gamma(2)}{\lambda} = \frac{1}{\lambda}$$

The fact that the expected waiting time until the first event is the reciprocal of the arrival rate coincides with what intuition would suggest.

DEFINITION 4.3.2. If $g(X)$ is a function of the continuous random variable X with pdf given by $f(x)$, then

$$E(g(X)) = \int_{-\infty}^\infty g(x)f(x) \, dx$$

Again we assume the latter integral converges absolutely. As in the discrete case the kth moments and kth central moments are defined, respectively, as

$$m_k = E(X^k) = \int_{-\infty}^\infty x^k f(x) \, dx$$

and

$$\mu_k = E(X - \mu)^k = \int_{-\infty}^\infty (x - \mu)^k f(x) \, dx$$

if these integrals are absolutely convergent. As an example of a
variable for which the expectation is not defined, consider X to
have pdf given by $f(x) = 1/x^2$ for $x \geq 1$ and $f(x) = 0$ for $x < 1$. We
see that

$$\int_1^\infty \frac{dx}{x^2} = -\frac{1}{x}\Big|_1^\infty = 1$$

but when we try to compute $E(X)$, we obtain $\int_1^\infty (1/x)\, dx$ which diverges.
The variance of a continuous variable X is defined by:

DEFINITION 4.3.3. The variance of a random variable X is
$\mu_2 = E(X - \mu)^2$ the second central moment of X.

The properties of random variables given in Theorem 3.2.1 carry
over to the continuous case, almost word for word. The proof is
omitted.

THEOREM 4.3.1. If X is a continuous random variable defined on a
continuous sample space S, then if $E(g(X))$ and $E(h(X))$ exist,

1. $E(kg(X)) = kE(g(X))$ for any constant k.
2. $E(g(X) + h(X)) = E(g(X)) + E(h(X))$.

COROLLARY 1. For any constant k, $E(k) = 0$.

COROLLARY 2. For any constant k, $Var(k) = 0$.

COROLLARY 3. $Var(kX) = k^2 Var(X)$.

COROLLARY 4. $Var(X) = E(X^2) - \mu^2 = m_2 - (m_1)^2$

The definition of the standard deviation of a continuous random variable X is analogous to the definition in the discrete case:

DEFINITION 4.3.4. If a continuous random variable X has variance σ^2_X, the standard deviation of X is defined to be $\sqrt{Var(X)} = \sigma_X$.

The Chebyshev inequality and its implications carry over in a straightforward manner from the discrete case. It is stated in the following:

THEOREM 4.3.2. If a continuous random variable X has variance, then for any $\varepsilon > 0$, the following inequality holds:

$$P(|X - \mu| \geq \varepsilon) \leq \frac{\sigma^2}{\varepsilon^2}$$

Again we obtain the result that $P(|X - \mu| \geq k\sigma) \leq 1/k^2$, yielding a bound on the probability of an observation k or more standard deviations from the theoretical expectation. The proof of the theorem is exactly analogous to the discrete case, with integration replacing summation and is left as an exercise.

As examples of variance calculations let us consider again the family of random variables with density functions $f_k(x) = kx^{k-1}$ on [0, 1] and $f_k(x) = 0$ elsewhere. We find

$$E(X^2_k) = \int_0^1 kx^{k+1}dx = \frac{k}{k + 2}$$

Hence, $\sigma^2_{X_k} = k/(k + 2) - (k/(k + 1))^2 = k/(k + 2)(k + 1)^2$. In the particular case k = 1, that is, $f(x) = 1$ on [0, 1], we obtain $\sigma^2_{X_1} = 1/12$.

For the example of the random variable T representing the waiting time until the first Poisson event, we found

$$f(t) = \lambda e^{-\lambda t} \qquad \text{for } t > 0 \text{ and } E(T) = \frac{1}{\lambda}$$

where λ represented the arrival rate of the Poisson events. We find

$$E(T^2) = \int_0^\infty t^2 e^{-\lambda t} dt = \int_0^\infty u^2 e^{-u} \frac{du}{2} = \frac{\Gamma(3)}{\lambda^2} = \frac{2}{\lambda^2}$$

using the substitution $u = \lambda t$ as before. Hence $\sigma^2_T = 2/\lambda^2 - 1/\lambda^2 = 1/\lambda^2$. The standard deviation of the first arrival time is $1/\lambda$ which is also the expected waiting time. This could not be predicted by intuition.

The response times in the program ANSWER have been generated from a continuous distribution with a probability density function

$$f(t) = \lambda e^{-\lambda t} \qquad \text{for } t > 0$$

The value of $E(t) = 1/\lambda$ is available as input in response to the request for the average response time. The value of $E(T) = 4$ has been used in the sample run of ANSWER. In the output of this sample run let us observe the frequency of observations of times equal to or greater than $16 = E(T) + 3\sigma_T$. We see that this frequency is 0.016. This is far less than the Chebyshev upper bound given by

$$P\left(\left|T - E(T)\right| \geq 3\sigma_T\right) \leq \frac{1}{9} = 0.111$$

However, using the knowledge that $f(t) = (1/4)e^{-t/4}$ for $t > 0$, we find $P(\left|T - 4\right| \geq 12) = P(T \geq 16) = \int_{16}^\infty (1/4)e^{-t/4} dt = -e^{-t/4}\Big|_{16}^\infty = e^{-4} = 0.0183$. This is in substantial agreement with the observed frequency of 0.016. Again it should be noted that the Chebyshev inequality, although a correct probability statement, is of little value for calculating probabilities. For this, knowledge of the probabilistic law generating the observations is required. The question of which probability law has generated a set of numerical observations is a statistical one, which will be addressed in the sequel.

Problems 4.3

1. Find E(X) for random variables with the following pdf's.
    (a) $f(x) = 1/4$ on $[0, 4]$, and $f(x) = 0$ elsewhere.
    (b) $f(x) = 6x(1 - x)$ on $[0, 1]$, and $f(x) = 0$ elsewhere.
    (c) $f(x) = 2e^{-x^2}/\sqrt{\pi}$ for $x \geq 0$, and $f(x) = 0$ elsewhere.

2. Find the Var(X) for the random variables with the pdf's given
    in parts (a) and (b) of Problem 4.3.1.

3. If X is a continuous random variable for which E(X) exists and
    the pdf of X is symmetric about 0 (that is, $f(-x) = f(x)$ for
    all x), prove E(X) = 0. Hint: $E(X) = \int_{-\infty}^{0} xf(x)\ dx + \int_{0}^{\infty} xf(x)\ dx.$

4. Let X have pdf $f(x) = 1$ on $[0, 1]$ and $f(x) = 0$ elsewhere. Find
    (a) $E(\sin\pi x)$.
    (b) m_k for k = 1, 2, 3, ...
    (c) μ_k for k = 1, 2, 3, ...

5. Prove Corollaries 1-4 of Theorem 4.3.1.

6. Let X have pdf $f(x) = x$ on $[0, 1]$ and $f(x) = 2 - x$ on $[1, 2]$.
    Find E(X) and Var(X).

7. Let X have pdf $f(x) = e^{-x}$ for $x \geq 0$ and $f(x) = 0$ elsewhere.
    $E(X) = \sigma_X = 1$. Find $P(|X - \mu| \geq k\sigma_X)$ for k = 1, 2, 3, 4 and
    compare these probabilities with the Chebyshev upper bounds.

8. Let X have pdf $f(x) = 1/\pi(1 + x^2)$.
    (a) Find F(0), where F is the cdf for X.
    (b) What can be said about E(X)? (See note to Definition
         4.5.1.)

9. If the pdf of a random variable X is $f(x) = 2/x^3$ for $x \geq 1$ and
    $f(x) = 0$ for $x < 1$, show that E(X) exists and that Var(X) does
    not exist. Find E(X).

10. Prove the Chebyshev inequality for a continuous random variable
    with variance σ^2. Hint: $\sigma^2 = \int_{|x-\mu|<\varepsilon}(x - \mu)^2 f(x)\ dx +$
    $\int_{|x-\mu|\geq\varepsilon}(x - \mu)^2 f(x)\ dx.$

4.4   SPECIAL CONTINUOUS RANDOM VARIABLES

In this section we consider several cases of particular families of random variables which have important applications in theory and practice.

4.4.1   *The Uniform Random Variable*

The uniform random variable takes on values on an interval [a, b] of the real line in such a way that each interval of [a, b] is assigned a probability in proportion to the length of the interval. The probability density function for this class of random variables is defined to be

$$f(x) = \begin{cases} \dfrac{1}{b - a} & \text{for x on } [a, b] \\ \\ 0 & \text{elsewhere} \end{cases}$$

By integration we find the cumulative distribution function to be given by

$$F(x) = \frac{x - a}{b - a} \quad \text{on } [a, b]$$

$$F(x) = \begin{cases} 0 & \text{for } x < a \\ \\ 1 & \text{for } x > b \end{cases}$$

In Fig. 4.4.1 graphs of $f(x)$ and $F(x)$ are presented. The expectation of this random variable is obtained in a straightforward manner by integration.

$$E(X) = \int_a^b \frac{x}{b - a} \, dx = \frac{a + b}{2}$$

Similarly we find $Var(X) = (b - a)^2/12$. The uniform distribution is important in describing observations which occur "randomly" on an interval of the real line, such as errors in rounding numbers, the value in radians of an angle selected "at random" on $[0, \pi/2]$, the value of a real number selected by a "spinner" on a circle of unit

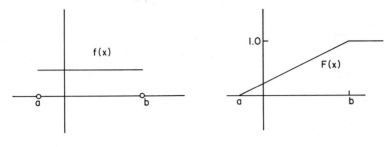

Fig. 4.4.1

circumference, and other similar selections. The uniform distribution on [0, 1] has particular practical importance which we shall see subsequently.

4.4.2   *The Gamma Distribution*

The probability density function for the family of gamma distributions is given by

$$f(x;\alpha,\beta) = \begin{cases} \dfrac{x^{\alpha-1}e^{-x/\beta}}{\Gamma(\alpha)\beta^{\alpha}} & \text{for } x > 0, \ \alpha > 0, \text{ and } \beta > 0 \\[2ex] 0 & \text{elsewhere} \end{cases}$$

By letting $v = x/\beta$, we see that

$$\int_0^\infty f(x;\alpha,\beta) \ dx = \frac{1}{\Gamma(\alpha)} \int_0^\infty v^{\alpha-1}e^{-v}dv = \frac{\Gamma(\alpha)}{\Gamma(\alpha)} = 1$$

The expectation and variance of the gamma distribution are easy to calculate.

$$E(X) = \frac{1}{\Gamma(\alpha)} \int_0^\infty x^{\alpha}e^{-x/\beta}dx = \frac{\beta}{\Gamma(\alpha)} \int_0^\infty v^{\alpha}e^{-v}dv$$

$$= \beta\frac{\Gamma(\alpha + 1)}{\Gamma(\alpha)} = \alpha\beta\frac{\Gamma(\alpha)}{\Gamma(\alpha)} = \alpha\beta$$

Similarly the variance can be shown to be $\alpha\beta^2$. The family of exponential distributions are gamma distributions with

$$\alpha = 1 \quad \text{and} \quad \beta = 1/\lambda$$

and hence pdf

$$f(x;\lambda) = \begin{cases} \lambda e^{-\lambda x} & \text{for } x > 0 \\ \\ 0 & \text{otherwise} \end{cases}$$

As an example of the usefulness of the gamma distribution let us consider the waiting time T until the second Poisson event, when the events occur with arrival rate λ.

$$P(T > t) = P(0 \text{ or } 1 \text{ events in } [0, t])$$
$$1 - F(t) = e^{-\lambda t} + (\lambda t)e^{-\lambda t}$$

and hence

$$F(t) = 1 - e^{-\lambda t} - (\lambda t)e^{-\lambda t}$$

By differentiation we find

$$f(t) = \lambda e^{-\lambda t} - \lambda e^{-\lambda t} + \lambda^2 t e^{-\lambda t} = \lambda^2 t e^{-\lambda t}$$

This is the pdf for a gamma random variable with $\beta = 1/\lambda$ and $\alpha = 2$. Hence the expected waiting time until the second Poisson event is $\alpha\beta = 2/\lambda$, which coincides with intuition. It can be shown that the pdf of the waiting time T_k until the kth Poisson event is

$$f_{T_k}(t) = \frac{\lambda^k t^{k-1} e^{-\lambda t}}{\Gamma(k)} \quad \text{for } t > 0$$

Hence T_k has the gamma distribution with $\alpha = k$ and $\beta = 1/\lambda$.

The program WAIT simulates N times until the kth Poisson event in the case that the Poisson events occur with arrival rate λ. The number of times N, the arrival rate λ, the value of k, and an upper bound on the interval on which the times are recorded are input in response to questions asked by the program. The program outputs the frequency table of the resulting times. It is only necessary to run the program and to answer the questions asked. A sample run is shown in the case that $\lambda = 1$, k = 2, and N = 1000. The actual distribution

WAIT

```
10 DIM F(25)
20 FOR I = 1 TO 20
30 LET F(I)=0
40 NEXT I
50 PRINT "HOW MANY ";"WAITING TIMES?"
60 INPUT N
70 PRINT "WHAT IS THE";" ARRIVAL RATE?"
80 INPUT L
90 LET M = 1/L
100 PRINT "WAIT UNTIL KTH "; "EVENT."; " WHAT IS K?"
110 INPUT K
120 PRINT "WHAT IS "; "MAX WAIT?"
130 INPUT T
140 FOR I = 1 TO N
150 LET X = 0
160 FOR J = 1 TO K
170 LET Z = RND
180 LET X = -LOG(1-Z)*M +X
190 NEXT J
200 FOR J = 1 TO 20
210 IF X > J*T/20 THEN 230
220 GO TO 240
230 NEXT J
240 LET F(J)=F(J)+1
250 NEXT I
260 LET A =0
270 LET B=0
280 PRINT "FREQUENCY ";"COUNTS FOR ";"INTERVALS"
290 PRINT
300 PRINT "INTERVAL NO.","LOWER END","UPPER END","REL FREQ","CUM FREQ"
310 FOR J = 1 TO 20
320 LET A = A +F(J)*(J-1/2)*T/(20*N)
330 LET B = B +F(J)/N
340 PRINT J,(J-1)*T/20,J*T/20,F(J)/N,B
350 NEXT J
360 PRINT
370 PRINT "AVERAGE=",A
380 END
*
```

```
RUN
HOW MANY WAITING TIMES?
? 1000
WHAT IS THE ARRIVAL RATE?
? 1
WAIT UNTIL KTH EVENT. WHAT IS K?
? 2
WHAT IS MAX WAIT?
? 20
FREQUENCY COUNTS FOR INTERVALS
```

INTERVAL NO.	LOWER END	UPPER END	REL FREQ	CUM FREQ
1	0	1	0.271	0.271
2	1	2	0.349	0.62
3	2	3	0.198	0.818
4	3	4	0.106	0.924
5	4	5	0.046	0.97
6	5	6	0.02	0.99
7	6	7	0.006	0.996
8	7	8	0.002	0.998
9	8	9	0.001	0.999
10	9	10	0	0.999
11	10	11	0	0.999
12	11	12	0.001	1
13	12	13	0	1
14	13	14	0	1
15	14	15	0	1
16	15	16	0	1
17	16	17	0	1
18	17	18	0	1
19	18	19	0	1
20	19	20	0	1

```
AVERAGE=        1.916
```

*

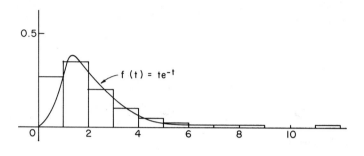

Fig. 4.4.2

is a gamma distribution with $\alpha = 2$ and $\beta = 1$. A histogram of the resulting times appears in Fig. 4.4.2. The corresponding probability density function for the gamma distribution $f(t) = te^{-t}$ for $t > 0$ is also graphed in Fig. 4.4.2. The correspondence between the histogram, which has area 1, and the probability density function $f(t;2,1)$ for the gamma distribution is evident from the figure.

4.4.3   The Normal Distribution

A very important distribution, widely used in practice, is the normal distribution. The normal distribution is not a single distribution, but a two-parameter family of distributions. The probability density function for the normal random variable is

$$f(x;\mu,\sigma^2) = \frac{1}{\sqrt{2\pi}\sigma} e^{-(x-\mu)^2/2\sigma^2} \qquad -\infty < x < \infty,\ \sigma > 0$$

In order to show that the area under this curve is 1, the transformation $z = (x - \mu)/\sigma$ is made and one obtains

$$\int_{-\infty}^{\infty} f(x;\mu,\sigma^2)\ dx = \frac{1}{\sqrt{2\pi}} \int_{-\infty}^{\infty} e^{-z^2/2} dz$$

The latter integral can be shown to be 1, but the proof requires the calculus of more than a single variable. We shall accept this fact

here (while including the method of proof in the problems). To
compute $E(X)$, the transformation above is made yielding

$$E(X) = \int_{-\infty}^{\infty} xf(x;\mu,\sigma^2)\ dx = \frac{1}{\sqrt{2\pi}} \int_{-\infty}^{\infty} (\mu + \sigma z)e^{-z^2/2}dz$$

$$= \mu - \sigma \frac{e^{-z^2/2}}{\sqrt{2\pi}} \Bigg|_{-\infty}^{\infty} = \mu$$

Similarly the variance is calculated as

$$Var(X) = \int_{-\infty}^{\infty} (x-\mu)^2 f(x;\mu,\sigma)\ dx = \frac{\sigma^2}{\sqrt{2\pi}} \int_{-\infty}^{\infty} z^2 e^{-z^2/2}dz$$

Letting $u = z$ and $dv = ze^{-z^2/2}$, we obtain

$$Var(X) = \frac{\sigma^2}{\sqrt{2\pi}} \int_{-\infty}^{\infty} z^2 e^{-z^2/2}dz = \sigma^2 \left[-z\frac{e^{-z^2/2}}{\sqrt{2\pi}} \Bigg|_{-\infty}^{\infty} + \int_{-\infty}^{\infty} \frac{e^{-z^2/2}}{\sqrt{2\pi}}\ dz \right]$$

The first term in parentheses in the last expression is 0 and the
last term is 1. Thus $Var(X) = \sigma^2$. The choice of the letters μ and
σ^2 as the names of the parameters is thus justified.

The moment generating function of the normal random variable
with parameters μ and σ^2 is given by

$$M_X(t) = e^{t\mu + t^2\sigma^2/2}$$

We leave the proof of this fact to the problems. The particular
normal distribution for which $\mu = 0$ and $\sigma^2 = 1$ is known as the
standard normal variable. It is denoted by Z and has pdf

$$f(z) = \frac{1}{\sqrt{2\pi}} e^{-z^2/2} \qquad -\infty < z < \infty$$

and

$$M_Z(t) = e^{t^2/2}$$

The following theorem indicates the reason for the importance of the standard normal variable.

THEOREM 4.4.1. If X is normal with expectation μ and variance σ^2, then $(X - \mu)/\sigma$ is a standard normal variable.

Proof: Using the properties of the moment generating function, we have

$$M_{\frac{X-\mu}{\sigma}}(t) = e^{-\mu t/\sigma} M_X\left(\frac{t}{\sigma}\right) = e^{-\mu t/\sigma} e^{\mu t/\sigma + t^2/2} = e^{t^2/2}$$

Hence $(X - \mu)/\sigma$ has the mgf of a standard normal variable and therefore is distributed as a standard normal variable. We write $(X - \mu)/\sigma \sim Z$.

The importance of the standard normal distribution lies in the fact that the probabilities associated with *any* normal distribution may be computed from the probabilities given by the standard normal distribution. For example, if $X \sim N(\mu, \sigma^2)$ (X is normally distributed with parameters μ and σ^2), consider the probability that X has a value on [a, b]:

$$P(a \leq X \leq b) = \int_a^b f(x; \mu, \sigma^2)\, dx$$

If we make the transformation $Z = (X - \mu)/\sigma$, we find

$$P(a \leq X \leq b) = \int_{\frac{(a-\mu)}{\sigma}}^{\frac{(b-\mu)}{\sigma}} \frac{e^{-z^2/2}}{\sqrt{2\pi}}\, dz = F_Z\left(\frac{b - \mu}{\sigma}\right) - F_Z\left(\frac{a - \mu}{\sigma}\right)$$

Here $F_Z(z) = \int_{-\infty}^z f(z; 0, 1)\, dz$ is the cumulative distribution function of the standard normal variable. Thus probabilities concerning any normal distribution can be reduced to corresponding probabilities associated with the standard normal distribution.

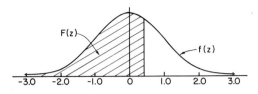

Fig. 4.4.3

The values of $F_Z(z)$, the cumulative distribution function for the standard normal distribution, are available in App. B.I. or can be obtained by using the program Z-PROB. It is only necessary to input the value of z requested and the computer will calculate $F_Z(z)$. There is no expression for $F_Z(z)$ in terms of elementary functions, but this fact does not make the interpretation of the normal distribution more difficult than any other continuous distribution. Figure 4.4.3 gives the graph of the probability density function for the standard normal variable. The area under this curve as a function of z, that is,

$$\int_{-\infty}^{z} \frac{e^{-z^2/2}}{\sqrt{2\pi}}\, dz = F_Z(z)$$

provides a geometric interpretation of $F_Z(z)$.

As an example of the use of the normal distribution, let us assume that scores on the Scholastic Aptitude Test (SAT) are normally distributed with $\mu = 500$ and $\sigma = 100$. What proportion of the population taking this exam have scores of 630 or less? We write

$$P(X \le 630) = P(\frac{(X - \mu)}{\sigma} \le \frac{630 - 500}{100})$$
$$= P(Z \le 1.3) = F_Z(1.3)$$

The value $F_Z(1.3) = 0.9032$, and hence 630 is roughly at the 90th percentile of scores on this exam. Suppose we wish to know the score x^* corresponding to the 67th percentile on this exam.

Z-PROB

```
10 REM PROGRAM GIVES LOWER TAILED NORMAL PROBABILITY FOR
20 REM Z VALUE INPUT BY USER
30 PRINT "WHAT IS THE ";"VALUE OF Z"
40 INPUT Z
50 LET Z1= ABS(Z)
60 LET C=1/SQR(2)
70 LET C1= .14112821
80 LET C2= .08864027
90 LET C3=.02743349
100 LET C4= -.00039446
110 LET C5=.00328975
120 DEF FNZ(X)=1-1/(1+C1*X+C2*X^2+C3*X^3+C4*X^4+C5*X^5)^8
130 LET P = .5+.5*FNZ(Z1*C)
140 LET P = 1E-4*(INT(1E4*P))
150 IF Z > 0 THEN 170
160 LET P = 1-P
170 PRINT "Z","PROB < Z ";"IN STANDARD ";"NORMAL DIST"
180 PRINT Z,P
190 END
*RUN
WHAT IS THE VALUE OF Z
 ?1
Z                   PROB < Z IN STANDARD NORMAL DIST
 1                      0.8413


*RUN
WHAT IS THE VALUE OF Z
 ?2
Z                   PROB < Z IN STANDARD NORMAL DIST
 2                      0.9772


*RUN
WHAT IS THE VALUE OF Z
 ?0
Z                   PROB < Z IN STANDARD NORMAL DIST
 0                      0.5


*
```

$$0.67 = P(X \leq x^*) = P(\frac{(X - 500)}{100} \leq \frac{(x^* - 500)}{100})$$

Thus

$$P(Z \leq \frac{(x^* - 500)}{100} = F_Z(\frac{(x^* - 500)}{100}) = 0.67$$

However as $F_Z(0.44) = 0.67$ (from App. B.I.) we must have $(x^* - 500)/100 = 0.44$ or $x^* = 544$.

Another important property of the normal distribution is given by the following equality for $X \sim N(\mu, \sigma^2)$.

$$P(|X - \mu| \leq k\sigma) = P(|Z| \leq k) \quad \text{for any } k \geq 0$$

Hence the probability of an observation of a normal variable falling within k standard deviations of μ is the same for all normal distributions. Table 4.4.1 gives these probabilities for $X \sim N(\mu, \sigma^2)$ for several values of k and the corresponding Chebyshev (lower) bounds.

An observation in a normal distribution will be outside the interval $[\mu - 3\sigma, \mu + 3\sigma]$ with probability only 0.0026. Such events are unusual. The program NORSAMP samples observations from a normal distribution with values of μ and σ requested by the program. A run with N = 10,000 observations (with $\mu = 0$ and $\sigma = 1$) produced 44 or a frequency of 0.0044 observations outside the interval $[-2.75, 2.75]$. This compares with a theoretical probability of 0.0066. This is additional evidence that observations more than 3σ units from μ are

Table 4.4.1 Probabilities for $X \sim N(\mu, \sigma^2)$

k	$P(|X - \mu| \leq k\sigma)$	Bound $(1 - 1/k^2)$
1	0.6826	0.0000
2	0.9544	0.7500
3	0.9974	0.8889
4	0.99994	0.9375

NORSAMP

```
10 REM SAMPLE FROM NORMAL
20 DIM F(20)
30 FOR I = 1 TO 18
40 LET F(I)=0
50 NEXT I
60 PRINT "WHAT IS SAMPLE";" SIZE?"
70 INPUT N
80 PRINT "WHAT IS TRUE";" MEAN?"
90 INPUT M
100 PRINT "WHAT IS ";"STD DEV?"
110 INPUT S
120 FOR I = 1 TO N
130 GOSUB 340
140 LET X = M+S*Z
150 FOR J= 1 TO 17
160 IF X > M+(2*J-1)/4*S-4*S THEN 180
170 GO TO 190
180 NEXT J
190 LET F(J)=F(J)+1
200 NEXT I
210 LET A = 0
220 LET B = 0
230 PRINT "FREQUENCY ";"COUNTS FOR ";"INTERVALS"
240 PRINT
250 PRINT "INTERVAL NO.","LOWER END","UPPER END","FREQENCY","CUM FREQ"
260 FOR J= 1 TO 17
270 LET A= A+(F(J)/N)*(M+(J-1)/2*S-4*S)
280 LET B = B+F(J)/N
290 LET L=M+(2*J-3)/4*S-4*S
300 LET R=L+S/2
310 PRINT J,L,R,F(J)/N,B
320 NEXT J
325 PRINT "AVERAGE=";A
330 GO TO 380
340 LET Z1=SQR(-2*LOG(RND))
350 LET Z2=6.2831853*RND
360 LET Z=Z1*COS(Z2)
370 RETURN
380 END
*
```

```
 RUN
WHAT IS SAMPLE SIZE?
? 10000
WHAT IS TRUE MEAN?
? 0
WHAT IS STD DEV?
? 1
FREQUENCY COUNTS FOR INTERVALS
```

INTERVAL NO.	LOWER END	UPPER END	FREQUENCY	CUM FREQ
1	-4.25	-3.75	0	0
2	-3.75	-3.25	0.0004	0.0004
3	-3.25	-2.75	0.0015	0.0019
4	-2.75	-2.25	0.0091	0.011
5	-2.25	-1.75	0.0295	0.0405
6	-1.75	-1.25	0.0642	0.1047
7	-1.25	-0.75	0.1248	0.2295
8	-0.75	-0.25	0.1751	0.4046
9	-0.25	0.25	0.1931	0.5977
10	0.25	0.75	0.1766	0.7743
11	0.75	1.25	0.1214	0.8957
12	1.25	1.75	0.0642	0.9599
13	1.75	2.25	0.0284	0.9883
14	2.25	2.75	0.0092	0.9975
15	2.75	3.25	0.0021	0.9996
16	3.25	3.75	0.0003	0.9999
17	3.75	4.25	0.0001	1.

```
AVERAGE= -0.00275
```

*

unusual in normal populations and observations more than 4σ units from μ are very rare indeed. (The probability is 0.00006.)

The importance of the normal distribution lies in the fact that many measurements of continuous random variables, such as crop yields, lengths, heights, and voltages, can be considered to be approximately normally distributed. Secondly, it is often possible to apply a real-valued transformation to measurements so that the resulting transformed values have a normal distribution. Additionally, the mathematical theory of the normal random variable has been highly developed, so that the properties of normal random variables are well understood. Also, as we shall see, averages of measurements of random variable can often be thought of as normal random variables, even though the measurements themselves are not normally distributed.

Problems 4.4

1. Prove that for X uniformly distributed on [a, b], Var(X) = $(b - a)^2/12$.

2. Suppose that θ is uniformly distributed on $[0, \pi/4]$. Find $E(\cos \theta)$.

3. Show that the variance of a gamma random variable is $\alpha\beta^2$. Hint: Find $E(X^2)$.

4. A particular example of the gamma distribution is the chi-square distribution with parameter n. The parameters $\alpha = n/2$ and $\beta = 2$. Find the expectation and variance of a chi-square variable with parameter n (called the *degrees of freedom*).

5. Graph the pdf for the gamma distribution for $\alpha = 1$ and $\beta = 1$, 2, 3 on the same axes.

6. Use the fact that $\int_0^\infty e^{-v^2} dv = \sqrt{\pi}/2$ to show that $\Gamma(1/2) = \sqrt{\pi}$.

7. Using the fact that $\Gamma(1/2) = \sqrt{\pi}$, show that $\int_{-\infty}^\infty (e^{-z^2/2}\sqrt{2\pi}) \, dz = 1$.

8. Show that the moment generating function of $X \sim N(\mu, \sigma^2)$ is given by $e^{\mu t + (\sigma t)^2/2}$

9. Verify that $E(X) = \mu$ and $Var(X) = \sigma^2$ for the normal distribution using the result in Problem 4.4.8.

10. For the standard normal variable Z, find the following probabilities: (a) $P(Z \leq 1.22)$, (b) $P(1.00 \leq Z \leq 2.00)$, (c) $P(-0.25 \leq Z \leq 0.5)$, and (d) $P(Z \geq 1.28)$.

11. Assume that heights of men are distributed as $X \sim N(70, \sigma^2 = 4)$.
    (a) What proportion of male heights exceed 6 ft?
    (b) What proportion of male heights are less than 66 in.?
    (c) What height is at the 80th percentile of these heights?

12. In the standard normal distribution find the values c satisfying
    (a) $P(Z \geq c) = 0.025$
    (b) $P(-c \leq Z \leq c) = 0.95$
    (c) $P(Z \leq c) = 0.05$

13. If IQ scores X are considered to be normally distributed with $\mu = 100$ and $\sigma = 15$, find the following probabilities:
    (a) $P(85 \leq X \leq 115)$
    (b) $P(100 \leq X \leq 130)$
    (c) $P(X \geq 100)$
    (d) $P(X \geq 137.5)$

14. Consider the integral

$$\int_0^\infty \int_0^\infty e^{-(x^2+y^2)/2} \, dx \, dy = I^2$$

where $I = \int_0^\infty e^{-x^2/2} \, dx$. Change the integral to polar coordinates to evaluate I^2. Prove that

$$\frac{1}{\sqrt{2\pi}} \int_{-\infty}^\infty e^{-x^2/2} \, dx = 1$$

(This problem requires the multivariate calculus.)

Exercises 4.4

1. Use the program NORSAMP to select a sample of the ages of legis-
    lators which are assumed to be normally distributed with $\mu = 50$
    and $\sigma = 4$. Use $N = 2000$ and plot a relative frequency histogram
    for the sample obtained.

2. Write a program to sample N observations from the uniform distri-
    bution on $[a, b]$. The values of a, b, and N should be requested
    by the program as input. The output should include a frequency
    table of the sample. The number of classes should also be input
    by the user. The critical command is LET X = A + (B - A)*RND.

3. Run the program WAIT with data $\lambda = 1$, $k = 3$, and $N = 1000$ to
    simulate the waiting times until the third Poisson event which
    occur with a rate of 1 per time period. Use a maximum wait of
    20. Construct a histogram of these times. What pdf describes
    the distribution of these waiting times?

4.5   TRANSFORMATIONS OF CONTINUOUS RANDOM VARIABLES

Suppose we consider a continuous random variable X defined on a
sample space S by the probability density function $f(x)$. Consider
$Y = r(X)$, a continuous real-valued function defined on the range of
X. Let us take R_Y to be the range of $Y = r(X(s))$. It is clear that
Y is a random variable defined on S with values in R_Y given by $Y =
r(X(s))$, which is again a real-valued function on S. If $[c, d]$ is
an interval contained in R_Y, then

$$P(c \leq Y \leq d) = P(c \leq r(X) \leq d)$$
$$= P(\omega | c \leq r(X(\omega)) \leq d) \qquad \text{for } \omega \in S$$

so that the probability distribution of the random variable Y is
defined via the original probability distribution on S. For example,
if X is the random length of the edge of a cube, then $Y = r(X) = X^3$
is a random variable giving the volume of the cube. We shall generally be interested in obtaining the probability density function and
cumulative distribution function of Y from a knowledge of the pdf and
cdf of X.

As an example, suppose that a cube has a side of length X, where X is a random variable with probability density function

$$f(x) = \begin{cases} 1 & \text{on } [0, 1] \\ 0 & \text{elsewhere} \end{cases}$$

It is clear that the pdf of $Y = X^3$ is positive only on $[0, 1]$. The cdf for X is

$$F(x) = \begin{cases} 0 & \text{for } x > 0 \\ 1 & \text{for } x > 1 \end{cases}$$

and

$$F(x) = x \qquad \text{for } 0 \leq x \leq 1$$

Hence the cdf for Y may be written as $G(y) = P(Y \leq y) = P(X^3 \leq y) = P(X \leq y^{1/3}) = y^{1/3}$, on $(0, 1)$, as the function x^3 is an increasing one-to-one function mapping $[0, 1]$ onto $[0, 1]$. Hence

$$G(y) = \begin{cases} 0 & \text{for } y < 0 \\ 1 & \text{for } y > 0 \end{cases}$$

and

$$G(y) = y^{1/3} \qquad \text{for } 0 \leq y \leq 1$$

By differentiation we obtain the pdf g(y) for Y. $G'(y) = g(y) = 0$ if $y < 0$ or $y > 1$, and $G'(y) = g(y) = (1/3)y^{-2/3}$ for $0 \leq y \leq 1$. Note that $E(Y) = 1/3 \int_0^1 y^{1/3} dy = 1/4$. This is not equal to $(E(X))^3 = (\int_0^1 x \, dx)^3 = 1/8$.

In order to emphasize this example, let us consider the program CUBE which simulates the selection of N cubes with edge X, where X is assumed to be uniformly distributed on $[0, 1]$. The program computes the edge x and volume x^3 of each cube. The average length of an edge and average volume of the N cubes is calculated. Notice that these values are close to 0.5 and 0.25 in the example for which $N = 10{,}000$. This coincides with our previous theoretical results.

CUBE

```
10 REM SIMULATES THE SELECTION OF N CUBES WITH EDGE UNIFORMLY
20 REM DISTRIBUTED (0,1)
25 PRINT "HOW MANY CUBES?"
30 INPUT N
40 LET A = 0
50 LET B =0
60 FOR I = 1 TO N
70 LET S = RND
80 LET V = S^3
90 LET A = A +S
100 LET B = B +V
105 NEXT I
110 PRINT "NO. OF CUBES","AVERAGE SIDE","AVERAGE VOL."
130 PRINT N,A/N,B/N
200 END
*
 RUN
HOW MANY CUBES?
 ?1000
NO. OF CUBES    AVERAGE SIDE    AVERAGE VOL.
 1000              0.49247         0.24788


*
 RUN
HOW MANY CUBES?
 ?10000
NO. OF CUBES    AVERAGE SIDE    AVERAGE VOL.
 10000             0.49555         0.24795


*
```

The method used in this example can be used to find the pdf of the random variable $Y = r(X)$ which is a continuous increasing function of X, where we suppose $f(x) > 0$ on $a < x < b$. The method is stated in the following theorem which is stated without proof.

THEOREM 4.5.1. Suppose a continuous random variable X has a pdf $f(x)$ which is positive on $a < x < b$ and that $r(x)$ is strictly increasing (or decreasing) on $a \leq x \leq b$. If the function $r(x)$ is differentiable on $a < x < b$, then the pdf for $Y = r(X)$ is given by

$$g(y) = f(r^{-1}(y)) \left| \frac{dr^{-1}(y)}{dy} \right| = f(x(y)) \left| \frac{dx}{dy} \right|$$

for $r(a) < y < r(b)$ [or for $r(a) > y > r(b)$].

As an example, consider the radius of a sphere to be uniformly distributed on $[0, a]$. We find the cdf and pdf of the surface area of the sphere. The radius X has pdf $f(x) = 1/a$ on $[0, a]$. The surface area $Y = r(X) = 4\pi X^2$, an increasing function on $[0, a]$. Hence the pdf of Y is $f(x(y)) \, d(\sqrt{y}/2\sqrt{\pi})/dy$ on $(0, 4\pi a^2]$, that is, $(1/a)(1/4\sqrt{\pi y})$ on $(0, 4\pi a^2]$. To see that this is a density function we integrate $g(y)$ on $(0, 4\pi a^2]$:

$$\frac{1}{4a\sqrt{\pi}} \int_0^{4\pi a^2} (\sqrt{y})^{-1} dy = \frac{1}{4a\sqrt{\pi}} \left. 2\sqrt{y} \right|_0^{4\pi a^2} = \frac{4\sqrt{\pi}a}{4\sqrt{\pi}a} = 1$$

We find

$$E(Y) = \frac{1}{4a\sqrt{\pi}} \int_0^{4\pi a^2} \sqrt{y} \, dy = \frac{4\pi a^2}{3}$$

Note that $E(Y) \neq 4\pi(E(X))^2 = 4\pi(a/2)^2 = \pi a^2$. However we can use the pdf of X to obtain $E(Y) = 4\pi E(X^2) = (4\pi/a) \int_0^a x^2 \, dx = 4\pi a^2/3$. We shall now turn to a general statement in this regard.

Suppose the random variable X is defined on a continuous sample space S. We have seen that $Y = r(X)$ for a continuous real-valued function is again a random variable. In several situations we have seen that the probability density function of Y, $g(y)$, can be obtained from the pdf of X, say $f(x)$. We can then calculate

$$E(Y) = \int_{R_Y} y g(y) \, dy$$

if this expectation exists. It is also possible to calculate $E(r(X))$ using the density function $f(x)$:

$$E(r(X)) = \int_{-\infty}^{\infty} r(x) f(x) \, dx$$

For example, in the case of a cube with edge uniformly distributed on $[0, 1]$ with volume $Y = X^3$ we have found $E(Y) = 1/4$. If we calculate $E(X^3) = \int_0^1 x^3 \, dx = x^4/4 \big|_0^1 = 1/4$, so we obtain the same result. In general we have the following statement. If X has a pdf $f(x)$ and $Y = r(X)$ is any function of the type met in practice, then $E(Y) = \int_{-\infty}^{\infty} r(x) f(x) \, dx = E(r(X))$. This result is often useful because it means that the pdf of Y need not be found in order to calculate $E(Y)$. For example, the cube described above has surface area $6X^2$ and thus the expected surface area is

$$E(6X^2) = \int_0^1 6x^2 \, dx = 2$$

Problems 4.5

1. Assume that the edge of a square has side X uniformly distributed on $[0, 1]$. Find the pdf for $Y = X^2$, the area of the square. Evaluate $E(Y)$ using the pdf of Y and also by finding $E(X^2)$ using the pdf of X.

2. Assume that an angle X is uniformly distributed on $[0, \pi/2]$. Find the pdf for $Y = \sin X$ on the interval $0 < y < 1$. Show that this pdf integrates to 1.

3. Let a circle have radius X, uniformly distributed on $[0, a]$. Find the pdf for $Y = \pi X^2$, the area of this circle. Find $E(Y)$ and $\text{Var}(Y)$.

4. If X is uniformly distributed on $[a, b]$, show that $Y = (X - a)/(b - a)$ is uniformly distributed on $[0, 1]$. Using the fact that $X = a + (b - a)Y$ and the expectation and variance of a uniformly distributed variable on $[0, 1]$, show that $E(X) = (a + b)/2$ and $\text{Var}(X) = (b - a)^2/12$.

5. If X is uniformly distributed on $[0, 1]$, show that $Y = -\ln X$ has the exponential distribution with $\lambda = 1$.

6. Assume X has a pdf satisfying $f(x) = f(-x)$ for all X. Consider the transformation $Y = X^2$ (which is not strictly decreasing or increasing). Using the equality

$$P(Y \le y) = P(X^2 \le y) = P(-\sqrt{y} \le X \le \sqrt{y})$$

show that $g(y) = f(\sqrt{y})/\sqrt{y}$ for $0 < y < \infty$, where $g(y)$ is the pdf of Y.

7. Using the result of Problem 4.5.6 show that if Z is standard normal, then Z^2 has the gamma distribution with $\alpha = 1/2$ and $\beta = 2$. (Note: This distribution is also the chi-square distribution with $n = 1$.)

8. Assume X has pdf $f(x)$ for all x and $E(X)$ and $\text{Var}(X)$ exist. Show that $Y = aX + b$ for $a > 0$ has pdf $f((y - b)/a)/a$. Show $E(Y) = aE(X) + b$ and $\text{Var}(Y) = a^2\text{Var}(X)$.

9. If X has a pdf such that $f(x) = f(-x)$ for all x, show that $Y = |X|$ has pdf $2f(y)$ for $y > 0$.

10. Using Problem 4.5.9 show that if $Z \sim N(0,1)$ then for $Y = |Z|$ we obtain $g(y) = \sqrt{2/\pi} \, e^{-y^2/2}$ for $y > 0$. Find $E(|Z|)$.

Exercise 4.5

1. Write a program which simulates the edge and the area of 2000
    squares with edge X uniformly distributed on [0, 1]. Find the
    average length of an edge of the squares and the average area.
    Do these results correspond to the theoretical expectations
    found in Problem 4.5.1?

4.6   GENERATING RANDOM SAMPLES FROM CONTINUOUS DISTRIBUTIONS

It is often desirable to generate a large number of observations
from a particular continuous distribution. A number of the programs
mentioned in the chapter, e.g., ANSWER, ERRORS, and WAIT perform
exactly this function. It is an interesting fact that this can be
accomplished numerically in a large number of cases by sampling from
the uniform distribution on [0, 1] and then transforming these observations by a particular function. This is based on the following:

THEOREM 4.6.1. Assume X is a continuous random variable with a
strictly increasing cumulative distribution function $F(x)$ on an
interval [a, b], $(-\infty, \infty)$, [a, ∞), or $(-\infty, b]$. Assume U has the
uniform distribuiton on [0, 1]. Then $F^{-1}(U)$ has the cumulative
distribution function $F(x)$ on the appropriate interval.

   Proof: We must show that $P(F^{-1}(U) \leq x) = F(x)$. As F is strictly
increasing, it has an inverse, say F^{-1}, on its range. On the one
hand, as U is uniform $P(U \leq F(x)) = F(x)$. On the other hand, as F
has an inverse, $P(U \leq F(x)) = P(F^{-1}(U) \leq F^{-1}(F(x))) = P(F^{-1}(U) \leq x)$.
Hence

$$P(F^{-1}(U) \leq x) = F(x)$$

and $F^{-1}(U)$ has the specified cdf.

Thus in order to take a random sample of size n from a distribution with strictly increasing cumulative distribution function, it is only necessary to sample n observations from the uniform distribution on [0, 1] and then obtain $x_i = F^{-1}(u_i)$, i = 1, 2, ..., n. The values x_i will be a random sample of size n from the distribution with cdf $F(x)$. As an example, consider sampling from the distribution with pdf

$$f(x) = \frac{1}{\pi(1 + x^2)} \qquad -\infty < x < \infty$$

We find

$$F(x) = \int_{-\infty}^{\infty} \frac{1}{\pi(1 + x^2)} \, dx = \frac{1}{\pi} \arctan x \Big|_{-\infty}^{x}$$

$$= \frac{1}{\pi} \arctan x + 0.5$$

The program CAUCHY samples n observations from this distribution by finding $x_i = \tan \pi(u_i - 0.5)$, where the u_i are uniformly distributed on [0, 1]. The u_i are easily obtained by using the function RND in BASIC.

Exercises 4.6

1. Write a program to sample N observations from an exponential distribution with expectation $1/\lambda$. The values of N and the expectation should be input in response to questions in the program. The output should contain a frequency table of the N observations and the observed average value, but not the observations themselves.

2. Run the program NORSAMP for the standard normal distribution with N = 100, N = 1000, and N = 2000. Delete lines 230, 240, 250, and 310 (which supresses the frequency table). Do the averages seem to approach 0 as N increases?

CAUCHY

```
10 REM SAMPLES N OBSERVATIONS FROM CAUCHY DISTRIBUTION
20 PRINT "HOW MANY ";"OBSERVATIONS ";"FROM CAUCHY ";"DISTRIBUTION"
30 INPUT N
40 LET S = 0
50 FOR I = 1 TO N
60 LET Z = RND
70 LET X = TAN(3.14159263*(Z-.5))
80 PRINT X,
90 LET S = S + X
100 NEXT I
105 PRINT
110 PRINT "AVERAGE = ";S/N
120 END
*
```

```
RUN
HOW MANY OBSERVATIONS FROM CAUCHY DISTRIBUTION
?20
-2.53199          -0.63602          -3.18786          -0.70455           2.58189
-1.54645          -4.23989          -0.70455          -0.2422           -1.35782
-0.83008          -0.40191          -0.69468          -0.37868          -0.49694
 0.64787           1.4014           2.76477E-2         0.56724           0.89034
AVERAGE = -0.59186
```

*

3. Run the program CAUCHY for values of N = 100, N = 1000, and
    N = 2000. *Delete the PRINT statement at line 80.* Do these
    averages seem to approach the true median value of 0? Can you
    explain this?

4. Write a program to calculate the average of N observations from
    the distribution with pdf $f(x) = 3x^2$ on [0, 1] in response to a
    program question. What is the theoretical expectation of this
    distribution? What average values do you obtain for N = 100,
    1000, and 2000?

5

SAMPLING

5.1   INTRODUCTION

In the previous chapters we have examined the mathematical properties
of certain probability models, which have been assumed to be appropriate to describe particular real-world situations. Random variables
such as the binomial, Poisson, hypergeometric, normal, and gamma
families are examples of such probability models. A user is generally
faced with several questions as he proceeds to use such probability
models. Each of the probability models depends upon a parameter
(such as p in the binomial case) or parameters (μ and σ^2 in the normal
case). Values have to be assigned to these parameters before probabilities can be computed. In the use of these models, how should the
values of the parameters be assigned? Secondly, it may not be clear
which probability model is appropriate to describe the situation at
hand. A user knows that a model will not exactly reflect reality,
but does the model provide a sufficiently close description of observed
data to permit its use? Additionally, if two models might possibly
be appropriate, then how is a choice made between the two?

The information available to answer such questions is almost
invariably incomplete. The value of parameters of a probability model
have to be estimated from a sample from the relevant population. For
example, a hospital association may wish to estimate the true proportion p of a state's population with blood type O. This value can

then be used in the binomial distribution to compute probabilities
of interest to a hospital in the administration of its blood trans-
fusions. As a second example, the normal distribution may be appropriate to describe the distribution of the yield (in bushels per acre)
of corn within a certain state. However, estimates of μ and σ^2 will
be required in order to use the normal distribution. As a third
example, the Poisson distribution may be appropriate to describe the
number of serious accidents in a large factory in a month. How can
we decide from the accident history at the factory whether the Poisson
distribution should be used?

Questions such as those posed in the previous paragraphs require
inferential reasoning. In order to estimate parameters of a distribution we must take a sample from the population in question and,
based on the properties of the sample, infer to properties of the
population from which the sample was taken. It is never possible to
be absolutely certain concerning the characteristics of a population
by observing only a part of it. For example, the true proportion of
individuals with type O blood can only be found with certainty by
typing the blood of all the state's residents. This process would
clearly be too costly, time consuming, and impractical to implement.
Nevertheless, by selecting a sample of the state's population, an
estimate of the true proportion p can be obtained. It is with properties of such inferential methods that statistics is concerned.

Before proceeding, let us define what is meant here by a population.

DEFINITION 5.1.1. A *population* is the aggregate or set of possible
observations of interest.

Although this idea seems simple enough, it is known from experience
that many studies are made without adequate definition of the population to which the study is directed. This is most glaringly pointed
out when inference is made to a population, based on the results of

a sample of a particular subset of the population. A classic example
was the 1936 poll of the *Literary Digest*, aimed at predicting the
winner of the Roosevelt-Landon presidential race of that year. The
poll obtained several million responses and the conclusion was made
that Landon would win easily. In that year Landon won in Maine and
Vermont while Roosevelt won in the other 46 states. The poll sampled
telephone directories and lists of magazine subscribers. This sub-
group of the electorate favored Landon, while those without telephones
and magazine subscriptions were strongly for Roosevelt. At the end
of a long period of depression, this latter group was a substantial
proportion of the electorate. It was not represented in the sampling
technique employed by the *Literary Digest*. This became very clear
in November 1936.

We also remark here that Definition 5.1.1 in no way requires
that a population be animate or that it be possible to observe all
elements of a population. For example, the following may all be
considered to be populations:

1. The tires produced on a given line at a particular factory
    during the year 1976.
2. The male undergraduate population at a large state university
    on a given date.
3. The form 1040 income tax returns received by the U. S. Government
    prior to a specific cutoff date in a given year.
4. All possible samples of a cubic centimeter of a liquid in a tank.

In general terms the population of interest corresponds to the
outcome space S which is defined in Chap. 2 as a basic component of
a probability model. With each element of the population we associate
a characteristic or characteristics, to which a numerical value is
assigned. For example, for the populations in the preceding paragraph,
the following characteristics might be appropriate:

1. A measurement of the quality of the tire: 1 if the tire is
    acceptable according to manufacturing standards, 2 if it is
    safe but has other production defects, and 3 if it is defective.

2. The IQ scores of the students.

3. The gross reported income on each 1040 form.

4. The density measured in g/cm^3 of a cubic centimeter of the liquid.

Such measurements correspond to observations of a random variable X, which assigns a numerical value to each element of S. It is for this reason that the distributions of random variables are so important, and why we have devoted substantial effort toward understanding the properties of random variables.

It is clear that one would like to have a method of sampling from a population which will yield samples representative of the population being sampled. It is also clear that for large populations it is impossible to guarantee that a sample will reflect the population characteristic or characteristics being measured. Unusual samples, e.g., samples of an electorate evenly balanced between Democrats and non-Democrats, may include either a high or low percentage of Democrats. We wish to design a method of sampling which will make small the probability of samples unrepresentative of the sampled population. The basic idea of the sampling process called *random sampling* is to avoid bias in favor or against any particular sample of a given size. For a finite population we give the following definition of random sample.

DEFINITION 5.1.2a. Suppose a finite population has N elements. *A random sample of size* n *with replacement* is one obtained in such a way that each element of the population has probability 1/N of selection at each of the n selections. A *random sample of size* n *without replacement* is a selection of n items in such a way that at each selection, the elements not already chosen have equal chance of selection.

The notation employed to describe a random sample is a vector of n random variables $(X_1, X_2, \ldots, X_n)$, where X_i is a random variable which describes the possible values that the ith sampled observation may have. A random sample is then a *vector* of random variables. A particular observation of size five from the population of IQ scores described above might be (120, 118, 125, 138, 110). The first student selected has IQ 120, the second has IQ 118, and the last has an IQ of 110. Notice that if a random sample is taken *with* replacement, the random variables X_1, X_2, ..., X_n will be independent and have the same distribution. This is because the item selected at each step is replaced, so that the next selection does not depend on the previous ones and the distribution of the random variable X_i is the same at each step, as the population is the same. It can be shown that in random sampling *without* replacement the distribution of each X_i is the *same*, but the random variables are *not* independent. For example, suppose a population contains N = 5 elements {1, 2, 3, 4, 5} and a random sample of size n = 2 is taken without replacement. For X_1, we have $P(X_1 = i) = 1/5$ for i = 1, 2, 3, 4, 5. Let us find $P(X_2 = 1) = P((2,1) + (3,1) + (4,1) + (5,1)) = 4/20 = 1/5$ because each of the 20 outcomes $(X_1 = i, X_2 = j)$ for i ≠ j, i = 1, 2, 3, 4, 5 and j = 1, 2, 3, 4, 5 has equal probability under random sampling without replacement. Similarly $P(X_2 = j) = 1/5$ for j = 1, 2, 3, 4, 5. However it is clear that X_1 and X_2 are not independent, for if $X_1 = 1$, X_2 cannot be equal to 1. Hence $P(X_1 = 1, X_2 = 1) = 0 \neq P(X_1 = 1)P(X_2 = 1) = 1/25$, showing clearly that X_1 and X_2 are not independent.

The idea of a random sample from a infinite population follows from the properties of random sampling with replacement for a finite population.

DEFINITION 5.1.2b. Suppose an infinite population has a characteristic described by a random variable X. A *random sample* from this population is a vector of random variables $(X_1, X_2, ..., X_n)$ where the X_i are independent random variables with the common distribution of X, called the *parent distribution*.

In a sense, the removal of an element from an infinite population corresponds to sampling with replacement, as the removal does not affect the distribution of the remaining population. A realization of such a random sample might be (1.02, 0.98, 1.10, 1.03, 1.04), where the values are the densities of 5 cm^3 of the liquid described in preceding example (4). In the sequel we will use the word *random sample* to refer to a vector of random variables $(X_1, X_2, \ldots, X_n)$ which are *independently* and *identically* distributed. If the sampling involves random samples from a finite population *without* replacement, it will be specifically called *random sampling without replacement*.

In the case of a finite population in which each of the elements may be numbered from 1 to N, it is easy to use the RND command to take a random sample of size M with or without replacement. The program RANSAM prints out M random integers from the set {1, 2, 3, ..., N}. The elements in the population labeled with the M numbers selected would constitute the random sample with replacement. If random sampling without replacement is required, the program RANDSAMO can be used to obtain a random sample without replacement. This means that no two of the M integers selected will be the same. Both programs can be utilized by running the program and answering the questions concerning the values of N and M. RANSAM and RANSAMO have been run for N = 1000 and M = 50, and the associated output appears following the program listing.

Table 5.1.1 gives the number of days required by 1000 students in a self-paced course to complete the required examinations. We can use the output of RANSAM to take a random sample of size 50 from this population. In this case 120 corresponds to 50, 320 to 62, and finally, 724 to 54. The 50 observations are

```
50  62  51  49  73  52  50  49  67  50
46  45  46  48  60  57  55  65  56  49
57  44  43  54  60  40  56  52  54  37
68  54  44  56  52  62  52  56  53  42
55  43  59  45  60  42  53  53  61  54
```

RANDSAM

```
10 REM RANDOM SAMPLE FROM FINITE POPULATION WITH REPLACEMENT
20 PRINT "WHAT IS ";"THE POPULATION ";"SIZE?"
30 INPUT N
40 PRINT "WHAT IS ";"THE SAMPLE ";"SIZE?"
50 INPUT M
60 PRINT
70 PRINT M;" RANDOM NUMBERS";" FROM POPULATION";" OF SIZE ";N
80 PRINT
90 FOR I = 1 TO M
100 LET K = 1+INT(RND*N)
110 PRINT K,
120 NEXT I
130 END
*


 RUN
WHAT IS THE POPULATION SIZE?
 ?1000
WHAT IS THE SAMPLE SIZE?
 ?50

 50   RANDOM NUMBERS FROM POPULATION OF SIZE   1000
```

120	320	97	305	883
183	74	305	425	203
280	379	307	385	354
683	803	509	665	732
110	894	18	415	535
619	558	749	644	512
576	164	284	262	70
325	928	774	645	488
608	189	844	970	535
963	489	457	577	724

*

RANDSAMO

```
5 DIM F(100)
10 REM RANDOM SAMPLE FROM FINITE POPULATION WITHOUT REPLACEMENT
20 PRINT "WHAT IS ";"THE POPULATION ";"SIZE?"
30 INPUT N
40 PRINT "WHAT IS ";"THE SAMPLE ";"SIZE?"
50 INPUT M
60 PRINT
70 PRINT M;" RANDOM NUMBERS";" FROM POPULATION";" OF SIZE ";N
80 PRINT
90 FOR I = 1 TO M
100 LET K = 1+INT(RND*N)
110 LET F(I)=K
120 FOR J= 1 TO I-1
130 IF F(J)= K THEN 100
140 NEXT J
150 PRINT K,
160 NEXT I
170 END
*


 RUN
WHAT IS THE POPULATION SIZE?
 ?1000
WHAT IS THE SAMPLE SIZE?
 ?50

 50   RANDOM NUMBERS FROM POPULATION OF SIZE   1000
```

120	320	97	305	883
183	74	425	203	280
379	307	385	354	683
803	509	665	732	110
894	18	415	535	619
558	749	644	512	576
164	284	262	70	325
928	774	645	488	608
189	844	970	963	489
457	577	724	807	458

*

Table 5.1.1   Number of Days Required by 1000 Students

Row number	Column number									
	1	2	3	4	5	6	7	8	9	10
00	61	51	45	53	54	52	54	41	51	51
01	25	32	54	44	26	42	47	43	35	49
02	38	62	59	60	44	57	46	53	51	50
03	59	52	67	48	54	38	47	42	47	50
04	46	47	52	46	50	35	40	47	38	35
05	45	57	45	53	32	48	51	54	57	47
06	53	59	55	50	44	45	38	54	49	52
07	52	63	56	50	50	58	64	57	52	49
08	62	50	50	54	53	54	66	58	52	55
09	42	55	51	50	38	41	51	53	58	49
10	53	59	49	50	56	49	66	49	46	57
11	46	65	63	51	47	56	59	37	50	50
12	54	38	36	44	25	25	29	32	40	35
13	60	57	53	57	42	45	55	50	55	57
14	50	59	51	52	53	51	48	47	56	47
15	49	48	59	50	51	35	38	41	46	49
16	46	67	48	54	55	51	68	61	51	56
17	46	53	54	45	49	48	52	54	55	54
18	50	61	52	45	40	45	60	40	43	46
19	51	44	50	44	51	59	54	53	52	49
20	39	50	50	54	36	50	38	43	46	48
21	60	52	50	50	48	53	52	50	45	54
22	59	64	53	63	50	58	50	44	45	55
23	52	58	57	49	49	46	49	48	59	45
24	46	42	60	40	48	42	45	49	56	46
25	50	56	53	54	41	48	54	51	61	52
26	52	56	46	47	44	51	47	50	48	52
27	41	56	55	54	49	46	40	56	60	46
28	40	53	50	44	50	52	43	55	43	56
29	54	54	50	52	42	54	58	53	55	46
30	55	54	47	68	49	50	46	41	49	55
31	45	48	58	56	52	47	61	42	54	62
32	53	69	56	50	62	51	60	52	52	55
33	56	56	53	48	41	50	57	54	57	54
34	56	53	43	44	56	54	57	49	46	47
35	50	53	54	60	31	36	49	57	58	45
36	44	61	50	43	58	45	34	26	48	49
37	46	37	38	37	40	43	41	46	45	52
38	54	49	49	41	48	48	50	50	56	60
39	44	41	51	47	55	44	49	49	61	49
40	47	54	53	51	44	48	51	52	51	53

Row number	Column number									
	1	2	3	4	5	6	7	8	9	10
41	55	47	55	40	54	46	42	48	65	61
42	54	43	56	63	67	53	58	59	55	53
43	56	55	58	57	57	57	53	49	51	58
44	58	53	56	65	55	61	54	52	61	58
45	43	52	53	52	51	47	53	55	55	49
46	60	50	57	49	51	49	47	53	51	52
47	55	59	54	48	53	47	51	48	54	47
48	42	32	39	48	48	53	47	42	53	44
49	52	57	55	49	54	53	60	48	50	53
50	48	48	50	48	47	48	37	51	65	50
51	43	37	44	35	54	46	36	28	43	46
52	57	52	58	48	51	57	61	55	50	57
53	57	51	58	49	60	55	61	62	49	47
54	46	59	54	56	54	61	52	60	48	53
55	58	61	56	61	57	63	72	56	57	58
56	64	50	52	50	57	65	57	65	55	57
57	61	48	50	55	64	68	61	57	63	52
58	59	66	61	59	61	61	51	70	52	62
59	52	50	50	52	50	54	44	49	50	47
60	52	48	44	40	50	56	49	55	57	46
61	50	48	47	45	40	52	51	51	40	44
62	49	48	41	49	44	48	46	41	46	40
63	57	57	51	46	49	59	50	58	53	48
64	60	59	52	54	53	53	50	49	45	55
65	61	50	50	52	56	51	55	59	55	43
66	55	57	55	57	56	58	59	55	55	52
67	56	49	50	49	52	48	50	47	57	54
68	48	49	57	56	46	49	49	63	51	51
69	50	48	44	51	47	47	39	52	47	50
70	45	44	39	45	45	41	45	37	41	41
71	61	53	49	53	51	49	51	54	55	48
72	41	53	57	54	50	51	48	48	48	42
73	58	49	52	55	52	48	47	49	55	57
74	60	56	55	52	57	56	54	49	52	45
75	57	55	48	52	46	49	43	55	36	45
76	53	61	55	53	49	43	53	45	47	45
77	55	51	53	56	50	49	48	54	51	43
78	44	53	49	44	50	59	53	42	50	38
79	63	54	45	51	57	48	59	57	53	44
80	57	51	55	56	60	56	52	53	49	43
81	48	55	44	55	53	46	50	49	48	43
82	49	50	48	55	58	53	46	50	49	58
83	55	57	56	43	40	54	59	47	55	44
84	74	55	45	59	53	53	53	57	56	47
85	53	48	47	43	45	48	59	53	52	39

Row number	Column number									
	1	2	3	4	5	6	7	8	9	10
86	45	45	31	46	61	52	41	30	36	45
87	51	41	55	58	58	45	53	45	47	33
88	55	58	73	55	65	50	48	46	54	43
89	68	68	51	44	51	57	52	55	54	39
90	58	49	71	51	54	57	55	58	55	40
91	45	45	52	57	56	53	52	50	61	45
92	50	56	49	49	53	49	57	52	55	41
93	55	53	40	48	57	51	50	48	56	59
94	56	50	51	63	52	51	60	46	52	47
95	53	46	61	51	49	50	48	55	45	37
96	50	50	42	54	51	55	47	48	48	45
97	48	53	57	53	53	49	52	49	66	46
98	56	54	49	51	58	46	50	50	51	40
99	46	46	51	45	61	54	50	48	58	51

Typically we wish to estimate the true average of the *population*. The average of the 50 sampled values is 52.82 days. The true average of the population is 50.87. We shall presently investigate the properties of such estimates.

We consider, in somewhat more detail, sampling without replacement from a finite population of size N. Assume that the N values are v_1, v_2, ..., v_N and define

$$E(V) = \sum_{i=1}^{N} \frac{v_i}{N} \quad \text{and} \quad \sigma_V^2 = \sum_{i=1}^{N} \frac{(v_i - E(V))^2}{N}$$

Assume that a random sample of size n is taken from this population *without* replacement. It can be shown that

$$\text{Var}\left(\sum_{i=1}^{n} X_i \right) = \sum_{i=1}^{n} \text{Var}(X_i) + \sum_{i \neq j}^{n} \text{Cov}(X_i, X_j)$$

where

$$\text{Cov}(X_i, X_j) = E(X_i X_j) - E(X_i)E(X_j)$$

As all X_i have the same distribution $\text{Var}(X_i) = \sigma_V^2$ for all i. It is also true that all the covariances are equal, although this is not so apparent. Hence

$$\text{Var}\left(\sum_{i=1}^{n} X_i\right) = \sigma_V^2 + n(n - 1)\gamma \tag{5.1.1}$$

where γ is the common covariance. Setting $n = N$ yields

$$N\sigma_V^2 + N(N - 1) = \text{Var}\left(\sum_{i=1}^{N} X_i\right) = 0$$

The latter equation follows because the sum of all the population values is a constant. Hence $\gamma = -\sigma_V^2/(N - 1)$. Substitution in Eq. (5.1.1) and simplification yields

$$\text{Var}\left(\sum_{i=1}^{n} X_i\right) = \frac{n\sigma_V^2(N - n)}{N - 1} \tag{5.1.2}$$

Furthermore we have

$$E\left(\sum_{i=1}^{n} X_i\right) = nE(V)$$

These yield

$$E\left(\sum_{i=1}^{n} \frac{X_i}{n}\right) = E(V)$$

and

$$\text{Var}\left(\sum_{i=1}^{n} \frac{X_i}{n}\right) = \frac{\sigma_V^2}{n} \left(\frac{N - n}{N - 1}\right)$$

Thus the sample average has expectation equal to $E(V)$ and variance $(\sigma_V^2/n)((N - n)/(n - 1))$. If the sampling is done with replacement, the corresponding values are $E(V)$ and σ_V^2/n. As the factor $(N - n)/(N - 1)$, referred to as the finite population correction factor, will be close to 1 when the sample size n is very much smaller than N, we have a strong indication that in this situation, sampling either with or without replacement will permit similar inferential statements about the population.

Problem 5.1

1. Using the result in the last paragraph, show that the variance of
    H, the hypergeometric random variable with finite population of
    size N, with R special items, from which a sample of size n is
    taken without replacement is given by

    $$Var(H) = n\frac{R}{N} \left(1 - \frac{R}{N}\right) \frac{N - n}{N - 1}$$

Exercises 5.1

1. Suppose that 10,000 lottery tickets numbered 1 to 10,000 are sold.
    Take a random sample of 25 such numbers without replacement using
    the appropriate program. What population is being sampled? What
    is the expected value and variance of the random variable X_i which
    gives the number on the lottery ticket drawn on the ith draw?
    Use RANDSAMO.

2. Write a program to compute the probability that for a sample of
    size n, at least two persons have the same birthday. Find this
    probability for n = 5, 10, 15, 20, 25, 30, 35, 40, 45, and 50.
    Use the program RANDSAM to select 40 birthdays at random from the
    365 birth dates. Are there any two birth dates alike?

3. The following is a list of the ages of 500 legislators. Take a
    sample of size 25 from this population without replacement.
    Average the ages of these 25 sampled individuals. Compare this
    average with 48 which is the average age of the population from
    which the 500 were chosen.

```
46   44   45   40   46   52   47   48   42   44
57   36   53   34   63   24   62   46   61   43
50   48   45   62   56   46   46   40   44   52
43   39   49   47   42   53   53   43   41   47
46   34   38   44   43   48   48   45   41   61
42   56   40   49   52   44   49   48   46   61
61   39   48   54   49   46   51   65   51   42
47   42   56   45   42   35   48   51   58   37
45   55   36   49   59   51   52   43   40   43
43   43   46   47   48   57   50   44   54   44
39   45   47   44   45   56   49   43   38   43
36   45   56   60   46   43   38   47   43   40
51   59   45   45   41   43   35   47   43   40
55   52   59   42   46   46   44   53   48   47
53   58   47   49   52   47   53   46   43   39
50   54   47   47   37   48   52   43   54   50
56   35   57   42   49   44   51   43   55   43
37   52   40   51   40   45   60   49   41   56
47   37   56   49   53   38   52   57   42   38
33   56   42   44   37   42   63   50   50   44
52   46   40   45   39   47   45   32   33   44
58   44   52   48   38   48   56   53   42   43
59   68   41   39   55   56   51   54   45   50
46   43   43   51   46   58   58   57   52   44
49   50   41   57   39   50   41   54   35   41
63   53   52   47   51   35   55   34   44   44
33   54   47   36   40   65   40   68   53   35
43   56   54   43   54   45   55   46   46   33
48   51   58   45   53   34   44   28   51   50
64   31   54   40   32   38   54   50   56   45
50   46   38   40   45   43   24   47   46   58
44   50   46   51   47   48   55   46   48   51
50   52   51   48   58   54   48   55   42   33
47   46   54   48   53   48   50   50   65   51
41   30   48   48   51   51   57   49   43   48
44   55   51   42   34   51   44   37   38   40
46   45   44   41   35   53   46   60   44   47
49   39   56   44   43   36   46   52   58   49
45   49   51   43   40   50   46   48   39   54
47   49   49   49   42   36   49   45   44   42
54   48   46   39   50   53   43   52   49   49
41   31   46   30   30   46   33   47   33   43
49   52   46   59   40   45   50   38   49   51
52   48   51   39   49   52   51   52   37   61
65   49   46   42   42   58   60   50   44   52
48   50   41   41   30   51   51   43   54   55
41   45   48   54   54   49   46   52   56   45
45   50   35   52   49   44   45   50   52   52
45   47   45   60   48   49   32   50   52   47
57   42   48   41   40   48   39   39   53   45
```

*

5.2   STATISTICS ESTIMATING LOCATION

We have seen in Chap. 4 that it is useful to employ some means of
summarizing the information in a random sample from a population.
In Sec. 4.1 random samples of size 1000 were taken from three populations. The populations sampled denoted S_1, S_2, and S_3 were response
times, net weights of a breakfast food, and the values of round-off
errors in rounding a number to the nearest integer, respectively.
The samples were summarized by means of relative frequency histograms
displayed in Figs. 4.1.1, 4.1.2, and 4.1.3. It is clear that the
ordinate measurement could be changed to frequency by multiplication
by n = 1000. Each bar would then represent the number of observations
for the appropriate interval. Histograms which display the properties
of a sample are a useful *graphical* means of summarizing data. In
Sec. 4.1 a frequency table was used to summarize the information in
the sample of round-off errors in *tabular* form. Such a frequency
table is often used to display the properties of a sample. In
statistics, however, we are mainly interested in *algebraic or mathe-*
matical ways of summarizing the information in a sample. The sample
characteristics will in turn be used inferentially to describe
characteristics of the population from which the sample was taken.
The first characteristic which we consider is "location." An important inferential question is how to estimate the parameter $\mu = E(X)$,
the expectation of the random variable X, which describes the
"location" of a population, using the information in a sample. We
will use certain statistics based on a random sample of size n from
the population. We now define the word statistic in this usage and
differentiate it from the body of knowledge referred to as statistics.

DEFINITION 5.2.1. A *statistic* is a calculation based on the values
of a random sample. More precisely, a statistic is a real-valued
function of the observations $(X_1, X_2, \ldots, X_n)$ in a random sample of
size n from a population.

The statistic is to be used to describe a characteristic of the random variable X, the parent distribution.

5.2.1 The Sample Mean

The sample mean is given by the statistic $\bar{X} = (X_1 + X_2 + \cdots + X_n)/n$. Notice that this is a real-valued function which, in a sense, summarizes certain information about the sample in a single number. Note also that since $X_1, X_2, \ldots, X_n$ are random variables, $\bar{X}$ is a random variable also. For different random samples of a fixed size n, $\bar{X}$ will have different values. In general, we observe only one realization of this random variable $\bar{X}$ and denote it as $\bar{x}$. On the other hand, $\bar{X}$ seems a reasonable statistic to use as an estimate of $\mu = E(X)$ for several reasons. The first is due to the fact that

$$E(\bar{X}) = E\left\{ \sum_{i=1}^{n} \frac{X_i}{n} \right\} = \sum_{i=1}^{n} \frac{E(X_i)}{n} = \frac{n\mu}{n} = \mu$$

because $E(X_i) = \mu$ for all i in a random sample. In words, the expectation of the sample mean is the expected value of X. We also have

$$Var(\bar{X}) = Var\left(\sum_{i=1}^{n} \frac{X_i}{n} \right) = \sum_{i=1}^{n} \frac{Var(X_i)}{n^2} = \frac{n\sigma^2}{n^2} = \frac{\sigma^2}{n}$$

where σ^2 is $Var(X)$. This follows from the fact that $Var(U + W) = Var(U) + Var(W)$, if U and W are independent random variables. The importance of the relation $Var(\bar{X}) = \sigma^2/n$ lies in the fact that as n increases the variance of $\bar{X}$ decreases, so roughly the probability that the sample mean $\bar{X}$ differs from μ becomes small as n increases. In fact, using the Chebyshev inequality in the form expressed in Theorem 4.3.2, we have

$$P(|\bar{X} - \mu| \geq \epsilon) \leq \frac{\sigma^2}{n\epsilon^2}$$

for any $\epsilon > 0$. Hence

$$\lim_{n \to \infty} P(|\bar{X} - \mu| \geq \epsilon) = 0$$

In words, the probability that the random variable $\bar{X}$ differs from μ by more than any $\varepsilon > 0$ approaches 0 as the sample size increases.

With the aid of the computer we now illustrate the properties of the random variable $\bar{X}$ by taking all possible samples of size 3 with replacement from the integer population $\{1, 2, 3, \ldots, 10\}$. We have seen before that if $P(X = i) = 1/10$ for $i = 1, 2, \ldots, 10$, then $E(X) = 5.5$. There are $10^3 = 1000$ samples of size 3 from this population. The program MVAL3 calculates the expected value of the *averages* of these 1000 samples of size 3 by assigning each a probability of 1/1000. In addition the probability distribution of these averages is printed out. This gives the probability of an average falling into the intervals $[i - 0.5, i + 0.5)$ for $i = 1, 2, \ldots, 10$. The probability histograms for the original population and the population of means $(n = 3)$ appear in Fig. 5.2.1. We see that although $E(X) = E(\bar{X}) = 5.5$, the averages have a distribution with much smaller variance. In fact, $\text{Var}(X) = 99/12 = 33/4$ and $\text{Var}(\bar{X}) = \text{Var}(X)/n = 11/4$.

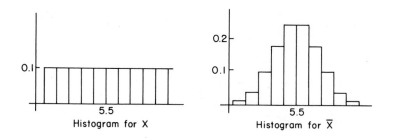

Fig. 5.2.1

5.2.2   *The Population and Sample Median*

We shall often use $E(X) = \mu$ as a measure of the population location. The statistic $\bar{X}$ will generally be used as the appropriate calculation from the sample data to estimate $\mu = E(X)$. In the next chapter we

MVAL3

```
10 REM MEANS OF SIZE 3 FROM INTEGER POPULATION 1-10
20 LET T=0
30 FOR I = 1 TO 10
40 FOR J = 1 TO 10
50 FOR K = 1 TO 10
60 LET S = (I+J+K)/3
70 FOR N= 1 TO 10
80 IF S > (N+.5) THEN 110
90 LET F(N)=F(N)+1
100 GO TO 120
110 NEXT N
120 LET T = T+S/1000
130 NEXT K
140 NEXT J
150 NEXT I
160 PRINT "PROBABILITY ";"DISTIBUTION ";"FOR MEANS"
170 PRINT
180 PRINT "LOWER BOUNDARY","UPPER BOUNDARY","PROBABILITY"
190 PRINT
200 FOR I = 1 TO 10
210 PRINT I-.5,I+.5,F(I)/1000
220 NEXT I
230 PRINT "AVERAGE MEAN = ";T
240 END
*
```

```
RUN
PROBABILITY DISTIBUTION FOR MEANS

LOWER BOUNDARY UPPER BOUNDARY PROBABILITY

0.5                1.5                4.00000E-3
1.5                2.5                3.10000E-2
2.5                3.5                8.50000E-2
3.5                4.5                0.163
4.5                5.5                0.217
5.5                6.5                0.217
6.5                7.5                0.163
7.5                8.5                8.50000E-2
8.5                9.5                3.10000E-2
9.5                10.5               4.00000E-3
AVERAGE MEAN =   5.5
```

*

shall note several properties of $\bar{X}$ which suggest that it is an efficient statistic. There are, on the other hand, populations for which
a measure other than $E(X)$, would be appropriate as a measure of the
population location. In highly skewed populations such as that
described by the probability density function $f(t) = te^{-t}$, $t > 0$,
presented in Fig. 5.2.2, the population median may be a more meaning-
ful measure of location.

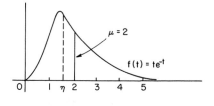

Fig. 5.2.2

DEFINITION 5.2.2. The *population median* for a continuous random
variable defined by the cdf $F(x)$ is a value of x, say η, for which
$F(\eta) = 1/2$.

The population median is then a value for which $P(X \le \eta) = 1/2$ and
$P(X > \eta) = 1/2$. In the probability sense, it divides the population
into two equal parts. Figure 5.2.2 shows the values of $E(T) =$
$\int_0^\infty t^2 e^{-t}\ dt = 2$ and $\eta = 1.678$ which satisfies the equation

$$e^{-\eta}(1 + \eta) = 0.5 \quad [\text{as } F(t) = 1 - e^{-t} - te^{-t} \text{ for } t > 0]$$

If the random variable T is thought of as describing response times
to a stimulus in minutes, then one-half of the population has a
response time of 1.678 min or less, while the expected response time
is $E(T) = 2$ min, which exceeds η because a sizable proportion of the
population requires a rather long time to respond.

The appropriate estimate of the population median based on a random sample of size μ from a population is the sample median, which is calculated as follows:

DEFINITION 5.2.3. Let $X_{(1)}$, $X_{(2)}$, ..., $X_{(n)}$ represent the values of a random sample $(X_1, X_2, ..., X_n)$ ordered by magnitude (that is, $X_{(1)} \leq X_{(2)} \leq \cdots \leq X_{(n)}$). The *sample median* M is defined to be

$$M = \begin{cases} X_{(n+1)/2} & \text{if } n \text{ is odd} \\ \\ (X_{(n/2)} + X_{(n/2+1)})/2 & \text{if } n \text{ is even} \end{cases}$$

Suppose we observe a random sample of five response times (3.5, 1.6, 2.3, 1.5, 1.7). Then the sample median is denoted $m = x_{(3)} = 1.7$. As in the case of the sample mean, the sample median M is a random variable. We generally have only one sample of size n and observe a single value of the median, denoted by m. However, we understand that different samples of size n from the same population will produce different sample medians and hence there is a distribution of such values. Generally the sample mean is preferred as a measure of location when the population can be thought of as symmetrically distributed about its expected value. For skewed populations such as incomes, response times, and city sizes, the sample median is often more acceptable as a measure of location.

Problems 5.2

1. Suppose that the following is a random sample of 10 heights of female students in inches: 64.5, 63.0, 67.5, 62.25, 64.0, 69.25, 62.0, 65.25, 66.0, 61.0. Find $\bar{x}$ and m for this sample.

2. Suppose that in Problem 5.2.1 that 69.25 is replaced by 78.0. What is the new value of $\bar{x}$ for the sample? The new value of m? Which of these measures of location is more sensitive to extreme observations.

3. Prove the following for any real numbers x_1, x_2, ..., x_n.

   (a) $\sum_{i=1}^{n} x_i = n\bar{x}$

   (b) $\sum_{i=1}^{n} (x_i - \bar{x}) = 0$

4. Show that $\bar{x}$ minimizes

$$f(a) = \sum_{i=1}^{n} (x_i - a)^2$$

   Hint: Expand

$$f(a) = \sum_{i=1}^{n} (x_i - \bar{x} + \bar{x} - a)^2$$

5. (a) Assume that $E(X) = \mu$ exists for a continuous random variable
   X. Show that if the probability density function for X satisfies
   $f(a - x) = f(a + x)$ for all X then $E(X) = a$ and the median of X
   is also a.
   (b) For the density function

$$f(z) = \frac{1}{\sqrt{2\pi}} e^{-z^2/2}$$

   for what value of a does the relation $f(a - z) = f(a + z)$ hold
   for all real z?
   (c) What are the expected value and theoretical median of the
   standard normal distribution?

6. Given $f(x) = (1/\pi)1/(1 + x^2)$ is the density function of a random
   variable X what can be said about $E(X)$. Why does this not contradict Problem 5.2.5(a). What is the median of the random
   variable X?

7. Let a population have values x = 1, 2, 3 with probabilities
   $f(x) = 1/3$ for x = 1, 2, 3.
   (a) Find $E(X)$ and $Var(X)$.

(b) Find the probability function for $\bar{X}$, the average of two observations taken from this population with replacement.

(c) Find $E(\bar{X})$ and $Var(\bar{X})$ from the probability function in part (b) and show that $E(\bar{X}) = E(X)$ and $Var(\bar{X}) = Var(X)/2$.

8. For the exponential function with pdf $f(x) = e^{-x}$ for $x > 0$ and $f(x) = 0$ elsewhere, find $E(X)$ and the median of X. Which is larger? Give a qualitative reason why this is plausible.

9. If all observations in a sample, x_1, x_2, $\ldots$, x_n, are replaced by $y_i = ax_i + b$ show that

(a) $\bar{y} = a\bar{x} + b$.

(b) If $x_{(1)}$, $x_{(2)}$, $\ldots$, $x_{(n)}$ are the ordered x_i values with median $m = x_{((n+1)/2)}$ (n assumed odd), then the median of the $y_{(i)}$ values is $ax_{((n+1)/2)} + b$.

10. Let p^* be the true probability a theoretical tack falls "point upward." Let f_n be the relative frequency of such a tack falling point upward in n tosses. Show that for any $\varepsilon > 0$,

$$\lim_{n\to\infty} P(|f_n - p^*| \geq \varepsilon) = 0$$

Exercises 5.2

1. Use the program MEANS to sample 400 means of samples of size 1, 9, 25, and 100 each from the exponential distribution with $E(X) = E(X) = 4$. Draw relative frequency histograms for each of the 400 means with boundaries 0, 1, 2, $\ldots$, 20.

2. Write a program to calculate the mean $\bar{x}$ and median m for a sample of size $n \leq 50$. Find mean and median of the following heights:

72.1 70.0 67.1 69.2 71.3 72.4 74.1 74.0 71.0 67.5
67.8 70.1 71.8 72.1 74.5 74.1 73.2 71.5 70.3 68.0
68.8 70.3 71.4 73.8 74.0 75.0 73.7 72.6 70.4 69.1
69.1 71.6 72.8 73.1 75.8 76.5 74.0 72.5 71.7 69.4
69.2 71.4 72.3 74.8 77.1 78.0 74.3 72.1 71.0 69.5

MEANS

```
10 REM FREQUENCY DISTRIBUTION FOR K MEANS FROM EXPONENTIAL
20 REM DISTRIBUTION (TRUE EXPECTATION 4). SAMPLE SIZE N
30 PRINT "HOW MANY MEANS?"
40 INPUT K
50 PRINT "WHAT SAMPLE SIZE?"
60 INPUT N
70 PRINT "CLASS NO.","LOWER","UPPER","REL FREQ"
80 DIM F(25)
90 FOR J = 1 TO K
100 FOR I = 1 TO N
110 LET Z = Z-4*(LOG(1-RND))
120 NEXT I
130 LET A= Z/N
140 LET Z=0
150 FOR I = 1 TO 20
160 IF A>I THEN 210
170 LET F(I)=F(I)+1/K
180 GO TO 220
190 IF I<20 THEN 210
200 LET F(20)=F(20)+1/K
210 NEXT I
220 NEXT J
230 FOR I = 1 TO 20
240 PRINT I,I-1,I,F(I)
250 LET V=(I-.5)*F(I)+V
260 LET T = F(I)+T
270 IF T>.999999 THEN 290
280 NEXT I
290 PRINT "AVERAGE MEAN= ";V
300 PRINT "TOTAL FREQ ";T
310 END
*
```

```
 RUN
HOW MANY MEANS?
 ?1000
WHAT SAMPLE SIZE?
 ?25
```

CLASS NO.	LOWER	UPPER	REL FREQ
1	0	1	0
2	1	2	1.00000E-3
3	2	3	9.60000E-2
4	3	4	0.451
5	4	5	0.347
6	5	6	9.30000E-2
7	6	7	1.10000E-2
8	7	8	1.00000E-3

```
AVERAGE MEAN= 3.972
TOTAL FREQ  1.
```

*

Plot a histogram for these data with a class interval of 1. Would you call these data symmetric or skewed?

3. Use the program written in Exercise 5.2.2 to find the mean and median of the following incomes in thousands of dollars. Which measure of location would be preferred? Why?

11.2	13.6	14.7	11.8	14.7
10.3	11.2	11.1	11.8	11.2
20.6	13.5	17.3	13.2	21.2
12.7	10.1	14.6	18.2	11.8
10.9	15.8	14.9	10.1	10
11.2	15.2	13.6	12.1	10.1
13.8	13.7	16.2	11.1	11.9
19.6	16.9	12.4	12.3	15.4
13.5	13.3	13	14.7	12.9
17.4	15.7	14.8	12.1	20

5.3 STATISTICS ESTIMATING VARIABILITY

The parameter σ^2, the variance of the random variable X, is a measure of variability of a population described by X. We often are able to assume that the observations which we sample come from a normal distribution, but the variance σ^2 is not known. The value of σ^2 must then be estimated before the normal model can be used. A statistic often used to estimate σ^2 is the *sample variance*, which we define as follows:

DEFINITION 5.3.1. The sample variance based on a random sample of size n is defined by

$$S^2 = \sum_{i=1}^{n} \frac{(X_i - \bar{X})^2}{n - 1}$$

We note here that S^2, like $\bar{X}$, is a random variable which will have different values for different random samples from a population. The value of a single observation of S^2 is denoted

$$s^2 = \sum_{i=1}^{n} \frac{(x_i - \bar{x})^2}{n - 1}$$

The population variance $\sigma^2 = E(X - \mu)^2$ is the expected squared difference between the values of X and $\mu = E(X)$, the population expectation. The sample variance is almost an average of the squared differences of the sample values from the sample mean, but the divisor $n - 1$ is used instead of n. The reason for this is that $E(S^2) = \sigma^2$, while

$$E(\frac{(n - 1)S^2}{n}) = E(\sum_{i=1}^{n} \frac{(X_i - \bar{X})^2}{n}) = \frac{\sigma^2(n - 1)}{n}$$

If we write

$$\sum_{i=1}^{n} (X_i - \mu)^2 = \sum_{i=1}^{n} (X_i - \bar{X} + \bar{X} - \mu)^2$$

$$= \sum_{i=1}^{n} (X_i - \bar{X})^2 + 2(\bar{X} - \mu) \sum_{i=1}^{n} (X_i - \bar{X}) + n(\bar{X} - \mu)^2$$

$$= \sum_{i=1}^{n} (X_i - \bar{X})^2 + n(\bar{X} - \mu)^2 \quad \text{as } \sum_{i=1}^{n} (X_i - \bar{X}) = 0$$

we have

$$E(\sum_{i=1}^{n} (X_i - \mu)^2) = E(\sum_{i=1}^{n} (X_i - \bar{X})^2) + E(n(\bar{X} - \mu)^2)$$

or

$$n\sigma^2 = E((n - 1)S^2) + \frac{n\sigma^2}{n}$$

Thus $E((n - 1)S^2) = (n - 1)\sigma^2$. Division by $n - 1$ yields $E(S^2) = \sigma^2$, and division by n yields $E((n - 1)S^2/n) = \sigma^2(n - 1)/n$. The property that $E(S^2) = \sigma^2$, i.e., the expected value of S^2 is equal to the parameter being estimated, seems to be a reasonable requirement for an estimate. The statistic $(n - 1)S^2/n$ would be an under estimate of σ^2 on the average because the factor $(n - 1)/n < 1$. For large n, clearly there would be little difference between the two statistics.

The program VARSAMP has been written to give an idea of the distribution of sample variances, taken from the standard normal distribution. In this case, of course, $\sigma^2 = 1$. For samples of size N = 5, 10, and 25, the program computes 1000 sample variances. A frequency table of the 1000 sample variances is printed out for N = 5, 10, and 25. As can be seen from these frequency tables, these variances cluster more closely about the population variance $\sigma^2 = 1$ as n increases. In fact, it may be shown that for a population for which $E(X^4)$ exists,

$$\lim_{n \to \infty} P(|S^2 - \sigma^2| > \varepsilon) = 0 \qquad \text{for any } \varepsilon > 0$$

This property corresponds to the analogous property of $\bar{X}$, namely that

$$\lim_{n \to \infty} P(|\bar{X} - \mu| \geq \varepsilon) = 0$$

and is also proved using the Chebyshev inequality.

We give an example of two samples of size 5 with the same mean $\bar{x} = 9$ to illustrate the calculation of s^2.

		Sample 1			Sample 2	
i	x_i	$x_i - \bar{x}$	$(x_i - \bar{x})^2$	x_i	$x_i - \bar{x}$	$(x_i - \bar{x})^2$
1	7	-2	4	1	-8	64
2	8	-1	1	8	-1	1
3	9	0	0	9	0	0
4	10	1	1	11	2	4
5	11	2	4	16	7	49
Totals	45	0	10	45	0	118

$$\bar{x}_1 = 9 \qquad s_1^2 = 2.5 \qquad \bar{x}_2 = 9 \qquad s_2^2 = 29.5$$

VARSAMP

```
10 REM FREQUENCY TABLE OF K VARIANCES OF SAMPLES
20 REM FROM NORMAL DISTRIBUTION
30 DIM F(35)
40 FOR I = 1 TO 35
50 LET F(I)=0
60 NEXT I
70 PRINT "WHAT IS SAMPLE";" SIZE?"
80 INPUT N
90 PRINT "HOW MANY";" REPETITIONS?"
100 INPUT K
110 FOR J = 1 TO K
120 FOR I = 1 TO N
130 GOSUB 410
140 LET S = S +Z
150 LET S2= S2+Z^2
160 NEXT I
170 LET X1=S/N
180 LET V =(S2-N*X1^2)/(N-1)
190 FOR C = 1 TO 30
200 IF V>C/5 THEN 230
210 LET F(C)=F(C)+1/K
220 GO TO 260
230 IF C <30 THEN 250
240 LET F(C)=F(C)+1/K
250 NEXT C
260 LET S=0
270 LET S2=0
280 NEXT J
290 PRINT
300 PRINT "FREQUENCY ";"DISTRIBUTION OF";" VARIANCES"
310 PRINT
320 PRINT "LOWER END","UPPER END","REL FREQ"
330 FOR C = 1 TO 30
340 PRINT (C-1)/5,C/5,F(C)
350 LET T = T+F(C)
360 IF T >= 0.999999 THEN 380
370 NEXT C
380 PRINT
390 PRINT "TOTAL FREQ= ";T
400 GO TO 450
410 LET Z1=SQR(-2*LOG(RND))
420 LET Z2=6.2831853*RND
430 LET Z=Z1*COS(Z2)
440 RETURN
450 END
*
```

```
   RUN
WHAT IS SAMPLE SIZE?
 ? 5
HOW MANY REPETITIONS?
 ?1000

FREQUENCY DISTRIBUTION OF VARIANCES

 LOWER END        UPPER END        REL FREQ
 0                0.2              4.90000E-2
 0.2              0.4              0.126
 0.4              0.6              0.16
 0.6              0.8              0.119
 0.8              1                0.119
 1                1.2              0.112
 1.2              1.4              8.40000E-2
 1.4              1.6              6.50000E-2
 1.6              1.8              3.40000E-2
 1.8              2                2.60000E-2
 2                2.2              2.20000E-2
 2.2              2.4              2.00000E-2
 2.4              2.6              2.00000E-2
 2.6              2.8              1.00000E-2
 2.8              3                1.40000E-2
 3                3.2              2.00000E-3
 3.2              3.4              6.00000E-3
 3.4              3.6              6.00000E-3
 3.6              3.8              1.00000E-3
 3.8              4                2.00000E-3
 4                4.2              1.00000E-3
 4.2              4.4              0
 4.4              4.6              0
 4.6              4.8              0
 4.8              5                0
 5                5.2              0
 5.2              5.4              0
 5.4              5.6              1.00000E-3
 5.6              5.8              0
 5.8              6                1.00000E-3

TOTAL FREQ=  1.


*
```

```
RUN
WHAT IS SAMPLE SIZE?
?10
HOW MANY REPETITIONS?
?1000

FREQUENCY DISTRIBUTION OF VARIANCES

LOWER END        UPPER END        REL FREQ
0                0.2              3.00000E-3
0.2              0.4              4.80000E-2
0.4              0.6              0.155
0.6              0.8              0.162
0.8              1                0.177
1                1.2              0.149
1.2              1.4              0.1
1.4              1.6              8.80000E-2
1.6              1.8              4.60000E-2
1.8              2                2.90000E-2
2                2.2              2.10000E-2
2.2              2.4              7.00000E-3
2.4              2.6              4.00000E-3
2.6              2.8              4.00000E-3
2.8              3                2.00000E-3
3                3.2              2.00000E-3
3.2              3.4              1.00000E-3
3.4              3.6              1.00000E-3
3.6              3.8              0
3.8              4                0
4                4.2              0
4.2              4.4              1.00000E-3

TOTAL FREQ=  1.


*RUN
WHAT IS SAMPLE SIZE?
?25
HOW MANY REPETITIONS?
?1000

FREQUENCY DISTRIBUTION OF VARIANCES

LOWER END        UPPER END        REL FREQ
0                0.2              0
0.2              0.4              1.00000E-3
0.4              0.6              5.20000E-2
0.6              0.8              0.209
0.8              1                0.273
1                1.2              0.234
1.2              1.4              0.142
1.4              1.6              5.20000E-2
1.6              1.8              2.50000E-2
1.8              2                5.00000E-3
2                2.2              3.00000E-3
2.2              2.4              2.00000E-3
2.4              2.6              1.00000E-3
2.6              2.8              1.00000E-3

TOTAL FREQ=  1.
```

```
    RUN
WHAT IS SAMPLE SIZE?
    ?40
HOW MANY REPETITIONS?
    ?1000

FREQUENCY DISTRIBUTION OF VARIANCES

LOWER END        UPPER END        REL FREQ
    0                0.2              0
    0.2              0.4              3.00000E-3
    0.4              0.6              1.70000E-2
    0.6              0.8              0.16
    0.8              1                0.344
    1                1.2              0.275
    1.2              1.4              0.139
    1.4              1.6              4.80000E-2
    1.6              1.8              9.00000E-3
    1.8              2                4.00000E-3
    2                2.2              1.00000E-3

TOTAL FREQ=  1.


*
```

Although both samples have the same mean of 9, it is clear that the variability of the first sample is less than that of the second. This is reflected in the values of s_1^2 and s_2^2.

Another measure of variation of a population is the population standard deviation $\sigma = \sqrt{\sigma^2}$. Similarly the sample standard deviation is defined as $S = \sqrt{S^2}$, or for an individual sample $s = \sqrt{s^2}$. Hence the sample standard deviations for the two samples given above would be $s_1 = \sqrt{2.5} = 1.58$ and $s_2 = \sqrt{29.5} = 5.43$. A useful property of the standard deviation is that its units are the same as the units of the measurements of the sample values x_1, x_2, ..., x_n.

The units of s^2, on the other hand, are in terms of these units squared. It is not true that $E(S) = \sigma$, but it is true for normal populations that $E(S) = k(n)\sigma$, where $k(n)$ is a function of n for which $\lim_{n\to\infty} k(n) = 1$; so for large n, $E(S) \doteq \sigma$. For these reasons the sample standard deviation is generally used to estimate σ, although other statistics are also used.

Problems 5.3

1. Using the heights of the 10 students given in Problem 5.2.1 find s^2, the sample variance and the sample standard deviation s.

2. Suppose that in Problem 5.3.1 69.25 is replaced by 78.0. What are the values of s^2 and s? Does the variance seem sensitive to the extreme observation?

3. Suppose, as in Problem 5.2.9, that all observations x_1, x_2, ..., x_n are replaced by $y_i = ax_i + b$.
    (a) Show that if a = 1, then
    $$s_y^2 = s_x^2$$
    This implies that if the same constant is added to a set of n observations, then the variance is not changed.
    (b) Show that if a is any real value, then
    $$s_y^2 = a^2 s_x^2 \quad \text{and} \quad s_y = |a|s_x$$

4. The average of 20 Fahrenheit temperatures is 77 with a standard deviation of 9. Using the relationship $C = (5/9)(F - 32)$ find the values of the mean, standard deviation and variance of these temperatures in degrees Celsius.

5. Suppose 10 "yardsticks" have a mean of $\bar{x} = 37.2$ in. and a standard deviation of $s = 1.44$ in. What is the mean and standard deviation of these measurements in yards?

6. The uniform distribution on the set $S = \{0, 1, 2, \ldots, 9\}$ has $E(X) = 4.5$ and $Var(X) = 8.25$. The following is a sample of 25 observations from this distribution (i.e., a sample of random digits): 6, 0, 6, 7, 0, 2, 0, 7, 7, 7, 3, 7, 7, 4, 1, 1, 1, 6, 0, 0, 1, 4, 4, 4, 0. Find the value of s^2 for this sample. Does it seem a reasonable estimate of $Var(X)$?

7. The coefficient of variation, defined as $s/\bar{x}$ (or $100s/\bar{x}$ in percentage terms), gives a measure of variability of the sample as a proportion or percentage of the observed mean. This measure is used for measurements which are naturally measured in positive units (such as heights). Find $s/\bar{x}$ for the heights in Problem 5.3.1.

8. Prove that

(a) $\displaystyle\sum_{i=1}^{n} (x_i - \bar{x})^2 = \sum_{i=1}^{n} x_i^2 - n(\bar{x})^2$

$\displaystyle\qquad\qquad\qquad = \sum_{i=1}^{n} x_i^2 - \frac{(\sum_{i=1}^{n} x_i)^2}{n}$

(b) Use the data of Sample 2 of this section to illustrate the equalities in part (a).

(c) Show that

$$s^2 = \frac{\sum_{i=1}^{n} x_i^2 - n(\bar{x})^2}{n - 1}$$

9. Use the Chebyshev inequality to show that in any sample of size n, $P(|x_i - \bar{x}| \geq ks) \leq (n - 1)/nk^2$, for any observation x_i in the sample.

Exercises 5.3

1. Use the program VARSAMP to sample 1000 variances from the standard
    normal distribution with sample sizes 5, 10, and 40. Draw rela-
    tive frequency histograms for each set of observations.

2. Expand the program written in Exercise 5.2.2 to calculate the
    sample variance and standard deviation. Use the height data of
    that exercise to calculate s^2, s, and $s/\bar{x}$.

3. Use the program in Exercise 5.3.2 together with the data of
    Exercise 5.2.3 to find the variance, standard deviation, and
    coefficient of variation of the income data given there. Which
    of the two data sets is more variable, this income data or the
    height data of the previous exercise?

5.4   THE CENTRAL LIMIT THEOREM

A crucial theorem in probability and statistics concerns the distribution of means, which have been calculated from random samples from
a population. As mentioned above such means, $\bar{X}$, are random variables.
We have seen that $E(\bar{X}) = \mu$ and $Var(\bar{X}) = \sigma^2/n$, where μ and σ^2 are the
population expectation and variance. In order to give an example,
to suggest the behavior of sample means, let us again look at the
histograms in Fig. 5.2.1. Here the underlying distribution is the
discrete uniform distribution, assigning equal probabilities to the
integers 1, 2, ..., 10. The distribution of means for n = 3 again
takes on values on the interval [1, 10], but we notice that the
histogram reflects a "bell-shaped" appearance, with small probability
in the tails and substantial probability near the center.

    We shall pursue this example further, but first we consider
standardizing a random variable X. If $E(X) = \mu$ and $Var(X) = \sigma^2$,
the standardized random variable

$$Z = \frac{X - \mu}{\sigma}$$

has two important properties. First, $E(Z) = E((X - \mu)/\sigma) =$
$E(X - \mu)/\sigma = (\mu - \mu)/\sigma = 0$. Secondly, it is straightforward to show
that $Var(Z) = 1$. Thus, as we have found the expectation and variance
of $\bar{X}$ to equal μ and σ^2/n, respectively, we see that

$$Z = \frac{\bar{X} - \mu}{\sigma/\sqrt{n}}$$

has expectation zero and variance one. Although we shall not prove
it here, the variable $Z = (\bar{X} - \mu)/\sigma/\sqrt{n}$ "behaves" like the standard
normal variable for large n, *regardless of the distribution from
which the means are sampled*, providing only $E(X) = \mu$ and $Var(X) = \sigma^2$
exist. This is formally stated in the following:

THEOREM 5.4.1. (The Central Limit Theorem). If X_1, X_2, $\ldots$, X_n are
independently and identically distributed random variables, with
common distribution given by the random variable X, for which $E(X) =$
μ and $Var(X) = \sigma^2$, then

$$\lim_{n \to \infty} P(\frac{\bar{X} - \mu}{\sigma/\sqrt{n}} \leq t) = F_Z(t)$$

for all t, where $F_Z(t)$ is the cdf for the standard normal distribution. We say $\bar{X}$ is asymptotically normally distributed, with mean μ
and variance σ^2/n.

We use the program CLT to simulate 10,000 means of samples of
size n = 5, 10, and 50 from the discrete uniform distribution on
1, 2, $\ldots$, 10. These means are transformed by the standardizing
transformation

$$\frac{\bar{X} - 5.5}{\sqrt{99/12n}}$$

A histogram for the case of n = 10 and the corresponding histogram
for the standard normal distribution are displayed in Fig. 5.4.1.
The close relationship between the two histograms is apparent. For

CLT

```
10 REM COMPUTES K STANDARDIZED MEANS OF SAMPLES OF SIZE R
20 REM FROM DISCRETE UNIFORM DISTRIBUTION ON 1,2,...,N
30 PRINT "WHAT VALUE OF N";" FOR DISCRETE";" UNIFORM DIST?"
40 INPUT N
50 DIM Z(18),F(18)
60 PRINT "HOW MANY MEANS?"
70 INPUT K
80 PRINT "WHAT SAMPLE SIZE?"
90 INPUT R
100 FOR I = 1 TO 16
110 READ Z(I)
120 NEXT I
130 FOR L = 1 TO K
140 FOR J= 1 TO R
150 LET M = 1 +INT(RND*N)
160 LET T =T+M/R
170 NEXT J
180 LET T = T-(N+1)/2
190 LET T =T/SQR((N^2-1)/(12*R))
200 FOR I = 1 TO 16
210 IF T>-4+I/2 THEN 240
220 LET F(I)=F(I)+1/K
230 GO TO 270
240 IF I<16 THEN 260
250 LET F(I)=F(I)+1/K
260 NEXT I
270 LET T=0
280 NEXT L
290 PRINT "FREQUENCY TABLE";" FOR TRANSFORMED";" MEANS"
300 PRINT
310 PRINT "LOWER END","UPPER END","FREQUENCY","NORMAL PROB"
320 PRINT
330 FOR I = 1 TO 16
340 PRINT -4+(I-1)/2,-4+I/2,F(I),Z(I)
350 LET A =(-4.25+I/2)*F(I)+A
360 LET B = B +F(I)
370 NEXT I
380 PRINT "AVERAGE ";"TRANSFORMED ";"MEAN =";A
390 PRINT "TOTAL FREQ= ";B
400 DATA .0002,.0011,.0049,.0166,.0440,.0919,.1498,.1915
410 DATA .1915,.1498,.0919,.0440,.0166,.0049,.0011,.0002
420 END
*
```

```
 RUN
WHAT VALUE OF N FOR DISCRETE UNIFORM DIST?
? 10
HOW MANY MEANS?
? 10000
WHAT SAMPLE SIZE?
? 5
FREQUENCY TABLE FOR TRANSFORMED MEANS
```

LOWER END	UPPER END	FREQUENCY	NORMAL PROB
-4	-3.5	0	0.0002
-3.5	-3	0.0004	0.0011
-3	-2.5	0.0037	0.0049
-2.5	-2	0.0159	0.0166
-2	-1.5	0.0411	0.044
-1.5	-1	0.1166	0.0919
-1	-0.5	0.1442	0.1498
-0.5	0	0.1783	0.1915
0	0.5	0.1823	0.1915
0.5	1	0.1423	0.1498
1	1.5	0.1166	0.0919
1.5	2	0.0387	0.044
2	2.5	0.0154	0.0166
2.5	3	0.0044	0.0049
3	3.5	0.0001	0.0011
3.5	4	0	0.0002

```
AVERAGE TRANSFORMED MEAN =-4.79999 E-3
TOTAL FREQ=  0.999999
```

*

```
 RUN
WHAT VALUE OF N FOR DISCRETE UNIFORM DIST?
? 10
HOW MANY MEANS?
? 10000
WHAT SAMPLE SIZE?
? 10
FREQUENCY TABLE FOR TRANSFORMED MEANS
```

LOWER END	UPPER END	FREQUENCY	NORMAL PROB
-4	-3.5	0	0.0002
-3.5	-3	0.0003	0.0011
-3	-2.5	0.0061	0.0049
-2.5	-2	0.0143	0.0166
-2	-1.5	0.0489	0.044
-1.5	-1	8.47999 E-2	0.0919
-1	-0.5	0.1593	0.1498
-0.5	0	0.2039	0.1915
0	0.5	0.174	0.1915
0.5	1	0.16	0.1498
1	1.5	8.15999 E-2	0.0919
1.5	2	0.0458	0.044
2	2.5	0.0156	0.0166
2.5	3	0.0044	0.0049
3	3.5	0.001	0.0011
3.5	4	0	0.0002

```
AVERAGE TRANSFORMED MEAN =-0.01585
TOTAL FREQ=  0.999999
```

```
*
```

```
 RUN
WHAT VALUE OF N FOR DISCRETE UNIFORM DIST?
? 10
HOW MANY MEANS?
? 10000
WHAT SAMPLE SIZE?
? 50

FREQUENCY TABLE FOR TRANSFORMED MEANS
```

LOWER END	UPPER END	FREQUENCY	NORMAL PROB
-4	-3.5	0.0002	0.0002
-3.5	-3	0.0014	0.0011
-3	-2.5	0.0051	0.0049
-2.5	-2	0.0184	0.0166
-2	-1.5	0.0437	0.044
-1.5	-1	9.10999 E-2	0.0919
-1	-0.5	0.1461	0.1498
-0.5	0	0.2074	0.1915
0	0.5	0.1947	0.1915
0.5	1	0.1416	0.1498
1	1.5	8.63999 E-2	0.0919
1.5	2	0.0437	0.044
2	2.5	0.0147	0.0166
2.5	3	0.0043	0.0049
3	3.5	0.001	0.0011
3.5	4	0.00002	0.0002

```
AVERAGE TRANSFORMED MEAN =-0.02425
TOTAL FREQ=  0.999999


*
```

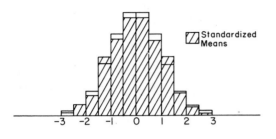

Fig. 5.4.1

n = 50, the simulation of the program CLT indicates that the standard
normal distribution and that of $\bar{X}$ are practically identical.

The central limit theorem has many important practical consequences in statistics. As the distribution of means of "large samples"
can be considered to be approximately normally distributed, questions
of inference concerning μ may be reduced to questions of inference
concerning μ in the case of a normal distribution. Additionally, as
many statistics and random variables can be thought of as sums of
independent and identically distributed (iid) random variables, the
distribution of such sums or an average of such sums can be approximated using the normal distribution.

As an example of this we shall consider X_1, X_2, ..., X_n to be
iid random variables, each with the exponential distribution. Thus
the random variable X, representing the common distribution has pdf
$f(x) = e^{-x}$ for x ≥ 0. However, $Y = X_1 + X_2 + \cdots + X_n$ would represent the waiting time until the nth Poisson event, when the events
have arrival rate 1/λ = 1. As stated in Sec. 4.4, this random variable has the gamma distribution with α = n and β = 1. Hence the pdf
for Y is

$$f_Y(y) = \frac{y^{n-1}}{\Gamma(n)} e^{-y} \qquad \text{for } y > 0$$

As the average waiting time for n arrivals is given by $Y/n = \bar{X}$ we
can use Theorem 4.5.1 to show that the pdf for $\bar{X}$ is given by

$$f_{\bar{X}_n}(x) = \frac{n(xn)^{n-1}}{\Gamma(n)} e^{-nx} \qquad \text{for } x > 0$$

The probability density functions for $\bar{X}_n$ are plotted in Fig. 5.4.2 for n = 1, 4, and 16. As is readily apparent, the exponential pdf is neither symmetric nor bell-shaped in appearance. However the averages of 16 independent observations from this population have a distribution which is decidedly bell-shaped in appearance. The expectations for each of these distributions is one, while the variance is 1/n. The approach to normality for fairly small n is quite remarkable, considering the shape of the parent distribution.

As a practical example of the use of the central limit theorem, let us assume that the dollar amount of losses due to theft at a clothing store per week is a uniformly distributed random variable on [50, 150] with the measurements made in dollars. Assuming independence, the total loss in a 50-week period is

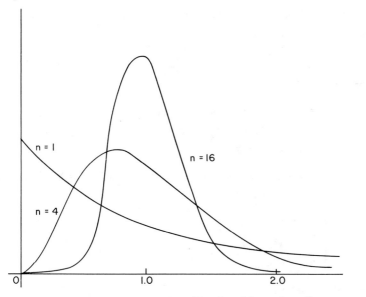

Fig. 5.4.2   Probability density functions for the average of n = 1, 4, and 16 observations.

$$L = \sum_{i=1}^{50} X_i$$

where X_i is the loss in the ith week. We find $E(X) = 100$ and $Var(X)$
$Var(X) = (100)^2/12$, where X is the common distribution of the X_i.
Thus $E(L) = 5000$ and $Var(L) = 50(100)^2/12$. Standardizing L, we find
that

$$\frac{L - 5000}{100\sqrt{50/12}} \sim Z$$

i.e., is approximately a standard normal variable. We may be interested in the probability L does not exceed \$6000. This is found as

$$P(L \leq 6000) = P(Z \leq 1000/100\sqrt{50/12}) = P(Z \leq 4.90)$$
$$= 1$$

Similarly the upper 95th percentile of the distribution of losses
over the 50-week period is $5000 + (100\sqrt{50/12})Z_{0.05} = 5335.78$. The
central limit theorem has allowed us to state that the sum of 50 iid
uniform variables, when properly standardized, has approximately a
standard normal distribution.

Problems 5.4

1. Assume that n = 25 observations are taken from a normal distri-
    bution X with $E(X) = 500$ and $\sigma_X = 100$. Find the expectation
    and standard deviation of $\bar{X}$.

2. Assume that the length of telephone calls T in minutes is described
    by the distribution with pdf

$$f(t) = \begin{cases} \frac{1}{2} e^{-t/2} & t > 0 \\ \\ 0 & t \leq 0 \end{cases}$$

    (a) Find $E(T)$ and $Var(T)$.

(b)   Let $S = T_1 + T_2 + \cdots + T_n$, where the T_i are independent
      random variables with the distribution given by $f(t)$. Find
      $E(S)$ and $Var(S)$. (These will be functions of n.)

(c)   Find the probability that the total length of 100 calls
      does not exceed 4 hr (T is assumed to be measured in minutes).
      Note that $(S - E(S))/\sqrt{Var(S)}$ is approximately a standard
      normal variable.

3. Assume that X is uniformly distributed on $[-0.5, 0.5]$.

(a)   Find $E(X)$ and $Var(X)$.

(b)   Find $E(\bar{X})$ and $Var(\bar{X})$ for $n = 48$.

(c)   Approximate the probability that an average of 48 indepen-
      dent observations from X satisfies

   $$-1/8 < \bar{X} < 1/8$$

(d)   What is $P(-1/8 < X < 1/8)$.

4. Assume X has pdf given by

$$f(x) = \begin{cases} 6x(1 - x) & \text{on } [0, 1] \\ \\ 0 & \text{elsewhere} \end{cases}$$

(a)   Find $P(0.25 < X < 0.75)$.

(b)   Approximate $P(0.25 < \bar{X} < 0.75)$ for $n = 30$.

Exercises 5.4

1. Use CLT to sample 1000 means of sample size 20 from the discrete
    uniform distribution on the integers 1, 2, 3, 4, 5. Construct
    a relative frequency histogram for the transformed means. How
    does this compare with the histogram for the uniform variable?

2. Use the DEF FNF(X) command in BASIC to evaluate the pdf of the
    average of 20 exponential random variables with $\lambda = 1$. Print
    the values for $X = (k - 1)/20$ for $k = 1, 2, \ldots, 41$. Plot a
    graph of this function. What general shape does the curve have?

3. Write a program to compute the averages of n = 100 observations from the distribution with cdf

$$F(x) = 1 - \frac{1}{x} \quad \text{for } x \geq 1$$

Find 1000 such means and plot a relative frequency histogram of these means. Does the resulting histogram appear symmetric? How can you explain this?

5.5 ESTIMATION OF μ USING LARGE SAMPLES

To illustrate the usefulness of the central limit theorem in questions of inference, we consider here the problem of estimating the expectation μ of a population with variance σ^2, using large samples. Questions of this type are common in practice, as we indicate in the following examples. What is the average net weight of packages of a breakfast food? What is the true average time required for a 6-year-old child to answer an arithmetic problem correctly? What is the average age of adult male voters in a city? To provide answers to such questions we assume that a random sample of size n is taken from a population described by the random variable X with $E(X) = \mu$ and $Var(X) = \sigma^2$. We assume for the moment that μ is unknown, but that σ^2 is known.

The random variable $\bar{X} = \sum_{i=1}^{n} X_i / n$ is, as mentioned above, a reasonable estimate for μ. However, this provides only a single number as an estimate of μ and gives no indication of the error associated with the estimate. Assuming large n, the asymptotic normality of $\bar{X}$ can be used to provide information of this kind. We denote by Z_α the value of Z in the standard normal distribution satisfying $P(Z \geq Z_\alpha) = \alpha$, as indicated by the tail area in Fig. 5.5.1. By the symmetry of the normal distribution we see that

$$P(|Z| \leq Z_{\alpha/2}) = 1 - \alpha$$

or, approximately, for large n

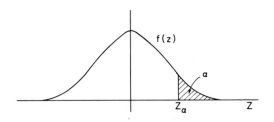

Fig. 5.5.1 The standard normal pdf

$$P(|\bar{X} - \mu/\sigma/\sqrt{n}| \leq Z_{\alpha/2}) = 1 - \alpha$$

or equivalently,

$$P(\bar{X} - (\alpha/\sqrt{n})Z_{\alpha/2} \leq \mu \leq \bar{X} + (\sigma/\sqrt{n})Z_{\alpha/2}) = 1 - \alpha$$

Hence for any preassigned probability α, the random interval $[\bar{X} - (\sigma/\sqrt{n})Z_{\alpha/2}, \bar{X} + (\sigma/\sqrt{n})Z_{\alpha/2}]$ covers μ with probability $1 - \alpha$. This means, in frequency terms, that if a large number of random samples of n are taken from a population and the interval centered at $\bar{X}$ of length $2(\sigma/\sqrt{n})Z_{\alpha/2}$ is marked off on the real-number axis, then $(1 - \alpha)100\%$ of these intervals will contain μ. Of course, we generally only obtain one such interval from a single sample. The interval is referred to as a confidence interval for μ with confidence coefficient $(1 - \alpha)100\%$. The probability $1 - \alpha$ is associated with the procedure which has led to the construction of the interval and not with the particular interval itself.

Suppose that we wish to obtain a confidence interval for the reaction time to a stimulus. We will assume that we have n = 100 individuals and that $\sigma = 2$. The sample yields an average reaction time $\bar{x} = 5.1$ min. A 95% confidence interval for μ is obtained as follows:

$$(1 - \alpha)100 = 95 \quad \text{yields} \quad \alpha = 0.05$$
$$Z_{\alpha/2} = Z_{0.025} = 1.96 \quad \text{(from App. B. I)}$$

The interval is

$$\bar{x} \pm \frac{\sigma}{\sqrt{n}} Z_{\alpha/2} = 5.1 \pm 0.2(1.96)$$
$$= 5.1 \pm 0.392 \text{ or } [4.708, 5.492]$$

Although we call this interval a 95% confidence interval for μ, in fact the probability that $[4.708, 5.492]$ contains μ is either 1 or 0 (as μ is in or not in this interval). However, it is the case that approximately 95% of intervals constructed in this way will contain μ.

To illustrate this point further we consider the program CONFINT which computes K 95% confidence intervals for samples of size N taken from the exponential distribution with mean μ. The program has been run for $\mu = 5$, K = 1000, and N = 20, 40, and 80. In each case the first 10 intervals are printed out. For N = 40, for example, all but the 10th interval contain the parameter $\mu = 5$. The program calculates the interval $\bar{X} \pm (s/\sqrt{n})Z_{0.025}$, using s as an estimate of σ, assuming a large sample size. The proportions of intervals containing μ in the 1000 samples are found to be 0.924, 0.94, and 0.928 for n = 20, 40, and 80, respectively. For large n, these confidence intervals become better for several reasons. First, the estimate s of σ improves. Second, the approximation to normality becomes better for larger n. Thirdly, the length of confidence intervals becomes shorter, on the average. This is most clearly seen by observing that if σ is known the length of the confidence interval is given by $2(\sigma/\sqrt{n})Z_{\alpha/2}$, which clearly decreases with n.

Problems 5.5

1. A sample of n = 64 college students yielded on average IQ score of $\bar{x} = 121.5$ with s = 18.

    (a) Find a 95% confidence interval for μ for the population from which these students were selected, assuming the population standard deviation is $\sigma = 15$.

    (b) Find a 95% confidence interval as in part (a) assuming σ is unknown.

CONFINT

```
10 REM 95 PERCENT CONFIDENCE INTERVALS FOR MU FOR EXPONENTIAL
20 PRINT "WHAT IS SAMPLE";" SIZE?"
30 INPUT N
40 PRINT "WHAT IS TRUE";" MEAN?"
50 INPUT M
60 PRINT "HOW MANY ";"CONFIDENCE";" INTERVALS?"
70 INPUT K
80 PRINT "ENDPOINTS OF ";"FIRST 10 INTERVALS"
90 PRINT "L","U"
100 FOR J = 1 TO K
110 FOR I = 1 TO N
120 LET X =-M*LOG(1-RND)
130 LET X1=X+X1
140 LET X2=X^2+X2
150 NEXT I
160 LET A = X1/N
170 LET B =SQR((X2-N*A^2)/(N-1))
180 LET B=B/SQR(N)
190 LET L = A-1.96*B
200 LET U = A+1.96*B
210 IF J>10 THEN 230
220 PRINT L,U
230 IF M<L THEN 270
240 IF M>U THEN 270
250 LET F(1)=F(1)+1/K
260 GO TO 280
270 LET F(2)=F(2)+1/K
280 LET X1=0
290 LET X2=0
300 NEXT J
310 PRINT
320 PRINT "PROPORTION COVERING","PROP NOT COVERING"
330 PRINT F(1),"",F(2)
340 GO TO 350
350 END
*
```

```
 RUN
WHAT IS SAMPLE SIZE?
? 20
WHAT IS TRUE MEAN?
? 5
HOW MANY CONFIDENCE INTERVALS?
? 1000
ENDPOINTS OF FIRST 10 INTERVALS
L                U
 2.28229         4.56973
 2.48621         6.51059
 2.07132         4.67738
 4.1221          6.66326
 3.68824         8.5829
 4.19414         9.57308
 2.46434         5.62246
 3.44853         7.69141
 2.69294         6.59116
 2.90931         7.04533


PROPORTION COVERING              PROP NOT COVERING
0.924                            0.076



*RUN
WHAT IS SAMPLE SIZE?
? 40
WHAT IS TRUE MEAN?
? 5
HOW MANY CONFIDENCE INTERVALS?
10000-
ENDPOINTS OF  FIRST 10 INTERVALS
L                U
 2.80755         5.11686
 3.43107         5.33596
 4.71107         8.30811
 3.47965         6.13373
 3.40615         6.21323
 3.5967          6.80164
 3.52781         5.79695
 4.65295         8.68237
 2.91102         5.59845
 2.61319         4.58944


PROPORTION COVERING              PROP NOT COVERING
0.94                             0.06


*
```

```
 RUN
WHAT IS SAMPLE SIZE?
? 80
WHAT IS TRUE MEAN?
? 5
HOW MANY CONFIDENCE INTERVALS?
? 1000
ENDPOINTS OF FIRST 10 INTERVALS
L                U
 3.42777        4.91795
 4.53192        6.78436
 3.94521        6.06365
 4.49517        6.83487
 3.09623        4.75982
 3.85123        6.10175
 3.23937        5.0517
 4.05192        6.58775
 3.99525        5.98547
 3.55178        5.32257

PROPORTION COVERING        PROP NOT COVERING
 0.928                      0.072


*
```

(c) Find a 99% confidence interval as in part (b).

2. The average GPA (grade point average) of 225 randomly selected
    students from a university was 3.12 with s = 0.31. Find 95 and
    99% confidence intervals for the true average GPA of students at
    this university.

3. Weight losses of 81 individuals on the first 30 days on a diet
    averaged 8.2 lb with sample variance of 4 lb^2.
    (a) Find a 99% confidence interval for μ, the true average loss
         of individuals on this diet.
    (b) The diet orginator claims that the true average weight loss
         in the first 30 days is 10 lb. Do the data support this
         assertion? Why or why not?

4. Suppose we wish to have a probability of at least $1 - \alpha$ that the
    error in estimating μ by $\bar{X}$ is no more than d in magnitude. For
    large n, we have

    $$P(|\bar{X} - \mu|/(\sigma/\sqrt{n}) < Z_{\alpha/2}) \doteq 1 - \alpha$$

    We wish the statement $P(|\bar{X} - \mu| < d) = 1 - \alpha$ to be true.
    (a) Assuming α, d, and σ are known, show that a sample of size
         $n = ((\sigma/d)Z_{\alpha/2})^2$ will satisfy the requirements.
    (b) Suppose IQ scores have $\sigma = 15$ and we wish to estimate the
         true average IQ of a large group using $\bar{X}$. If we wish a
         maximum absolute error of 3 with probability 0.95, what
         sample size is required?
    (c) If we wish the maximum error to be 1.5 with probability
         0.95, what sample size is required?

5. An assembly-line process is used to fill packages with flour.
    It is desired to estimate the true average net weight. Long
    experience indicates that the true variance of the net weight is
    0.25 oz^2. What sample size should be used in order to ensure
    the estimate $\bar{X}$ differs from μ by no more than 0.05 oz with prob-
    ability of 0.95? (Use Problem 5.5.4.)

Exercises 5.5

1. Use the program CONFINT with true mean $\mu = 4$, n = 25, and k = 100. Omit statement 210 so that the endpoints of the 100 intervals are printed out. Identify those which fail to cover μ.

2. Use the program written in Exercise 5.3.2 to find a 95% confidence interval for the true average height μ of the population from which the data of Exercise 5.2.2 were sampled.

3. Use the program in Exercise 5.5.2 to find a 95% confidence interval for μ for the income data in Exercise 5.2.3.

4. Alter CONFINT to provide 80, 90, 95, or 99% confidence intervals, where the program requests the confidence coefficient $1 - \alpha$.

5.6   APPROXIMATION OF BINOMIAL PROBABILITIES BY THE NORMAL
      DISTRIBUTION

As another example of the importance of the central limit theorem we condider approximating binomial probabilities in the case that the number of trials n is large. We have seen before that the Poisson distribution is a good approximation to the binomial distribution for large n and small p. The Poisson approximation to the binomial distribution has been described as satisfactory if $n \geq 20$ and $p \leq 0.05$ in several texts. It is very good for $n \geq 100$ and $p \leq 0.1$. However, for large values of n and $0.1 \leq p \leq 0.9$, it is often very tedious to compute binomial probabilities, and the Poisson approximation does not apply. However, in Sec. 3.3 we have seen that the binomial variable X could be written as $X = \Sigma_{k=1}^{n} I_k$, where the I_k were independently and identically distributed indicator variables with $P[I_k = 0] = 1 - p$ and $P[I_k = 1] = p$.

The central limit theorem allows us to assert that for $X \sim B(n,p)$,

$$\frac{X - np}{\sqrt{np(1 - p)}} = \frac{\bar{X} - p}{\sqrt{p(1 - p)/n}}$$

has approximately the standard normal distribution. This allows use of the normal distribution to approximate binomial probabilities for

large n. Suppose a basketball player makes 80% of his foul shots.
What is the probability he makes between 75 and 85 of his next 100
shots.

We have

$$P(75 \le X \le 85) = \sum_{k=75}^{85} \binom{100}{k}\left(\frac{4}{5}\right)^k\left(\frac{1}{5}\right)^{100-k}$$

This is an impossible calculation by hand, and even a somewhat annoy-
ing calculation using the computer. However, we can use the normal
approximation as follows:

$$
\begin{aligned}
P(75 \le X \le 85) &= P(75 - np \le X - np \le 85 - np) \\
&= P\left(\frac{75 - np}{\sqrt{np(1 - p)}} \le \frac{X - np}{\sqrt{np(1 - p)}} \le \frac{85 - np}{\sqrt{np(1 - p)}}\right) \\
&= P\left(\frac{75 - 80}{\sqrt{16}} \le Z \le \frac{85 - 80}{\sqrt{16}}\right) \\
&= P(-1.25 \le Z \le 1.25)
\end{aligned}
$$

where Z has the standard normal distribution. From Table B.I in
Appendix B, we find $P(Z \le 1.25) = 0.7888$.

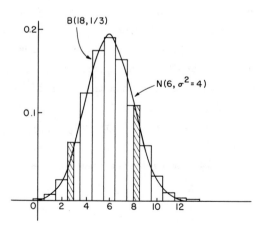

Fig. 5.6.1

In order to see how this approximation may be improved, we consider the case n = 18 and p = 1/3. Hence E(X) = np = 6 and Var(X) = 4. In Fig. 5.6.1 the probability histogram for this binomial distribution and the normal density function with $\mu = 6$ and $\sigma^2 = 4$ are graphed. We are estimating the area under the histogram, by the area under the corresponding normal probability density function. If we consider finding P(3 ≤ X ≤ 8), we can run the program BINOM and find the exact probability to be 0.859750. The normal approximation yields

$$P\left(\frac{3-6}{2} \leq Z \leq \frac{8-6}{2}\right) = P(-1.5 \leq Z \leq 1)$$

$$= F_Z(1) - F_Z(-1.5)$$

$$= 0.8413 - 0.0668 = 0.7745$$

It is clear from Fig. 5.6.1 that the approximation would be improved if we find the area under the normal density function from 2.5 to 8.5, as this will give a better approximation to the area under the histogram. The shaded area has not been approximated. Hence we apply a "continuity correction" and find

$$P\left(\frac{2.5-6}{2} \leq Z \leq \frac{8.5-6}{2}\right) = F_Z(1.25) - F_Z(-1.75)$$

$$= 0.8944 - 0.0401 = 0.8543$$

This result is much closer to the true probability.

In general, if we approximate $P(k_1 \leq X \leq k_2)$, we use the following approximation:

$$P\left(\frac{k_1 - 0.5 - np}{\sqrt{np(1-p)}} \leq Z \leq \frac{k_2 + 0.5 - np}{\sqrt{np(1-p)}}\right)$$

$$= F_Z\left(\frac{k_2 + 0.5 - np}{\sqrt{np(1-p)}}\right) - F_Z\left(\frac{k_1 - 0.5 - np}{\sqrt{np(1-p)}}\right)$$

Hence for the basketball player we would estimate P(75 ≤ X ≤ 85) as

$$F_Z\left(\frac{85.5 - 80}{4}\right) - F_Z\left(\frac{74.5 - 80}{4}\right) = 0.8310$$

This will be an improvement over the previous estimate. If we wish
to estimate P(X = k), say P(X = 6) for n = 18, and p = 1/3, we use
the estimate

$$F_Z\left(\frac{5.5 - 6}{2}\right) - F_Z\left(\frac{5.5 - 6}{2}\right) = F_Z(0.25) - F_Z(-0.25) = 0.1974$$

The actual probability is 0.1963. This is quite a good approximation
considering that n is only 18. We turn to a more detailed discussion
of the question of the estimation of parameters in the next chapter.

Problems 5.6

1. Genetic theory predicts that certain flowers will be white with
    probability 1/4 and yellow with probability 3/4. Approximate
    the probability that of 300 such flowers between 210 and 240
    flowers inclusive will be yellow. (Make two calculations with-
    out and with the continuity correction.)

2. In a triangle test, persons unable to detect the specified char-
    acteristic will be correct by chance with probability 1/3. For
    n = 72 persons, find the probability that between 20 and 30 will
    identify the characteristic correctly. Again approximate this
    probability without and with the continuity correction.

3. A fair coin is tossed n times.
    (a) If n = 16, find the exact probability of 8 heads and the
        corresponding normal approximation.
    (b) If n = 100, use the normal approximation to estimate the
        probability of exactly 50 heads.
    (c) For even n, find an approximate expression for exactly n/2
        heads in n tosses.
    (d) What is the limit of the probability in [part (c)] as n → ∞?

4. An operation is successful with probability 0.9. What is the
    probability that in the next 100 operations,
    (a) Ninety-three or more will be successful.
    (b) Eighty-five or more will be successful.
    (Use the continuity correction.)

5. A batter gets a hit with probability p = 1/4. Compute his prob-
    ability of exactly 4 hits in his next 12 at bats.
    (a) Using the binomial distribution with n = 12.
    (b) Using the normal approximation to the binomial distribution
         with the continuity correction.

6. For large n, assume we estimate an unknown probability p by the
    estimator $\hat{p}$ defined as the frequency of the event in n trials.
    (a) Show that

    $$P(|\hat{p} - p| \leq 2\sqrt{\frac{p(1 - p)}{n}} \doteq 0.9544$$

    and

    $$P(|\hat{p} - p| \leq \frac{1}{\sqrt{n}}) \geq 0.9544$$

    (Note: $p(1 - p) \leq 1/4$.)
    (b) Show that by choosing $n = 1/d^2$ we can achieve, for a pre-
         determined value of d, the result that the estimate $\hat{p}$ will
         not differ in magnitude from p by more than d with probabil-
         ity at least 0.9544.

7. In order to estimate the percent of an electorate who prefer the
    Republican candidate to all others, how large a sample size should
    be used in order to have a probability at least 0.95 that the
    error of the estimate will not exceed 2.5 percentage points?
    What sample size is required if the error is not to exceed 2 per-
    centage points? Use Problem 5.6.6.

8. (a) We wish to estimate the proportion of the registered voters
         in a city in favor of a bond issue. How large a sample of
         these voters do we require in order that the probability is
         at least 0.95 that the error in the estimate of this propor-
         tion does not exceed 5 percentage points?
    (b) If the allowable error is cut in half (2.5 percentage points),
         what happens to the sample size?
    (c) If the allowable error is divided by k (5/k), what happens
         to the result in parts (a)?

Exercise 5.6

1. Write a BASIC program to evaluate the probability density function
    of the normal density function used to approximate the binomial
    probabilities for n = 25 and p = 1/2. Evaluate the pdf on the
    interval [0, 25] with step size 0.5. Plot this function and the
    probability histogram for B(25,0.5) using the same axes.

ESTIMATION OF PARAMETERS

6.1   INTRODUCTION AND DEFINITIONS

As has been mentioned in the previous chapter, one of the important inferential questions considered in statistics is the estimation of parameters of a theoretical distribution. Here we shall consider some important properties of statistical estimators. It is conventional to denote an estimator of the parameter θ by $\hat{\theta}_n$. We recall that $\hat{\theta}_n$ is a real-valued function of the observations in a random sample of size n. The statistics $\bar{X}$ and S^2 are particular examples of such estimators (of the parameters μ and σ^2, respectively). The estimator $\hat{\theta}_n$ is called a *point estimate* of θ because a random sample provides a single real number estimating θ. On the other hand, it should be kept clearly in mind that $\hat{\theta}_n$ is a random variable, as different random samples of size n will yield different values of $\hat{\theta}_n$.

We next define certain properties of an estimator $\hat{\theta}_n$ of a parameter θ, which are considered to be important.

DEFINITION 6.1.1.   The estimator $\hat{\theta}_n$ is an *unbiased estimator* of θ if $E(\hat{\theta}_n) = \theta$.

In words, the expected value of the random variable $\hat{\theta}_n$ is equal to the parameter being estimated. We have seen in the previous chapter that $\bar{X}$ is an unbiased estimator of μ, the expectation of the population from which the sample has been taken, as $E(\bar{X}) = \mu$. More colloquially, a statistic $\hat{\theta}_n$ is an unbiased estimator of a parameter θ if it is equal to θ "on the average." We have seen that

$$S_*^2 = \sum_{i=1}^{n} \frac{(X_i - \bar{X})^2}{n}$$

is not an unbiased estimator of σ^2 as $E(S_*^2) = ((n - 1)/n)\sigma^2$, while S^2 is an unbiased estimator of σ^2.

The next property of an estimator concerns the behavior of the estimator $\hat{\theta}_n$ as n becomes large. Such a property is called an *asymptotic property* of $\hat{\theta}_n$.

DEFINITION 6.1.2. As estimator $\hat{\theta}_n$ is called a *consistent estimator* of θ if for any $\varepsilon > 0$,

$$\lim_{n \to \infty} P(|\hat{\theta}_n - \theta| > \varepsilon) = 0.$$

This property means that the *probability* that $\hat{\theta}_n$ has a value more than ε away from θ becomes small (approaches zero) as the sample size n is increased. Using the Chebyshev inequality, we have seen in the previous chapter that $\bar{X}$ is a consistent estimator of μ if the population from which we are sampling has a finite theoretical variance σ^2.

If we write

$$E(\hat{\theta}_n - \theta)^2 = E(\hat{\theta}_n - E(\hat{\theta}_n) + E(\hat{\theta}_n) - \theta)^2$$

$$= E(\hat{\theta}_n - E(\hat{\theta}_n))^2 + (\theta - E(\hat{\theta}_n))^2$$

we see that the expression $E(\hat{\theta}_n - \theta)^2$, called the *expected mean square* (EMS) *error*, can be written as

$$\text{EMS}(\hat{\theta}_n) = \text{Var}(\hat{\theta}_n) + (\text{Bias}(\hat{\theta}_n))^2 \qquad (6.1.1)$$

where $\theta - E(\hat{\theta}_n)$ is called the *bias* of $\hat{\theta}_n$. It is straightforward to show that if $Var(\hat{\theta}_n) \to 0$ and $Bias(\hat{\theta}_n) \to 0$ as $n \to \infty$, then $\hat{\theta}_n$ is a consistent estimator of θ. For example consider $\tilde{X} = (X_1 + X_2 + \cdots + X_n)/(n - 1)$ as an estimate of μ for a population with theoretical variance σ^2.

$$Var(\tilde{X}) = Var\left(\frac{n\bar{X}}{n - 1}\right) = \left(\frac{n}{n - 1}\right)^2 Var(\bar{X})$$

$$= \left(\frac{n}{n - 1}\right)^2 \frac{\sigma^2}{n}$$

and

$$Bias(\tilde{X}) = \mu - E(\tilde{X}) = \mu - \frac{n\mu}{n - 1}$$

$$= -\frac{\mu}{n - 1}$$

Clearly $Var(\tilde{X}) \to 0$ and $Bias(\tilde{X}) \to 0$ as $n \to \infty$, so that $\tilde{X}$ is a *consistent* estimator of μ but not an unbiased estimator of μ.

A third property of $\hat{\theta}_n$ which we shall consider is the efficiency of one estimator $\hat{\theta}_n$ with respect to another estimator $\tilde{\theta}_n$, both estimating θ. The efficiency is measured in terms of the variance of an estimator when it is unbiased for θ. It is clear that for two unbiased estimators, the one with the smaller variance is to be preferred, as roughly, the smaller the variance the higher the probability that an estimator will have a value close to θ. (See Fig. 6.1.1.)

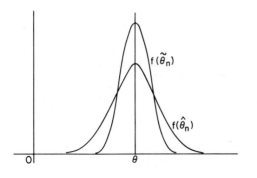

Fig. 6.1.1

DEFINITION 6.1.3. Let $\hat{\theta}_n$ and $\tilde{\theta}_n$ be two estimators of θ. The *efficiency* of $\hat{\theta}_n$ with respect to $\tilde{\theta}_n$ is given by

$$e(\hat{\theta}_n, \tilde{\theta}_n) = \frac{\text{EMS}(\tilde{\theta}_n)}{\text{EMS}(\hat{\theta}_n)}$$

The efficiency equals $\text{Var}(\tilde{\theta}_n)/\text{Var}(\hat{\theta}_n)$ if both estimators are unbiased.

DEFINITION 6.1.4. The *asymptotic efficiency* of $\hat{\theta}_n$ with respect to $\tilde{\theta}_n$ is given by

$$e = \lim_{n \to \infty} \frac{\text{EMS}(\tilde{\theta}_n)}{\text{EMS}(\hat{\theta}_n)}$$

if this limit exists.

For example consider $\tilde{X}$ and $\bar{X}$ mentioned above. From Fig. 6.1.1 we see that $\text{EMS}(\tilde{X}) = (n/(n-1))^2 \sigma^2/n + (\mu/(n-1))^2$ and $\text{EMS}(\bar{X}) = \sigma^2/n$. Hence the efficiency $e(\tilde{X}, \bar{X})$ is always *less* than 1 as it is equal to the ratio of the latter expected mean square to the former. The limit e is clearly 1 here.

Let us consider finding the best linear unbiased estimator (BLUE) of μ, based on a random sample of size n from a population with expectation μ and variance σ^2. The term *linear* refers to restricting our attention to estimates of the form

$$\sum_{i=1}^{n} a_i X_i$$

where $(X_1, X_2, \ldots, X_n)$ represents the observations in the random sample. As $E(X_i) = \mu$ and $\text{Var}(X_i) = \sigma^2$ for all i and the observations are independent, we find

$$E\left(\sum_{i=1}^{n} a_i X_i\right) = \left(\sum_{i=1}^{n} a_i\right)\mu \quad \text{and} \quad \text{Var}\left(\sum_{i=1}^{n} a_i X_i\right) = \left(\sum_{i=1}^{n} a_i^2\right)\sigma^2$$

As the estimate is to be unbiased, $\Sigma_{i=1}^n a_i = 1$, and as the best estimate has the smallest variance, we must minimize $\Sigma_{i=1}^n a_i^2$ subject to
this condition. Let us write $a_i = 1/n + e_i$, where from $\Sigma_{i=1}^n a_i = 1$,
it follows that $\Sigma_{i=1}^n e_i = 0$. Then

$$\sum_{i=1}^n a_i^2 = \sum_{i=1}^n (\frac{1}{n} + e_i)^2 = \frac{1}{n} + \sum_{i=1}^n e_i^2$$

Clearly this expression is minimized by setting $e_i = 0$ for all i, so
that $a_i = 1/n$ for $i = 1, 2, \ldots, n$. Thus $\bar{X} = \Sigma_{i=1}^n X_i/n$ is the best
linear unbiased estimator (BLUE) for μ.

Problems 6.1

1. Assume $(X_1, X_2, \ldots, X_n)$ represents a random sample of size n
   from a population X with $E(X) = \mu$ and $Var(X) = \sigma^2$. Assume also
   that n is even. Consider the estimators X^* and $\bar{X}$ where

   $$X^* = \sum_{i=1}^{n/2} \frac{X_i}{n/2}$$

   and $\bar{X}$ is the sample mean.
   (a) Show that X^* is an unbiased estimator of μ
   (b) Find $Var(X^*)$.
   (c) Show that the efficiency $e(X^*, \bar{X})$ is 1/2 for any even
       integer n.

2. Parallel the proof of Theorem 3.6.1 (the Chebyshev inequality)
   to show that
   (a) $P(|\hat{\theta}_n - \theta| \geq \epsilon) \leq EMS(\hat{\theta}_n)/\epsilon^2$ for any $\epsilon > 0$, in both the
       discrete and continuous cases, assuming $EMS(\hat{\theta}_n)$ exists.
   (b) Using Eq. (6.1.1) show that if $\lim_{n\to\infty} Var(\hat{\theta}_n) = 0$ and
       $\lim_{n\to\infty} Bias(\hat{\theta}_n) = 0$, then $\hat{\theta}_n$ is a consistent estimator of θ.

3. Let $(X_1, X_2, \ldots, X_n)$ be a random example of size n from the
   population described in Problem 6.1.1
   (a) Find $Bias(S_*^2)$.
   (b) Assuming $Var(S_*^2) \to 0$ as $n \to \infty$, show that S_*^2 is a
       consistent estimator of σ^2.

4. Assume that a random variable X is defined by the pdf

$$f(x;\beta) = \begin{cases} \dfrac{1}{\beta}e^{-x/\beta} & \text{for } x > 0 \\[2mm] 0 & \text{elsewhere} \end{cases}$$

(a) Prove that $\bar{X}$ is an unbiased estimator of β for a sample of size n.

(b) Find $\text{Var}(\bar{X})$ for a sample of size n.

5. Assume that a random variable X is defined by the pdf

$$f(x;\theta) = \begin{cases} \dfrac{1}{\theta} & \text{on } [0,\,\theta] \\[2mm] 0 & \text{elsewhere} \end{cases}$$

(a) Show that $2\bar{X}$ is an unbiased estimator of θ for a sample of size n.

(b) Find $\text{Var}(2\bar{X})$ for a sample of size n.

(c) Show that $2\bar{X}$ is a consistent estimator of θ.

6. (Continuation of Problem 6.1.5.) Let $X_{(n)} = \max_i(X_1, X_2, \ldots, X_n)$, the maximum of a sample of size n from the distribution described in Problem 6.1.5.

(a) Show that $P(X_{(n)} \le x) = (x/\theta)^n$ for x $[0,\,\theta]$.

(b) Show that $f_{X_{(n)}}(x) = nx^{n-1}/\theta^n$ on $[0,\,\theta]$ and 0 elsewhere.

(c) Find $\text{Var}(X_{(n)})$ and Bias $(X_{(n)})$.

(d) Use Problem 6.1.2 to demonstrate that $X_{(n)}$ is a consistent estimator of θ.

(e) Find $e(2\bar{X}, X_{(n)})$ and the limit of this efficiency as $n \to \infty$.

7. Consider random sampling from a population with probability function

$$f(x;p) = p^x(1 - p)^{1-x} \qquad x = 0,\, 1$$

In other words, we observe 1 with probability p and 0 with probability 1 - p. These are called *Bernoulli trials*.

(a) What is the distribution of $X = \sum_{i=1}^{n} X_i$?

(b) Find an unbiased estimate of p and the variance of this
estimate.

(c) Is your estimate in part (b) BLUE for p? Why or why not?

8. Assume that we are sampling from a population of tobacco seedlings, of which a proportion p have green stems and a proportion
1 - p have white stems. Out of 110 germinated seedlings 76 have
green stems. What is an estimate of p? Is the estimator which
you have used unbiased and consistent? (Use Problem 6.1.7.)

6.2   ESTIMATION OF LOCATION IN SMALL SAMPLES

In the previous chapter we have considered the estimation of $\mu = E(X)$
in the case that a large sample is taken from a population. Via the
central limit theorem, we have been able to find confidence intervals
for μ, based on the normal approximation for the distribution of $\bar{X}$
in large samples. Of course, it is often the case that for reasons
of cost, time, or ethical considerations that sample sizes cannot be
large. For example, medical experiments using primates to discover
characteristics which may be useful in the practice of medicine are
very expensive. Consideration of humane treatment of animals encour-
ages the practice of seeking the most information from "small" samples.

We consider first the estimation of the median η of a continuous
random variable. We recall that η satisfies $F(\eta) = 0.5$, where $F(x)$
is the cumulative distribution function of a random variable X. We
assume here that X has a unique median η. The natural point estimate
of η is given by the sample median M given in Definition 5.2.3. It
can be show that

$$\lim_{n \to \infty} P(|M - \eta| \geq \epsilon) = 0$$

for any $\epsilon > 0$, so that M is a consistent estimator of η. Here we are
interested in finding a confidence interval for η based on the order
statistics $X_{(1)}, X_{(2)}, \ldots, X_{(n)}$ of a random sample of size n. A
natural confidence interval for μ is given by $[X_{(1)}, X_{(n)}]$. The
corresponding confidence coefficient is

$$P(X_{(1)} \leq \eta \leq X_{(n)}) = 1 - P(X_{(1)} > \eta) - P(X_{(n)} < \eta)$$

However $P(X_{(1)} > \eta)$ is the probability that *all* observations in the sample exceed the population median. As the probability that any observation in a random sample exceeds η is $1/2$, the probability that n independent observations exceed η is the binomial probability $(1/2)^n$. Similarly $P(X_{(n)} < \eta)$ is the probability that all observations are less than the population median, which is also $(1/2)^n$. Hence the confidence coefficient associated with $[X_{(1)}, X_{(n)}]$ is $1 - 2(1/2)^n$.

The confidence interval which we shall use for η is given by $[X_{(r)}, X_{(n-r+1)}]$, i.e., the closed interval having the rth smallest and the rth largest observations in the random sample as endpoints. Again the corresponding confidence coefficient may be found from the binomial distribution by observing that

$$P(X_{(r)} \leq \eta \leq X_{(n-r+1)}) = 1 - 2P(X_{(r)} > \eta)$$

$$= 1 - 2 \sum_{k=0}^{r-1} \binom{n}{k}(\tfrac{1}{2})^n$$

where $P(X_{(r)} > \eta)$ is the probability that exactly $0, 1, 2, \ldots,$ or $r - 1$ observations of the sample of size n are less than η, which is given by the indicated binomial probability.

The program MEDIAN produces a confidence interval for the true population median based on a sample of size n. The number of observations n, followed by the n observations should be typed in DATA statements beginning at line 500. The program requests the value of r in response to an INPUT statement and then prints $X_{(r)}, X_{(n-r+1)}$ and the corresponding confidence coefficient. The observations need not be typed in order of magnitude, as the program will order them. Consider the batters in the National League who had the most hits in the years 1952-1971. These are listed in Table 6.2.1. Find a confidence interval for η, the theoretical median winning number of hits per year. As indicated by the output from the program MEDIAN, [200, 218] is a confidence interval with confidence coefficient 0.988, while [204, 215] is a confidence interval with coefficient 0.959.

MEDIAN

```
10 REM COMPUTES A CONFIDENCE INTERVAL FOR THE MEDIAN OF A
20 REM CONTINUOUS POPULATION OF FORM X(R),X(N-R+1) WHERE
30 REM X(I) IS THE ITH ORDER STATISTIC OF A RANDOM SAMPLE
40 REM OF SIZE N. THE CONFIDENCE COEFFICIENT IS ALSO PRINTED.
50 REM FIRST DATA POINT IS N, THE SAMPLE SIZE, FOLLOWED BY THE
60 REM OBSERVATIONS THEMSELVES BEGINNING AT LINE 500
70 REM THE SAMPLE SIZE IS LIMITED TO 100 OBSERVATIONS
80 DIM X(100)
90 READ N
100 FOR I = 1 TO N
110 READ X(I)
120 NEXT I
130 FOR J = 1 TO N-1
140 FOR I = 1 TO N-J
150 IF X(I) <= X(I+1) THEN 190
160 LET T=X(I+1)
170 LET X(I+1)=X(I)
180 LET X(I)=T
190 NEXT I
200 NEXT J
210 PRINT "WHAT IS THE ";"VALUE OF R?"
220 INPUT R
230 LET Q= (0.5)^N
240 LET C=Q
250 FOR I = 1 TO R-1
260 LET Q=(N-I+1)/I*Q
270 LET C = C+Q
280 NEXT I
290 PRINT "THE ENDPOINTS ";"OF THE CONFIDENCE";" INTERVAL ARE"
300 PRINT "X(R)=";X(R),"X(N-R+1)=";X(N-R+1)
310 PRINT "THE CONFIDENCE ";"COEFFICIENT=";(1-2*C)
500 DATA   20, 194, 205, 212, 192, 200, 200, 215, 223, 190, 208
510 DATA   230, 204, 211, 209, 218, 209, 210, 231, 205, 230
1000 END
*RUN
WHAT IS THE VALUE OF R?
 ?5
THE ENDPOINTS OF THE CONFIDENCE INTERVAL ARE
X(R)= 200      X(N-R+1)= 218
THE CONFIDENCE COEFFICIENT= 0.98818


*RUN
WHAT IS THE VALUE OF R?
 ?6
THE ENDPOINTS OF THE CONFIDENCE INTERVAL ARE
X(R)= 204      X(N-R+1)= 215'
THE CONFIDENCE COEFFICIENT= 0.95861


*
```

```
 LIST 490-520

00490 DATA 25
00500 DATA 42.52,41.82,41.94,40.25,42.37,44.6,42.74,43.14,41.17,41.9,46.2
00510 DATA 38.89,45.02,38.1,48.59,34.54,48.07,42.45,47.97,41.5,43.9,43.2
00520 DATA 42.02,48.3,45.9

*RUN
WHAT IS THE VALUE OF R?
 ?8
THE ENDPOINTS OF THE CONFIDENCE INTERVAL ARE
X(R)= 41.9      X(N-R+1)= 44.6
THE CONFIDENCE COEFFICIENT= 0.95671


*RUN
WHAT IS THE VALUE OF R?
 ?7
THE ENDPOINTS OF THE CONFIDENCE INTERVAL ARE
X(R)= 41.82     X(N-R+1)= 45.02
THE CONFIDENCE COEFFICIENT= 0.98537


*
```

Table 6.2.1   Total Number of Hits for the National League Leaders[a]

Year	Name	Hits	Year	Name	Hits
1952	Musial	194	1962	Davis, H.	230
1953	Ashburn	205	1963	Pinson	204
1954	Mueller	212	1964	Clemente/Flood	211
1955	Kluszewski	192	1965	Rose	209
1956	Aaron	200	1966	Alou, F.	218
1957	Schoendienst	200	1967	Clemente	209
1958	Ashburn	215	1968	Rose	210
1959	Aaron	223	1969	Alou, M.	231
1960	Mays	190	1970	Rose/Williams	205
1961	Pinson	208	1971	Torre	230

[a]The *Official Encyclopedia of Baseball*, 6th Ed. (revised).

If it is known that a population is normally distributed, then
a substantial improvement can be made in estimating μ in small samples
by using a point estimate based on $\bar{X}$. It can be shown that the
asymptotic efficiency of the median M in relation to $\bar{X}$ is given by
$e = 2/\pi = 0.64$. The asymptotic efficiency of M is thus rather low
if the population is normally distributed, but it may exceed 1 for
other distributions. Additionally, in skewed populations the population median η may be a more informative measure of location.

The determination of the endpoints of confidence intervals for
μ in small samples from a population assumed to be normally distributed $[X \sim N(\mu,\sigma^2)]$ with σ^2 unknown is based on the knowledge of the
distribution of the random variable

$$t = \frac{\bar{X} - \mu}{S/\sqrt{n}} \tag{6.2.1}$$

The distribution of this random variable was found by Gosset who
published his work under the pseudonym of Student in 1908. The random variable is referred to as Student's t distribution. The

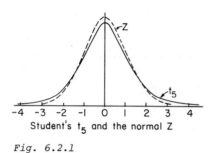

Student's t_5 and the normal Z

Fig. 6.2.1

distribution depends upon the sample size n. If the sample is of
size n the corresponding random variable t is said to have $\nu = n - 1$
degrees of freedom. In Fig. 6.2.1 we present the density functions
for the t distribution with $\nu = 5$ degrees of freedom and that of the
standard normal variable. Note that the t statistic in Eq. (6.2.1)
is similar to a standardized mean, where the unknown parameter σ
has been replaced by the estimator S. Roughly one can say that the
t distribution has greater probability in the "tails" than the standard normal Z, but less probability near zero. The probability
density function is symmetric about 0, as in the case of the standard
normal variable Z. As the sample size becomes arbitrarily large
$(n \rightarrow \infty)$, the percentiles of the Student t distribution approach those
of the standard normal distribution. This is not surprising as S is
a consistent estimator of σ.

We will denote by $t_{\alpha,\nu}$ the number exceeded with probability α
in the t distribution with ν degrees of freedom. From Table B.II we
see that $t_{0.05,4} = 2.132$ [that is, $P(t_4 > 2.132) = 0.05$], $t_{0.05,10} = 1.812$, and $t_{0.05,\infty} = Z_{.05} = 1.645$. The t distribution is used to
find the endpoints of confidence intervals in the same way that was
described in Sec. 5.5 for large samples, where the standard normal
distribution was used. We assume that we observe n independent
observations $(X_1, X_2, \ldots, X_n)$ from a parent population $X \sim N(\mu,\sigma^2)$.
Then

$$P(|t_{n-1}| < t_{\alpha/2,n-1}) = 1 - \alpha$$

or

$$P(|(X - \mu)/S/\sqrt{n}| < t_{\alpha/2,n-1}) = 1 - \alpha$$

This statement can be rewritten as

$$P(\bar{X} - (S/\sqrt{n})t_{\alpha/2,n-1} < \mu < \bar{X} + (S/\sqrt{n})t_{\alpha/2,n-1}) = 1 - \alpha$$

Hence a confidence interval for μ is given by the interval

$$(\bar{X} - (S/\sqrt{n})t_{\alpha/2,n-1}, \bar{X} + (S/\sqrt{n})t_{\alpha/2,n-1})$$

Again the probability $1 - \alpha$ is associated with the method of constructing the interval and not with a single observed interval. Note that the length of the interval $2(S/\sqrt{n})t_{\alpha/2,n-1}$ is a random variable since S is random, so that two intervals with the same confidence coefficient and the same sample size will not have the same length as in Sec. 5.5.

The program TCONFINF produces a confidence interval for μ, where it is assumed that one obtains a random sample of size n from a population $X \sim N(\mu,\sigma^2)$. It is assumed that σ^2 is not known. The sample observations are placed in DATA statements in line 600 and following. The program requests the sample size N as input and the value of the confidence coefficient (90, 95, or 99%). The program prints as output the endpoints of the confidence interval, the mean of the sample data, and the standard deviation of the sample data. The maximum sample size is 50. For larger sample sizes, the program CONFINT may be used.

As an example, suppose that the following are 25 measurements of the heights in inches of kindergarten boys in September of the school year: 42.52, 41.82, 41.94, 40.25, 42.37, 44.6, 42.74, 43.14, 41.17, 41.9, 46.2, 38.89, 45.02, 38.1, 48.59, 34.54, 48.07, 42.45, 47.97, 41.5, 43.9, 43.2, 42.02, 48.3, 45.9. The program TCONFINT has been run to find a 95% confidence interval for μ, the true average height of the population of such boys. The interval is found to be (41.71, 44.46), with $\bar{x} = 43.09$ and $s = 3.34$. The actual normal population sampled had a true mean of $\mu = 43$ and $\sigma = 2.5$. The program MEDIAN has also been run to find an approximate 95% confidence interval for

TCONFINT

```
10 REM PROGRAM CALCULATES A 90,95 OR 99 PERCENT CONFIDENCE
20 REM INTERVAL FOR THE TRUE MEAN OF A NORMAL DISTRIBUTION
30 REM ASSUMING THAT THE VARIANCE OF THE POPULATION IS
40 REM UNKNOWN. PUT DATA IN 600 AND FOLLOWING. THE MAXIMUM
50 REM NUMBER OF OBSERVATIONS IS 50. OTHERWISE USE CONFINT.
60 PRINT "WHAT IS THE ";"VALUE OF N?"
70 INPUT N
80 PRINT "WHAT IS CONFIDENCE";" COEFFICIENT? ";"90,95 OR 99?"
90 INPUT C
92 IF C = 90 THEN 105
94 IF C = 95 THEN 105
96 IF C = 99 THEN 105
98 PRINT "INCOREECT FORMAT";" TRY AGAIN"
100 GO TO 90
105 GOSUB 300
110 DIM X(50),T(50),U(50),V(50)
120 FOR I = 1 TO N
130 READ X(I)
140 LET S = S+X(I)
150 LET S1=S1+X(I)^2
160 NEXT I
170 LET X1=S/N
180 LET S2=SQR((S1-N*X1^2)/(N-1))
190 PRINT "A ";C;" PERCENT CONFIDENCE";" INTERVAL ";"FOR MU IS"
200 IF C =90 THEN 230
210 IF C=95 THEN 240
220 IF C=99 THEN 250
230 LET L =X1-S2/SQR(N)*T(N-1)
232 LET U = X1+S2/SQR(N)*T(N-1)
234 PRINT "L= ";L,"U= ";U
236 PRINT "SAMPLE MEAN= ";X1,"SAMPLE STD DEV = ";S2
238 GO TO 1000
240 LET L=X1-S2/SQR(N)*U(N-1)
242 LET U=X1+S2/SQR(N)*U(N-1)
244 PRINT "L= ";L,"U= ";U
246 PRINT "SAMPLE MEAN= ";X1,"SAMPLE STD DEV = ";S2
248 GO TO 1000
250 LET L=X1-S2/SQR(N)*V(N-1)
252 LET U=X1+S2/SQR(N)*V(N-1)
254 PRINT "L= ";L,"U= ";U
256 PRINT "SAMPLE MEAN= ";X1,"SAMPLE STD DEV = ";S2
258 GO TO 1000
300 FOR I = 1 TO 50
310 READ T(I)
320 NEXT I
330 FOR I = 1 TO 50
340 READ U(I)
350 NEXT I
360 FOR I = 1
```
.

```
   LIS 360-1000

00360FOR I = 1 TO 50
00370READ V(I)
00380NEXT I
00390RETURN
00400DATA 6.3138,2.9200,2.3534,2.1318,2.0150
00405DATA 1.9432,1.8946,1.8595,1.8331,1.8125
00410DATA 1.7959,1.7823,1.7709,1.7613,1.7531
00415DATA 1.7459,1.7396,1.7341,1.7291,1.7247
00420DATA 1.7202,1.7171,1.7139,1.7109,1.7081
00425DATA 1.7056,1.7033,1.7011,1.6991,1.6973
00430DATA 1.6955,1.6939,1.6924,1.6909,1.6896
00435DATA 1.6883,1.6871,1.6860,1.6849,1.6839
00440DATA 1.6829,1.6820,1.6811,1.6802,1.6794
00445DATA 1.6787,1.6779,1.6772,1.6766,1.6759
00450DATA 12.7062,4.3027,3.1824,2.7764,2.5706
00455DATA 2.4469,2.3646,2.3060,2.2622,2.2281
00460DATA 2.2010,2.1788,2.1604,2.1448,2.1315
00465DATA 2.1199,2.1098,2.1009,2.0930,2.0860
00470DATA 2.0796,2.0739,2.0687,2.0639,2.0595
00475DATA 2.0555,2.0518,2.0484,2.0452,2.0423
00480DATA 2.0395,2.0369,2.0345,2.0322,2.0301
00485DATA 2.0281,2.0262,2.0244,2.0227,2.0211
00490DATA 2.0195,2.0181,2.0167,2.0154,2.0141
00495DATA 2.0129,2.0117,2.0106,2.0096,2.0086
00500DATA 63.6574,9.9248,5.8409,4.6041,4.0322
00505DATA 3.7074,3.4995,3.3554,3.2498,3.1693
00510DATA 3.1058,3.0545,3.0123,2.9768,2.9467
00515DATA 2.9208,2.8982,2.8784,2.8609,2.8453
00520DATA 2.8314,2.8188,2.8073,2.7969,2.7874
00525DATA 2.7787,2.7707,2.7633,2.7564,2.7500
00530DATA 2.7440,2.7385,2.7333,2.7284,2.7238
00535DATA 2.7195,2.7154,2.7116,2.7079,2.7045
00540DATA 2.7012,2.6981,2.6951,2.6923,2.6896
00545DATA 2.6870,2.6846,2.6822,2.6800,2.6778
00600DATA 42.58,41.82,41.94,40.25,42.37,44.6,42.74,43.14,41.17,41.9,46.2
00610DATA 38.89,45.02,38.1,48.59,34.54,48.07,42.45,47.97,41.5,43.9,43.2
00620DATA 42.02,48.3,45.9
01000END

*RUN
WHAT IS THE VALUE OF N?
 ?25
WHAT IS CONFIDENCE COEFFICIENT? 90,95 OR 99?
 ?95
A  95  PERCENT CONFIDENCE INTERVAL FOR MU IS
L=  41.7097    U=  44.4631
SAMPLE MEAN=  43.0864         SAMPLE STD DEV =  3.33531


*
```

$\eta = \mu$. The interval is found to be $[41.9, 44.6]$ with confidence
coefficient 0.957. In this case, this interval compares favorably
with that obtained from TCONFINT. Its length is 2.7 compared with
2.75 for the interval based on the t distribution.

Problems 6.2

1. (a) Verify that for $X \sim B(n;1/2)$ that $P(X < r - 1) =$
         $P(X > n - r + 1)$.

    (b) For a random sample of size 25 from a continuous population,
         find the confidence coefficient associated with the interval
         $[X_{(8)}, X_{(18)}]$ used as a confidence interval for η

2. The following is a random sample of the March kilowatt usage of
    10 households in Florida: 780, 820, 960, 750, 850, 1040, 620,
    930, 2400, 800.
    (a) Find a 97.86% confidence interval for the true median house-
         hold power consumption.
    (b) Find a 89.06% confidence interval for this median.

3. Find the values of the following:
    (a) $t_{0.05,17}, \ t_{0.95,16}, \ t_{0.01,5}, \ t_{0.01,20}, \ t_{0.01,\infty}$
    (b) $P(t_9 \geq 1.833)$, $P(t_3 \leq -3.182)$, $P(t_{11} \leq 2.718)$,
         $P(t_\infty \geq 1.960)$, $P(t_\infty \geq 1.282)$, $P(t_\infty \leq 1.645)$

4. The homerun leaders in the National League over a 20-year period
    are given in Table 6.2.2.
    (a) Find a 95.9% confidence interval for η, the true median
         winning number of homeruns.
    (b) Assuming normality of these observations find a 95% confi-
         dence interval for μ, the true average winning number.
    (c) Which of these intervals do you prefer. Why?

Table 6.2.2   Homerun Leaders 1952-1971[a]

Year	Name	Number	Year	Name	Number
1952	Kiner/Sauer	37	1962	Mays	49
1953	Mathews	47	1963	Aaron/McCovey	44
1954	Kluszewski	49	1964	Mays	47
1955	Mays	51	1965	Mays	52
1956	Snider	43	1966	Aaron	44
1957	Aaron	44	1967	Aaron	39
1958	Banks	47	1968	McCovey	36
1959	Mathews	46	1969	McCovey	45
1960	Banks	41	1970	Bench	45
1961	Cepeda	46	1971	Stargell	48

[a]*The Official Encyclopedia of Baseball,* 6th Ed. (revised).

5. The average hourly earnings of production workers in manufacturing industries in 15 eastern and southeastern states as of November 1975 are given below*:

State	Earnings	State	Earnings
Connecticut	4.89	New Jersey	5.06
Deleware	5.40	New York	4.99
Florida	4.11	North Carolina	3.62
Georgia	4.00	Pennsylvania	5.10
Maine	3.95	Rhode Island	3.90
Maryland	5.16	South Carolina	3.72
Massachusetts	4.59	Vermont	4.18
New Hampshire	4.05		

(a) Find a 90% confidence interval for μ, the true average hourly earnings in these states assuming normality.

*Source: *The Handbook of Economic Statistics*, January 1976.

(b) Find an approximate 90% confidence interval for η, the true
     median hourly earnings in these states.

(c) Which interval would you prefer. Why?

6. Suppose that the IQ scores of a large group of children can be
    considered to be normally distributed. A sample of 25 such
    children yielded a sample IQ of 114.2 and a sample standard de-
    viation of 12.1.

    (a) Find 95 and 99% confidence intervals for the true average
         IQ of children in this group.

    (b) Is it reasonable to suppose that the true average IQ of this
         group of children is 100?

7. The following data give per capita income in 12 north central
    states in 1972*:

State	Income	State	Income
Illinois	5140	Missouri	4293
Indiana	4366	Nebraska	4355
Iowa	4300	North Dakota	3738
Kansas	4455	Ohio	4534
Michigan	4881	South Dakota	3699
Minnesota	4298	Wisconsin	4255

    Find an approximate 95% confidence interval for η, the true
    median per capita income of states in this region in 1972.

8. The periods of 29 comets due to return in the period 1975-1978
    are given below*:

Name	Period in years	Name	Period in years
Arend	7.98	Grigg-Skjellerup	5.12
Perrine-Mrkos	6.72	Encke	3.30
Gunn	6.80	Temple I	5.50

---

*Source: *The World Almanac and Book of Facts*, 1975.

Wolf	8.43	Arend-Rigauk	6.84
Churyumoa-Gerasimenko	6.55	Temple II	5.26
Harrington-Abell	7.19	Wolf-Harrington	6.55
Schaumasse	8.18	Whipple	7.47
Kiemola	11.00	Tsuchinshan I	6.64
d'Arrest	6.23	Comas-Sola	8.55
Pons-Winnecke	6.34	Daniel	7.09
Kojima	6.19	Ashbrook-Jackson	7.43
Johnson	6.77	Tsuchinshan II	6.80
Dutoit-Neujmin	6.31	Jackson-Neumin	8.39
Kopff	6.42	VanBiesbroeck	12.41
Faye	7.39		

Assuming normality of the observations, find a 95% confidence interval for μ, the true average period of such comets.

Exercises 6.2

1. Use the program MEDIAN to find an approximate 95% confidence interval for the median income for the population sampled in Exercise 5.2.3. (Note: n = 50.)

2. (a) Use the program TCONFINT to find a 95% confidence interval for the true average height of the population sampled in Exercise 5.2.2. (Note: n = 50.)

    (b) Use MEDIAN and the above sample data to find a 95% confidence interval for this true average height. Compare with the interval in part (a).

3. For large n, if $X \sim B(n;0.5)$, then

$$P(X \leq r) \doteq P\left[\frac{X - n/2}{\sqrt{n}/2} \leq \frac{r + 0.5 - n/2}{\sqrt{n}/2}\right]$$

Hence to attain $P(X \leq r) \doteq \alpha/2$ we use $(r + 0.5 - n/2)/\sqrt{n}/2 = -Z_{\alpha/2}$ or $r = [n/2 - 0.5 - (\sqrt{n}/2)Z_{\alpha/2}]$, where $[\ ]$ represents the greatest integer function. Use this fact to write a program to obtain an approximate 90, 95, 98, or 99% confidence interval for η in the large sample case. The value of $1 - \alpha$ and n should be input. The observations should appear in DATA statements.

6.3 ESTIMATING σ^2, THE POPULATION VARIANCE

We have seen in Chap. 5 that σ^2 is estimated unbiasedly by the statistic S^2. Additionally, it has been stated that in sampling from a population described by the random variable X, that for any $\epsilon > 0$

$$\lim_{n \to \infty} P(|S^2 - \sigma^2| \geq \epsilon) = 0$$

providing $E(X^4)$ is finite. In this case S^2 is a consistent estimator of σ^2. In large samples one can reasonably use S^2 as an estimator of σ^2 appealing to these properties. It is possible if we make additional assumptions about X to obtain confidence intervals for σ^2. We treat this question here.

Confidence intervals can be obtained only if we have knowledge of the distribution of the estimating statistic. In the case that we sample from $X \sim N(\mu, \sigma^2)$, $\bar{X}$ is an estimate of μ. It was stated that $(\bar{X} - \mu)/S/\sqrt{n}$ had Student's t distribution with $n - 1$ degrees of freedom. Here we introduce an important random variable known as the χ^2 or chi-square random variable, which will permit us to obtain confidence intervals for σ^2 based on a random sample from a normal population.

DEFINITION 6.3.1. A random variable having a gamma distribution with parameters $\alpha = \nu/2$ and $\beta = 2$ is called a *chi-square variable* with ν degrees of freedom. Such a random variable is denoted by χ_ν^2.

From the properties of the gamma random variable we see that $E(\chi_\nu^2) = \alpha\beta = \nu$ and $Var(\chi_\nu^2) = \alpha\beta^2 = 2\nu$. Figure 6.3.1 gives an example of the pdf for the chi-square random variable with $\nu = 4$

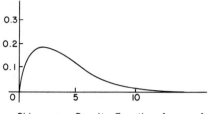

Chi-square Density Function for $\nu = 4$

Fig. 6.3.1

degrees of freedom. The chi-square variable takes on only nonnegative values as it is a gamma variable. We denote by $\chi^2_{\alpha,\nu}$ the number
satisfying

$$P(\chi^2_\nu \geq \chi^2_{\alpha,\nu}) = \alpha$$

Table B.III (Appendix B) gives values of these percentiles. For
example, we see that $\chi^2_{0.05,4} = 9.49$ and $\chi^2_{0.05,20} = 31.41$.

For large ν it is true that $(\chi^2_\nu - \nu)/\sqrt{2\nu}$, the standardized chi-
square variable, is approximately distributed as the standard normal
variable. It turns out that $\sqrt{2\chi^2_\nu} - \sqrt{2\nu - 1}$ is also approximately
distributed as the standard normal variable for large ν and that the
normal approximation is better. Hence for large ν we can find values
of $\chi^2_{\alpha,\nu}$ as follows:

$$\alpha = P(Z \geq Z_\alpha) = P(\sqrt{2\chi^2_\nu} - \sqrt{2\nu - 1} \geq Z_\alpha)$$

Hence we obtain

$$P(\chi^2_\nu \geq (\sqrt{2\nu - 1} + Z_\alpha)^2/2) \doteq \alpha$$

or

$$\chi_{\alpha,\nu} = (\sqrt{2\nu - 1} + Z_\alpha)^2/2 \qquad\qquad (6.3.1)$$

For example,

$$\chi^2_{0.05,60} \doteq (\sqrt{119} + 1.645)^2/2 = 78.798$$

The following theorem, stated without proof, can be used to find a confidence interval for σ^2.

THEOREM 6.3.1. If a random sample of size n is taken from $\chi \sim N(\mu, \sigma^2)$, then

$$\frac{(n-1)S^2}{\sigma^2} \sim \chi^2_{n-1}$$

i.e., this random variable has the chi-square distribution with n - 1 degrees of freedom.

This theorem permits the determination of the endpoints of a confidence interval for χ^2, assuming we are sampling from a normal distribution. Consider the statement

$$P(\chi^2_{1-\alpha/2,n-1} \leq \chi^2_{n-1} \leq \chi^2_{\alpha/2,n-1}) = 1 - \alpha$$

or

$$P(\chi^2_{1-\alpha/2,n-1} \leq \frac{(n-1)S^2}{\sigma^2} \leq \chi^2_{\alpha/2,n-1}) = 1 - \alpha$$

This may be rewritten as

$$P(\frac{(n-1)S^2}{\chi^2_{\alpha/2,n-1}} \leq \sigma^2 \leq \frac{(n-1)S^2}{\chi^2_{1-\alpha/2,n-1}}) = 1 - \alpha$$

This interval provides a $(1 - \alpha)100\%$ confidence interval for σ^2. If one takes the square roots of the endpoints of this interval, it is clear that we will obtain a $(1 - \alpha)100\%$ confidence interval for σ.

The program VCONFINT computes a 95% confidence interval for σ^2, assuming the population sampled is normally distributed. The endpoints of the confidence interval are based on the chi-square distribution using the expressions given in the previous paragraph. If n > 46, the approximation given in Eq. (6.3.1) is used to find $\chi^2_{0.025,n-1}$ and $\chi^2_{0.975,n-1}$. The observations in a sample should be

VCONFINT

```
10 REM PROGRAM CALCULATES A 95 PERCENT CONFIDENCE
20 REM INTERVAL FOR THE VARIANCE AND STANDARD DEVIATION
30 REM OF A NORMAL POPULATION. PUT DATA IN 500 AND FOLLOWING
40 PRINT "WHAT IS THE ";"VALUE OF N?"
50 INPUT N
60 DIM C(50),D(50)
70 GOSUB 350
80 FOR I = 1 TO N
90 READ X
100 LET S1=S1+X^2
110 LET S=S+X
120 NEXT I
130 LET S2= (S1-S^2/N)/(N-1)
140 LET S1=SQR(S2)
150 PRINT "A 95 PERCENT ";"CONFIDENCE INTERVAL ";"FOR THE POPULATION"
160 PRINT "VARIANCE IS"
170 IF N > 46 THEN 290
180 LET L = (N-1)*S2/D(N-1)
190 LET U = (N-1)*S2/C(N-1)
200 PRINT "L = ";L,"U = ";U
210 PRINT "A 95 PERCENT ";"CONFIDENCE INTERVAL ";"FOR THE POPULATION"
220 PRINT "STANDARD DEVIATION IS"
230 LET L = SQR(L)
240 LET U = SQR(U)
250 PRINT "L = ";L,"U = ";U
255 PRINT
260 PRINT "SAMPLE MEAN = ";S/N,"SAMPLE VARIANCE = ";S2
270 PRINT "SAMPLE STD. ";"DEVIATION = ";S1
280 GO TO 1000
290 LET C1=(SQR(2*N-3)-1.96)^2/2
300 LET C2=(SQR(2*N-3)+1.96)^2/2
310 LET L = (N-1)*S2/C2
320 LET U =(N-1)*S2/C1
330 GO TO 200
350 FOR I = 1 TO 45
360 READ C(I)
370 NEXT I
380 FOR I = 1 TO 45
390 READ D(I)
400 NEXT I
410 RETURN
420 DATA 0.001,0.051,0.216,0.484,0.831
425 DATA 1.237,1.690,2.180,2.700,3.247
430 DATA 3.816,4.404,5.009,5.629,6.262
435 DATA 6.908,7.564,8.231,8.907,9.591
440 DATA 10.283,10.982,11.689,12.401,13.120
```

```
00450DATA  17.539,18.291,19.047,19.806,20.569
00455DATA  21.336,22.106,22.878,23.654,24.433
00460DATA  25.215,25.999,26.785,27.575,28.366
00465DATA  5.024,7.378,9.348,11.143,12.833
00470DATA  14.449,16.013,17.535,19.023,20.483
00475DATA  21.920,23.337,24.736,26.119,27.488
00480DATA  28.845,30.191,31.526,32.852,34.170
00485DATA  35.479,36.781,38.076,39.364,40.646
00490DATA  41.923,43.194,44.461,45.722,46.979
00495DATA  48.232,49.480,50.725,51.966,53.203
00500DATA  54.437,55.668,56.896,58.120,59.342
00505DATA  60.561,61.777,62.990,64.201,65.410
00600DATA  16.28,15.45,15.38,16.04,16.35,16.08,16.12,16.58
00610DATA  16.42,16.46,16.42,15.41,15.54,15.67,15.04,16.62
00620DATA  15.9,15.95,15.67,16.04
01000END

*RUN
WHAT IS THE VALUE OF N?
 ?20
A 95 PERCENT CONFIDENCE INTERVAL FOR THE POPULATION
VARIANCE IS
L =  0.11846   U =  0.43692
A 95 PERCENT CONFIDENCE INTERVAL FOR THE POPULATION
STANDARD DEVIATION IS
L =  0.34418   U =  0.661

SAMPLE MEAN =  15.971        SAMPLE VARIANCE =  0.20482
SAMPLE STD. DEVIATION =  0.45258


*
```

entered in DATA statements beginning at line 600.   The sample size
is requested when the program is run.   Let us assume that the following are 20 weights of "1-lb" bags of flour in ounces: 16.28, 15.45,
15.38, 16.04, 16.35, 16.08, 16.12, 16.58, 16.42, 16.46, 16.42, 15.41,
15.54, 15.67, 15.04, 16.62, 15.90, 15.95, 15.67, 16.04.   As the output shows, a 95% confidence interval for σ^2 is given by (0.118, 0.437)
and for σ by (0.344, 0.661).   In this case, the true value of σ^2 =
0.25, which is covered by the first interval.

Problems 6.3

1. Find the values of the following: (a) $\chi^2_{0.05,1}$; (b) $\chi^2_{0.95,4}$; and
    (c) $\chi^2_{0.99,8}$.

2. Find the following probabilities:
    (a) $P(\chi^2_4 \geq 9.488)$
    (b) $P(\chi^2_6 \geq 1.635)$
    (c) $P(\chi^2_8 \leq 15.507)$
    (d) $P(\chi^2_{10} \geq 20.483)$

3. Use Eq. (6.3.1) to approximate (a) $\chi^2_{0.95,85}$; (b) $\chi^2_{0.05,85}$; and
    (c) $\chi^2_{0.50,113}$.

4. Assuming the validity of Theorem 6.3.1, show that this implies
    $E(S^2) = \sigma^2$ in the case of a random sample of size n from a normal
    distribution.

5. Assume that the observations given in Problem 6.2.4 of the number
    of homeruns by the National League leader from 1952-1971 inclusive
    can be assumed to be normally distributed.
    (a) Find a 95% confidence interval for σ^2, the true variance of
         the number of homeruns of the leader.
    (b) Find a 95% confidence interval for σ.

6. Assume the observations in Problem 6.2.5 of the hourly earnings
    of manufacturing workers in 15 eastern and southeastern states
    are normally distributed with expectation μ and variance σ^2.
    (a) Find a 90% confidence interval for σ^2.
    (b) Find a 90% confidence interval for σ.

7. Take the observations of the periods of comets in Problem 6.2.8
    to be normally distributed, that is, $X \sim N(\mu, \sigma^2)$. Find 95%
    confidence intervals for σ^2 and σ.

8. Assume that the lengths of legal bass caught in a certain lake
    are distributed as $X \sim N(\mu, \sigma^2)$. Suppose a random sample of 17
    bass yields $\bar{x} = 14.1$ in. and $s = 1.20$ in. Find 95% confidence
    intervals for μ, σ^2, and σ.

Exercises 6.3

1. Use VCONFINT with the height data for the 25 boys given in Sec.
    6.2 to find 95% confidence intervals for σ^2 and σ. Does the
    latter interval contain the true value of $\sigma = 2.5$?

2. Use VCONFINT with the height data given in Exercise 5.2.2 to
    find 95% confidence intervals for σ^2 and σ.

6.4   ESTIMATION OF σ IN A NORMAL POPULATION

A number of statistics have been suggested as estimators of σ in a
normal population. The estimator S is not an unbiased estimator of
σ, but it is a consistent estimator of σ. The sample range, properly
standardized, is used in practice to estimate σ unbiasedly. This
estimator is of the form $R/c(n)$, where $c(n)$ is a function of n satisfying $E(R/c(n)) = \sigma$. Its efficiency in relation to the standard
deviation decreases rapidly with n. For $n = 2$, $e_{R,S} = 1$; for $n = 10$,
$e_{R,S} = 0.850$; while for $n = 20$, $e_{R,S} = 0.700$ (Dixon and Massey, 1969).
Additionally, this estimator is quite sensitive to the normality
assumption.

Many recent papers have been devoted to estimates of σ based on linear combinations of the order statistics $(X_{(1)}, X_{(2)}, \ldots, X_{(n)})$. The estimate based on the range is of this form. Downton (1966) has suggested the estimator

$$\hat{\sigma} = \frac{2\sqrt{\pi}}{n(n-1)} \sum_{i=1}^{n} (i - 0.5(n+1))X_{(i)}$$

which is an unbiased estimator of σ with variance $V(\hat{\sigma}) = \sigma^2 A(n)$ where

$$A(n) = \frac{1}{n(n+1)} \{n(\frac{\pi}{3} + 2\sqrt{3} - 4) + (6 - 4\sqrt{3} + \frac{\pi}{3})\}$$

Using the measure of efficiency $EMS(S)/EMS(\hat{\sigma}) = e_{\hat{\sigma},S}$, the efficiency of this estimator with respect to S is never less than 0.97, for any sample size, and hence it is a reasonable estimate of σ. Additionally, since $(\hat{\sigma} - \sigma)/\sqrt{A(n)}\sigma$ is approximately distributed as the standard normal distribution for moderate and large n, we find

$$P(-Z_{\alpha/2} \leq \frac{\hat{\sigma} - \sigma}{\sqrt{A(n)}\sigma} \leq Z_{\alpha/2}) \doteq 1 - \alpha$$

or equivalently,

$$P(\frac{\hat{\sigma}}{1 + Z_{\alpha/2}\sqrt{A(n)}} \leq \sigma \leq \frac{\hat{\sigma}}{1 - Z_{\alpha/2}\sqrt{A(n)}}) \doteq 1 - \alpha$$

Thus we can obtain a confidence interval for σ using this estimator.

The program SCONFINT, which is similar to VCONFINT, provides the value of the Downton estimator $\hat{\sigma}$ and for $n \geq 20$ provides a 90, 95, or 99% confidence interval for σ, assuming the sample is from a normal distribution. The observations should be entered in DATA statements beginning at line 500, again the sample size is requested when the program is run. Using the weights of the "1-lb" bags of flour in the previous section we obtain $\hat{\sigma} = 0.468$ with a 95% confidence interval of (0.354, 0.692). This compares with s = 0.453 and the interval (0.344, 0.661). As the true value of σ = 0.5, the estimator $\hat{\sigma}$ compares favorably with S in this case.

SCONFINT

```
10 REM THIS PROGRAM COMPUTES DOWNTON:S ESTIMATOR OF
20 REM THE STANDARD DEVIATION OF A NORMAL POPULATION BASED
30 REM ON A SAMPLE OF SIZE N AND A CONFIDENCE INTERVAL IF N>=20
40 REM DATA SHOULD BE ENTERED IN DATA STATEMENTS BEGINNING
50 REM AT LINE 500. IF N>100 CHANGE DIMEN STATEMENT AT 80
60 PRINT "WHAT IS THE ";"VALUE OF N?"
70 INPUT N
80 DIM X(100)
90 FOR I = 1 TO N
100 READ X(I)
110 NEXT I
120 FOR J = 1 TO N-1
130 FOR I = 1 TO N-J
140 IF X(I) <= X(I+1) THEN 180
150 LET T = X(I+1)
160 LET X(I+1)=X(I)
170 LET X(I)=T
180 NEXT I
190 NEXT J
200 FOR I = 1 TO N
210 LET C = (I-(N+1)/2)*X(I)+C
220 NEXT I
230 LET C = 2/(N*(N-1))*SQR(3.14159263)*C
240 PRINT "THE VALUE OF ";"THE DOWNTON ESTIMATOR IS ";C
250 LET A =N*(3.14159263/3+2*SQR(3)-4)+(6-4*SQR(3)+3.14159263/3)
260 LET A = A/(N*(N-1))
270 IF N < 20 THEN 1000
280 PRINT "DO YOU WISH ";"A CONFIDENCE INTERVAL";" FOR SIGMA?"
285 PRINT "PLS ANSWER 0 IF NO AND 1 IF YES"
290 INPUT A1
300 IF A1=1 THEN 320
310 IF A1=0 THEN 1000
315 PRINT "INPUT ERROR";" PLEASE ANSWER 1 IF YES AND 0 IF NO"
318 GO TO 280
320 PRINT "CONFIDENCE COEFFICIENT ";"90,95 OR 99?"
330 INPUT C1
340 IF C1=90 THEN 390
350 IF C1=95 THEN 410
360 IF C1 = 99 THEN 430
370 PRINT "INPUT ERROR";" TRY AGAIN"
380 GO TO 320
390 LET Z = 1.645
400 GO TO 440
410 LET Z =1.960
420 GO TO 440
430 LET Z = 2.576
440 LET L = C/(1+Z*SQR(A))
450 LET U = C/(1-Z*SQR(A))
460 PRINT "A ";C1;"PERCENT ";"CONFIDENCE INTERVAL"
465 PRINT "FOR SIGMA IS"
470 PRINT
480 PRINT "L= ";L,"U= ";U
500 DATA 16.28,15.45,15.38,16.04,16.35,16.08,16.12,16.58
510 DATA 16.42,16.46,16.42,15.41,15.54,15.67,15.04,16.62
520 DATA 15.90,15.95,15.67,16.04
1000 END
*
```

```
 RUN
WHAT IS THE VALUE OF N?
 ?20
THE VALUE OF THE DOWNTON ESTIMATOR IS  0.46802
DO YOU WISH A CONFIDENCE INTERVAL FOR SIGMA?
PLS ANSWER 0 IF NO AND 1 IF YES
 ?1
CONFIDENCE COEFFICIENT 90,95 OR 99?
 ?90
A  90 PERCENT CONFIDENCE INTERVAL
FOR SIGMA IS

L=  0.36811    U=  0.64237


*RUN
WHAT IS THE VALUE OF N?
 ?20
THE VALUE OF THE DOWNTON ESTIMATOR IS  0.46802
DO YOU WISH A CONFIDENCE INTERVAL FOR SIGMA?
PLS ANSWER 0 IF NO AND 1 IF YES
 ?1
CONFIDENCE COEFFICIENT 90,95 OR 99?
 ?95
A  95 PERCENT CONFIDENCE INTERVAL
FOR SIGMA IS

L=  0.35365    U=  0.69172
```

Problems 6.4

1. Assume that the observations in Problem 6.2.5 of the hourly
   earnings in 15 states are normally distributed with expectation
   μ and variance σ^2. Compute the Downton estimator of σ.

2. Assume the per capita income in the 12 north central states
   given in Problem 6.2.7 is normally distributed. Compute the
   Downton estimator of σ.

Exercises 6.4

1. Use SCONFINT with the height data for the 25 boys given in Sec.
   6.2 to find a 95% confidence interval for σ and the Downton
   estimator $\hat{\sigma}$. Compare this interval with the corresponding output from VCONFINT. Compare the values of $\hat{\sigma}$ and s. Does the
   interval computed here contain the true value $\sigma = 2.5$?

2. Use SCONFINT with the height data given in Exercise 5.2.2 to find
   a 95% confidence interval for σ and the Downton estimator $\hat{\sigma}$.
   Compare the output of SCONFINT with that of VCONFINT.

3. Assume that the numbers of hits gotten by the National League
   leaders in the period 1952-1971 can be considered to be normally
   distributed (Table 6.2.1). Estimate σ and find a 95% confidence
   interval for σ using SCONFINT.

6.5   ESTIMATION OF A POPULATION PROPORTION p

It is often of interest to estimate the proportion of a population
having a certain characteristic. We may wish to know the proportion
p of a population in favor of a certain governmental policy. For
example, the proportion of voters in a state in favor of a state
lottery. In manufacturing, the proportion of defective items produced by a machine is important. We generally assume that a sample
of size n can be considered to constitute independent trials. We
observe $X_i = 1$ if the ith selected item has the characteristic and

X_i = 0 if it does not. The vector $(X_1, X_2, \ldots, X_n)$ of zeros and ones is observed, where the common distribution of the X_i is $P(X_i = 1)$ = p and $P(X_i = 0) = 1 - p = q$. This is, of course, the observation of n binomial trials. A natural estimator of p is $\hat{p} = \Sigma_{i=1}^{n} X_i/n$. This is just a particular case of observing $\bar{X}$ so that $\hat{p}$ is an unbiased and consistent estimator of p:

$$E(\hat{p}) = p \qquad Var(\hat{p}) = Var\left(\frac{X}{n}\right) = \frac{p(1 - p)}{n}$$

as $X \sim B(n;p)$. Additionally, by the central limit theorem

$$\frac{\hat{p} - p}{\sqrt{p(1 - p)/n}}$$

is approximately distributed as the standard normal variable for large n.

This latter fact is important in finding confidence intervals for p in large samples. We have

$$P(|Z| \leq Z_{\alpha/2}) = 1 - \alpha$$

so that

$$P\left(\left|\frac{\hat{p} - p}{\sqrt{p(1 - p)/n}}\right| \leq Z_{\alpha/2}\right) \doteq 1 - \alpha \qquad (6.5.1)$$

which may be rewritten in the form

$$P\left(\hat{p} - \frac{\sqrt{p(1 - p)}}{n} Z_{\alpha/2} \leq p \leq \hat{p} + \frac{\sqrt{p(1 - p)}}{n} Z_{\alpha/2}\right) \doteq 1 - \alpha$$

In the last expression, we obtain endpoints of a confidence interval for p with confidence coefficient $1 - \alpha$. However these expressions still depend upon p. We define $s_{\hat{p}} = \sqrt{\hat{p}(1 - \hat{p})/n}$ and obtain an approximate $(1 - \alpha)100\%$ confidence interval

$$\hat{p} \pm s_{\hat{p}} Z_{\alpha/2} \qquad \text{or} \qquad \hat{p} \pm \frac{\sqrt{\hat{p}(1 - \hat{p})}}{\sqrt{n}} Z_{\alpha/2}$$

For example, suppose that we wish to estimate the proportion of voters in favor of a state lottery. We find 144 or 400 randomly selected voters in favor of the lottery. Here $\hat{p} = 144/400 = 0.36$ and $s_{\hat{p}} =$ $\sqrt{0.36(0.64)/400} = 0.024$. Hence an approximate 90% confidence interval for p is given by

$$0.36 \pm 0.024 Z_{0.05} = 0.36 \pm 0.024(1.645)$$

or by $[0.321, 0.399]$. Apparently the chances are high that the lottery would not be approved in a referendum.

In the 1976 presidential election, the final Gallup interviews were made from October 28 to October 30. From a sampling of 3439 registered voters, 1926 of those who were "most likely to vote" were selected. The estimated proportion of these favoring President Ford was 0.49 and favoring Jimmy Carter was 0.48. Ninety-five percent confidence intervals fro these population proportions are

$$0.49 \pm \sqrt{\frac{0.49(0.51)}{1926}} \, (1.96) = 0.490 \pm 0.022$$

that is, $(0.468, 0.512)$

and

$$0.48 \pm \sqrt{\frac{0.48(0.52)}{1926}} \, (1.96) = 0.48 \pm 0.022$$

that is, $(0.458, 0.502)$

As can be seen from these confidence intervals it was not possible to predict the election winner in advance in that year. Of course in the final tabulation Jimmy Carter had a higher percentage of the popular vote than President Ford.

In the case of intermediate size n, say $(20 \le n \le 100)$, the expression (6.5.1) can be used to find a more exact confidence interval for p. We have from Eq. (6.5.1)

$$(\hat{p} - p)^2 \le \frac{(Z_{\alpha/2})^2 p(1 - p)}{n}$$

or

$$p^2 \left(1 + \frac{Z_{\alpha/2}^2}{n}\right) - \left(2\hat{p} + \frac{Z_{\alpha/2}^2}{n}\right)p + \hat{p}^2 \le 0$$

Hence the roots of this quadratic in p are endpoints of an approximate
$(1 - \alpha)100\%$ confidence interval for p. These are found to be

$$\frac{\hat{p} + z_{\alpha/2}^2/2n \pm z_{\alpha/2} \sqrt{\hat{p}(1 - \hat{p})/n + z_{\alpha/2}^2/4n^2}}{1 + z_{\alpha/2}^2/n} \tag{6.5.2}$$

Notice that for large n, if we neglect the terms $z_{\alpha/2}^2/n$, $z_{\alpha/2}^2/2n$,
and $z_{\alpha/2}^2/4n^2$, we obtain the interval

$$\hat{p} \pm z_{\alpha/2}s_{\hat{p}} \tag{6.5.3}$$

as before.

The program PCONFINT has been written to calculate the endpoints
of a confidence interval for p using the expression (6.5.2). An
elementary statistics class of 105 students was asked to respond to
the question: "Do you believe that students majoring in your major
subject should be required to take a statistics course?" The number
of "Yes" responses was 38. A 95% confidence interval for p, the
population proportion of affirmative responses is given by the program PCONFINT to be [0.276, 0.457]. The corresponding 95% confidence
interval using expression (6.5.3) is [0.270, 0.454], so that the
intervals differ slightly in this case.

Another question of importance in estimating population proportions is the necessary sample size to achieve a good estimate. We
shall require that $P(|p - \hat{p}| \leq \delta) \geq 1 - \alpha$, where δ is a preassigned
maximum tolerance of the absolute error in estimating p by $\hat{p}$ that we
are willing to accept. The probability $1 - \alpha$ is a preassigned (high)
probability that the absolute error of the estimate be no more than
δ. From Eq. (6.5.1) we have

$$1 - \alpha = P\left(|p - \hat{p}| \leq \sqrt{\frac{p(1 - p)}{n}}z_{\alpha/2}\right)$$

Hence by choosing n to satisfy

$$\delta = \sqrt{\frac{p(1 - p)}{n}}z_{\alpha/2} \qquad \text{or} \qquad n^* = [\frac{p(1 - p)}{\delta^2}z_{\alpha/2}^2] + 1$$

PCONFINT

```
10 REM THIS PROGRAM COMPUTES AN 80,90,95,98 OR 99 PERCENT
20 REM CONFIDENCE INTERVAL FOR P, A POPULATION PROPORTION
30 REM BASED ON A SAMPLE OF SIZE N WITH X "SUCCESSES"
40 PRINT "WHAT IS SAMPLE ";"SIZE?"
50 INPUT N
60 PRINT "HOW MANY SUCCESSES?"
70 INPUT X
80 PRINT "WHAT IS ";"CONFIDENCE COEFFICIENT   ";"80,90,95,98,OR 99?"
90 INPUT C
100 IF C = 80 THEN 160
110 IF C =90 THEN 170
120 IF C =95 THEN 180
130 IF C =98 THEN 190
140 IF C =99 THEN 200
150 PRINT "INCORRECT FORMAT";" TRY AGAIN"
155 GO TO 90
160 LET K =1.282
165 GO TO 220
170 LET K=1.645
175 GO TO 220
180 LET K = 1.96
185 GO TO 220
190 LET K =2.236
195 GO TO 220
200 LET K= 2.576
220 LET P =X/N
230 LET L =(P+K^2/(2*N))-K*SQR(P*(1-P)/N+K^2/(4*N^2))
240 LET U =(P+K^2/(2*N))+K*SQR(P*(1-P)/N+K^2/(4*N^2))
250 LET B = (1+K^2/N)
260 LET L =L/B
270 LET U= U/B
280 PRINT "PROPORTION OF ";"SUCCESSES= ";P
290 PRINT "A ";C;" PERCENT CONFIDENCE ";"INTERVAL FOR P IS"
300 PRINT
310 PRINT "L= ";L,"U= ";U
320 END
*RUN
WHAT IS SAMPLE SIZE?
 ?105
HOW MANY SUCCESSES?
 ?38
WHAT IS CONFIDENCE COEFFICIENT  80,90,95,98,OR 99?
 ?95
PROPORTION OF SUCCESSES=  0.3619
A  95   PERCENT CONFIDENCE INTERVAL FOR P IS

L=  0.27637    U=  0.45719
```

where [ ] represents the greatest integer function, we will achieve the desired result. If we choose $n_1 > n^*$, then defining $\delta_1 = \sqrt{p(1-p)/n_1}\,Z_{\alpha/2}$, we have $\delta_1 < \delta$. Hence

$$1 - \alpha = P(|p - \hat{p}| \le \delta_1) \le P(|p - \hat{p}| \le \delta).$$

Hence for $n \ge n^*$, we achieve the desired accuracy. The expression for n^* depends upon knowledge of p. If nothing is known about p, then as $\sqrt{p(1-p)} \le 1/2$, we may take $n^* \doteq (Z_{\alpha/2}/2\delta)^2$. On the other hand, if we know that $p \in [0, p_1]$, for $0 < p_1 < 1/2$, we may use the expression for n^* given above, evaluated at p_1, as $p(1-p)$ is an increasing function of p for $0 < p < 1/2$. Hence the largest sample size required so that $P(|p - \hat{p}| \le \delta)$ for any $p \in [0, p_1]$ will be found by evaluating n^* at $p = p_1$. For $p \in [p_1, 1]$, $0.5 < p_1 < 1$, similar reasoning implies that the value of n^* for $p = p_1$ is the maximum required sample size.

For example, assume that we wish that the error in the estimate of p by $\hat{p}$ not exceed 0.02 with probability at least 0.95. Without any knowledge of p we use $n^* = (1.96/2(0.02))^2 = 49^2 = 2401$. If it were known that $p \in [0, 0.20]$, then we find

$$n^* = [(0.2)(0.8)(\frac{1.96}{0.02})^2] + 1 = 1537$$

Thus the additional knowledge permits reduction of the sample size by about 1/3 without a sacrifice in the accuracy of the estimator.

Problems 6.5

1. In 1974, Carl Yastrzemski of the Boston Red Sox batted 0.301, getting 155 hits in 515 at bats. Find a 90% confidence interval for p, Yastrzemski's lifetime batting average. Why might this not be a good method of estimating p? (Source: *The World Almanac and Book of Facts*, 1975.)

2. In 1973-1974, Kareem Abdul-Jabbar attempted 1759 shots of which he made 948 for a proportion of success of 0.539. Find a 95% confidence interval for p, Jabbar's long-run shooting percentage. (Source: *The World Almanac and Book of Facts*, 1975.)

Table 6.5.1   Effect of Gasoline Shortage on Travel Frequency of
               Families with Automobiles[a]

Type of change	Number of families
No Change	237
Change frequency	
Switched to other modes	120
Reduce auto travel	84
Miscellaneous	16
Total	457

3. A city has a population equally divided between the races (black
    and white). A sample of 400 truck drivers reveals 350 whites
    and 50 blacks. Find a 95% confidence interval for p, the true
    proportion of truck drivers in the city who are white. Does
    there appear to be evidence that blacks are underrepresented
    among truck drivers?

4. In 1974, the Oregon Department of Transportation collected data
    concerning the effect of the energy (gasoline) shortage on the
    behavior of drivers in Portland, Oregon. The data are given in
    Table 6.5.1.
    (a) Estimate the proportion indicating no change and find a
         95% confidence interval for this population proportion.
    (b) Among those who altered their behavior, estimate the pro-
         portion who reduced automobile travel, and find a 95%
         confidence interval as in part (a).

5. A beer manufacturer, Lightdraft, Inc., wishes to estimate its
    proportion of the market in an urban area. A grocery chain
    agrees to supply the company with the proportion of six-packs

---

[a]*Behavior of Car Owners During the Gasoline Shortage,* B. W. Becker,
D. J. Brown, and Philip B. Schary, Traffic Quarterly, July 1976.

of Lightdraft sold during a given week among all six-packs sold. Of 1600 sold, 320 are Lightdraft. Find a 95 and 99% confidence interval for the true market share.

6. Suppose in Problem 6.5.5 that the manufacturer wishes the error in the estimate of the market share to be no more than 0.02 with probability at least 0.99.

(a) Find the minimum required sample size assuming no knowledge about p.

(b) Find the minimum required sample size assuming $p \leq 0.25$.

7. For a maximum error of δ with probability at least $1 - \alpha$ without knowledge of p, we use $n = (Z_{\alpha/2}/2\delta)^2$.

(a) Show that as α decreases (that is, $1 - \alpha$ increases), n increases for fixed δ.

(b) Show that if δ is replaced by $\delta/2$ (i.e., the maximum tolerable error is halved), then n is increased by a factor of 4 for fixed α.

8. If we choose $1 - \alpha = 0.9544$ in Problem 6.5.7, verify the relationship

$$n = \frac{1}{\delta^2}$$

Evaluate this function of δ for $\delta = 0.01(0.01)(0.10)$. How large a sample is required to estimate a population proportion with a maximum error of 0.01 with probability at least 0.9544?

9. Verify that the values in Eq. (6.5.2) are the roots of the appropriate polynomial. Sketch a graph of the polynomial for fixed $\hat{p}$ and $Z_{\alpha/2}$.

10. (a) We wish to estimate the proportion of the female population (16 years and older) with Rh negative blood in a given city. Suppose 900 women are typed and 90 are Rh negative. Find a 95% confidence interval for p.

(b) Find the minimum required sample size to estimate p with a maximum error of 0.005 if it can be assumed that $p \leq 0.2$. Let $1 - \alpha = 0.98$.

Exercises 6.5

1. Use PCONFINT with the data of Problem 6.5.4 to calculate 95% confidence intervals. Compare these intervals with your previous results in Problem 6.5.4.

2. Use PCONFINT with the data of Problem 6.5.5 to find 95 and 99% confidence intervals for the true market share. Compare these intervals with those found in Problem 6.5.5.

3. Use PCONFINT with the data of Problem 6.5.10 to find a 95% confidence interval for the proportion of Rh negative women. Compare this interval with that found in Problem 6.5.10(a).

4. Of 64 students, 37 rank a teacher as "good" or better. Using PCONFINT, find a 95% confidence interval for the corresponding population proportion p.

HYPOTHESIS TESTING

7.1   INTRODUCTION

In Chap. 6 we have considered the estimation of parameters, an important topic in statistics. Here we consider another important inferential area in statistics called *hypothesis testing*. We first give
a definition of what is meant by a statistical hypothesis.

DEFINITION 7.1.1. A *statistical hypothesis* is a statement about the
distribution of a random variable.

Such a statement is generally made in terms of a set of values
that a parameter may take on, where the parameter is used in the
probability law describing the distribution of a random variable.
Let us suppose that an election is to be held for mayor in a large
city and that there are two candidates, A and B. We decide to ask
100 randomly sampled voters to answer the question: "Do you prefer
candidate A to candidate B in the election for mayor?" We may consider the answers to be 100 independent trials in which the number
of "Yes" replies is distributed as $X \sim B(100;p)$. A statistical
hypothesis of interest is

$$H_0: \; p \leq \frac{1}{2}$$

where H_0 is referred to as the null hypothesis. We see that this is
a statement about the possible values of the parameter p. In statistical hypothesis testing we wish to find a reasonable method of
deciding between the hypothesis H_0 and an alternative

$$H_1: \quad p > \frac{1}{2}$$

Clearly a decision for H_1 should please A, but not B.

Before we consider a method of deciding between H_0 and H_1, let
us consider two possible errors to which the decision process may
lead. These are indicated in Table 7.1.1. On the one hand, we may
decide p > 1/2 when in fact p ≤ 1/2, i.e., we may decide that the
proportion of voters preferring A to B exceeds 1/2, when in fact, it
does not. This error, rejecting the null hypothesis when it is true,
is called a *type I error*. On the other hand, we may decide p ≤ 1/2,
when in fact, p > 1/2. This error, accepting H_0 when it is false,
is called a *type II error*. Clearly these errors may result in different responses. For example, if a type I error is made, candidate
A may relax his efforts when he should not. If a type II error is
made, candidate A may spend additional time and money to win an
election in which victory is already assured.

A decision between H_0 and H_1 described in Table 7.1.1 should,
of course, depend on the observed value of the random variable X,
the number of those of the 100 voters selected answering "Yes" to
the question. Clearly large values of X indicate that H_1 should be
chosen, while small values of X indicate that the decision should
be made in favor of H_0. A statistical test is determined by deciding

Table 7.1.1   Type I and Type II Errors

Decision	True state of nature	
	p ≤ 1/2	p > 1/2
H_0 is correct	No error	Type II error
H_1 is correct	Type I error	No error

exactly *which values of the test statistic lead to rejection of* H_0.
These values are referred to as the *critical region* for the test.
Let us consider the probability of a type I error. The standard
notation is to define

$$\alpha = \Pr(\text{rejection of } H_0 \,|\, H_0 \text{ is true})$$

The testing strategy which is traditionally adopted is to construct
a critical region so that α is not greater than a specific value,
say 0.05 or 0.01. The maximum value that α may attain is called the
significance level for the test. In the case at hand, critical
regions are of the form $X \geq j$ for some integer $0 \leq j \leq 100$. Now

$$\alpha = P(X \geq j \,|\, p \leq \tfrac{1}{2}) = \sum_{k=j}^{100} \binom{100}{k} p^k (1 - p)^{100-k}$$

It can be shown that this probability may be written as

$$c \int_0^p x^{j-1}(1 - x)^{100-j} \, dx \qquad \text{for } c > 0$$

so that α clearly increases with p. Thus, to achieve $\alpha \leq 0.05$, for
example, we must choose j so that

$$\sum_{k=j}^{100} \binom{100}{k} \left(\tfrac{1}{2}\right)^{100} = P(X \geq j \,|\, p = \tfrac{1}{2}) \leq 0.05$$

The program BINOM has been run for n = 100 and p = 0.5. We see that
$P(X \geq 59) = 1 - p(X \leq 58) = 0.0443$, while $P(X \geq 58) = 0.0666$. Hence
we would choose $X \geq 59$ as critical region for the test. If 59 or
more of the 100 persons answer "Yes," we decide that p > 1/2. The
significance level for the test is 0.0443.

One might ask why α should not be set equal to 0. This could
be achieved by using the empty set as the critical region. The
decision to accept $p \leq 1/2$ would always be made (regardless of the
outcome of the sampling process). However, clearly the probability
of a type II error, the probability of accepting H_0 given p > 1/2,
would be 1 (as H_0 is always accepted). In general, the probability
of a type II error depends on the critical region and the value of
p assumed to be correct. The standard notation is

```
RUN
BINOMIAL DISTRIBUTION

WHAT IS THE VALUE OF N
?100

WHAT IS THE VALUE OF P
?.5
```

X SUCCESSES	PROB OF X	CUM PROB OF X
39	7.11073E-3	1.76001E-2
40	1.08439E-2	2.84440E-2
41	1.58691E-2	4.43130E-2
42	2.22923E-2	6.66053E-2
43	3.00686E-2	9.66740E-2
44	3.89526E-2	0.13563
45	4.84743E-2	0.1841
46	5.79584E-2	0.24206
47	6.65905E-2	0.30865
48	7.35270E-2	0.38218
49	7.80287E-2	0.46021
50	7.95892E-2	0.53979
51	7.80287E-2	0.61782
52	7.35270E-2	0.69135
53	6.65905E-2	0.75794
54	5.79584E-2	0.8159
55	4.84743E-2	0.86437
56	3.89526E-2	0.90333
57	3.00686E-2	0.93339
58	2.22923E-2	0.95569
59	1.58691E-2	0.97156
60	1.08439E-2	0.9824
61	7.11073E-3	0.98951
62	4.47288E-3	0.99398
63	2.69793E-3	0.99668
64	1.55974E-3	0.99824
65	8.63856E-4	0.99911
66	4.58105E-4	0.99956
67	2.32471E-4	0.9998
68	1.12817E-4	0.99991
69	5.23209E-5	0.99996
70	2.31707E-5	0.99998
71	9.79043E-6	0.99999
72	3.94337E-6	1.
73	1.51252E-6	1.
74	5.51867E-7	1.
75	1.91314E-7	1.

```
*
```

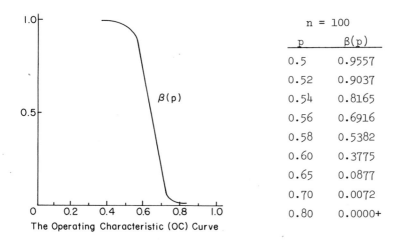

	n = 100
p	β(p)
0.5	0.9557
0.52	0.9037
0.54	0.8165
0.56	0.6916
0.58	0.5382
0.60	0.3775
0.65	0.0877
0.70	0.0072
0.80	0.0000+

The Operating Characteristic (OC) Curve

Fig. 7.1.1

$$\beta(p) = P(\text{accepting } H_0 | p \text{ is true value})$$

In the example

$$\beta(p) = P(X \leq 58 | p) = 1 - P(X \geq 59 | p)$$

the probabilities of a type II error for several values of p have
been calculated using the program BINOM. A graph of $\beta(p)$ as a function of p appears in Fig. 7.1.1. This graph is called the *operating
characteristic* (OC) curve for the test.

The function $\pi(p) = 1 - \beta(p)$ is called the *power function* of the
test. The graph of the power function for this test appears in Fig.
7.1.2. For p > 0.5, this function gives the probability of rejecting
H_0, the correct decision, as a function of p. As the probability of
rejection of H_0 for p = 0.5 is α, $\pi(0.5) = \alpha = 0.0443$. For values
of p slightly greater than 0.5, it is clear that $\pi(p)$ must be relatively small as $\pi(p) = c \int_0^p x^{58}(1 - x)^{41} dx$, is a continuous function
of p. It is difficult for any decision procedure to distinguish
between $p \leq 0.5$ and p > 0.5 if p is only slightly larger than 0.5.
For example, $\pi(0.55) = 0.2415$ and $\pi(0.60) = 0.6225$. However for
candidate A it may be the case that distinguishing between alternatives

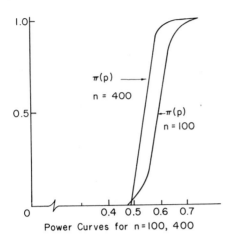

	n = 100	n = 400
p	$\pi(p)$	$\pi(p)$
0.50	0.0443	0.0495
0.52	0.0963	0.1975
0.54	0.1835	0.4800
0.56	0.3084	0.7750
0.58	0.4618	0.9418
0.60	0.6225	0.9917
0.65	0.9123	0.9999
0.70	0.9928	1.0000

Power Curves for n=100, 400

Fig. 7.1.2

such as $p \leq 0.5$ and $p = 0.6$ is exactly the question of importance. If this is the case, the power function can be improved for $\alpha = 0.05$, only by increasing the sample size n.

Suppose, for example, that 400 voters can be sampled. For this sample size we may use the normal approximation to the binomial distribution to find a test with significance level no greater than 0.05. Using the continuity correction, we must have

$$P(\frac{j - 0.5 - 200}{10} \leq \frac{X - 200}{10} \leq \frac{400.5 - 200}{10}) \doteq 0.05$$

or

$$P(\frac{j - 200.5}{10} \leq Z \leq 20.05) = 0.05$$

Hence $(j - 200.5)/10 \doteq 1.645$ and $j = 216.95$, and the critical region $X \geq 217$ would be used. The values $\beta(p)$ and $\pi(p)$ for $p > 0.5$ can also be found using the normal approximation. For example, as $400(0.6)(0.4) = 96$,

$$\pi(0.6) = P(\frac{217 - 200.5}{\sqrt{96}} \leq Z) = P(-2.398 \leq Z)$$

$$= 1 - F_Z(-2.398) = 0.9917$$

The graph of $\pi(p)$ for this new test also appears in Fig. 7.1.2. The power curve for this test (with approximately the same value of α) lies above that for the previous test for all $p > 0.5$. The value $\pi(0.60) = 0.9917$ for $n = 400$. In other words, if $p = 0.6$, the test has a probability of $0.99+$ of yielding a decision for H_1 with $\alpha = 0.0495$. For $n = 100$, $\pi(0.60) = 0.6225$, so the probability of a correct decision if $p = 0.6$ is only 0.6225. Clearly the increased sample size has brought an important improvement in the properties of the decision procedure.

It should be remarked that there is a substantial difference between rejecting the hypothesis H_0: $p \leq 1/2$ and retaining (or accepting) H_0. For the case $n = 100$, we reject H_0 if 59 or more persons in the sample answer "Yes." Suppose, in fact, 62 answer "Yes." The probability that such an event occurs if $p \leq 1/2$ is less than 0.05. Hence if 62 affirmative answers are recorded, we conclude that H_0 is false and that $p > 1/2$. If we observe 55 "Yes" replies, we retain H_0. We have not observed an event highly inconsistent with $p \leq 1/2$. However 55 "Yes" replies is consistent with values of p both greater than and less than 1/2. Indeed a 95% confidence interval for p is given by PCONFINT as $[0.452, 0.644]$. Hence in retaining H_0 we can only state that no evidence inconsistent with the null hypothesis has been found. If we decide to reject H_0, there is statistical evidence inconsistent with the truth of H_0. As it is stronger to reject H_0 in favor of H_1, studies seeking evidence for a hypothesis generally state the hypothesis as an alternative H_1 to a null hypothesis H_0.

Problems 7.1

1. Statistical testing methods are employed in the detection of aircraft by radar. The detector records voltage measurements at n times, $t = 0 + (i - 1)\Delta t$ for $i = 1, 2, \ldots, n$. The recorded values are denoted $X_1, X_2, \ldots, X_n$. If the signal is not present,

then $X \sim N(0, \sigma^2)$ is assumed to be the parent distribution. If the signal is present, then $X \sim N(\mu, \sigma^2)$, where $\mu > 0$. The hypotheses may be stated in terms of the parameter μ as

$$H_0: \ \mu = 0 \qquad \text{versus} \qquad H_1: \ \mu > 0$$

(a) State in words the meaning of a type I error.

(b) State in words the meaning of a type II error.

2. A jury in a criminal case makes a decision between

$$H_0: \ \text{the defendent is not guilty.}$$
$$H_1: \ \text{the defendent is guilty.}$$

(a) State in words the meaning of a type I and type II error.

(b) In English jurisprudence, which error probability, α or β, is the jury required to make small?

3. An individual claims to have ESP powers. He states that he can call the suit of a card before it is revealed with a frequency greater than chance. Let p represent the probability that a card's suit is correctly called by this individual. Assume that after each drawing the card drawn is replaced and the deck reshuffled. We wish to test

$$H_0: \ p = \frac{1}{4} \qquad \text{versus} \qquad H_1: \ > \frac{1}{4}$$

We record the outcome, success or failure, of 10 draws and call the number of successes X. Suppose that the critical region is taken to be $X \geq 6$.

(a) Why is H_0 as stated above the reasonable null hypothesis?

(b) Use BINOM or a table of binomial probabilities to find the significance level α for this test.

(c) Use BINOM or a table of binomial probabilities to find $\beta(p)$ for $p = 0.3(0.1)(1.0)$. Graph the OC curve.

(d) Find $\pi(p)$ for $p = 0.25$ and $p = 0.3(0.1)(1.0)$ and graph the power function.

4. We wish to have a test to determine which basketball players require practice in foul shooting. A player will be considered "acceptable" if he makes at least 60% of his foul shots. Let p represent the proportion of foul shots made by a player. We wish to test

$$H_0: \quad p \geq 0.6 \quad \text{versus} \quad H_1: \quad p < 0.6$$

   (a) Using the number of successes in 10 independent trials X as the appropriate statistic determine a critical region for a test so that $\alpha \doteq 0.05$ (but ≤ 0.05).

   (b) Evaluate $\beta(p)$ for $p = 0(0.1)(0.6)$ and graph the power function for the test in part (a).

5. A manufacturer claims that a gallon of his paint will cover more than 600 ft^2. We assume that the number of square feet covered by a gallon of this paint to be a random variable $X \sim N(\mu, \sigma^2)$. Assuming σ^2 is known, state the null and alternative hypotheses in terms of μ. The hypotheses should be stated so that the burden of proof falls on the manufacturer.

6. Let p represent the proportion of a population which does not get the flu during a winter. We test a vaccine's effectiveness by giving it to 100 volunteers and observing X, the number who do not get the flu. It is known that 3/4 of the population survives the winter without flu if no vaccine is given.

   (a) State the hypotheses H_0 and H_1, so that we place the statistical burden on the vaccine to demonstrate its effectiveness.

   (b) Using the normal approximation to the binomial distribution, find the critical region for this test using $\alpha = 0.05$.

   (c) Again using the normal approximation to the binomial distribution, find the probability of a type II error given $p = 0.9$.

7. A survey by Dunn's Review, October 1976, indicated that, of 300 top business executives polled, 85% favored President Ford over Jimmy Carter for President. Let p be the proportion of business executives favoring President Ford.

(a) Find a critical region for testing H_0: $p = 3/4$ versus
     H_1: $p > 3/4$ using the normal approximation to the binomial
     and $\alpha = 0.05$.

(b) Based on the poll in Dunn's Review, is there evidence that
     the percentage of business executive preferring President
     Ford exceeded 75%?

7.2   TESTS OF A POPULATION PROPORTION p FOR LARGE n

In this section we shall consider tests of the hypothesis H_0: $p = p_0$
for a fixed value of p versus various alternatives. We assume the
sample size n is sufficiently large that the normal approximation
applies. The three alternatives which we consider are

1. H_1: $p > p_0$

2. H_1: $p < p_0$

3. H_1: $p \neq p_0$

will require different critical regions for a test with significance
level α. For H_1: $p > p_0$, we have seen, assuming X is the random
variable representing the number of successes in n trials, that the
appropriate critical region is of form $X \geq j$. The exact value of j
is determined by requiring

$$P(j \leq X | p = p_0) = P\left(\frac{j - 0.5 - np_0}{\sqrt{np_0(1 - p_0)}} \leq Z\right) \leq \alpha$$

The critical value of j is found from the equation

$$j = [0.5 + np_0 + Z_\alpha\sqrt{np_0(1 - p_0)}] + 1$$

where [ ] is the greatest integer function. For the alternative
H_1: $p < p_0$ similar reasoning leads to the rejection region $X \leq j$,
where

$$j = [np_0 - 0.5 - Z_\alpha\sqrt{np_0(1 - p_0)}]$$

For the case in which the alternative is written as H_1: $p \neq p_0$, we reject H_0 if $X \geq j_1$ or $X \leq j_2$, where

$$j_1 = [np_0 + 0.5 + Z_{\alpha/2}\sqrt{np_0(1 - p_0)}] + 1$$

and

$$j_2 = [np_0 - 0.5 - Z_{\alpha/2}\sqrt{np_0(1 - p_0)}]$$

The program PTEST has been written to test the null hypothesis H_0: $p = p_0$ versus the three alternatives described above. The program requests the values of n, p_0, and the alternative. The value of the significance level may be 0.10, 0.05, or 0.01 and is entered in response to an INPUT statement. The number of successes X actually observed is entered in the same way. The program computes the rejection region according to the formulas given above. The power function for the test is also computed for appropriate values of p.

Suppose that passage of a bond issue requires a 60% favorable vote. In order to test the chance of passage we sample 180 voters randomly from the registration list. The hypotheses may be stated

$$H_0: \quad p = 0.60 \qquad H_1: \quad p > 0.60$$

Note that the alternative for which we are seeking evidence, namely the passage of the issue, is H_1. Also note that if H_0 is written $p \leq 0.60$, the same critical region is appropriate, as indicated previously. Suppose that we observe 130 voters in favor of the bond issue. The program PTEST, using $\alpha = 0.05$ indicates the rejection region is $X \geq 120$. Hence the null hypothesis would be rejected. There is good reason to believe that more than 60% of the registered voters favor the bond issue.

The values of the power function $\pi(p)$ are calculated for p = 0.6 to p = 0.95 with a step size of 0.05. For p = 0.75, we find $\pi(0.75) = 0.9961$, so that if the true proportion of the registered voters favor the bond issue is actually 75%, the probability is very high that p = 0.60 will be rejected in favor of H_1: p > 0.60. Note that $\pi(0.60) = 0.0401$, so that although the value of α is chosen to

PTEST

```
10 REM THIS PROGRAM TESTS THE HYPOTHESIS P = PO AT SIGNIFICANCE
20 REM LEVEL ALPHA AGAINST ONE OF THE ALTERNATIVES
30 REM 1) P > PO
40 REM 2) P < PO
50 REM 3) P UNEQUAL PO
100 PRINT "WHAT ARE THE VALUES OF N AND PO?"
110 INPUT N, P4
120 PRINT "WHICH ALTERNATIVE ";"1, 2 OR 3?"
130 INPUT A
132 IF A = 1 THEN 150
134 IF A = 2 THEN 150
136 IF A = 3 THEN 600
138 PRINT "INCORRECT FORMAT ";"TRY AGAIN"
140 GO TO 120
150 PRINT "WHAT IS THE ";"SIGNIFICANCE LEVEL, ";"0.10,0.05, OR 0.01?"
160 INPUT A1
170 IF A1= 0.10 THEN 220
180 IF A1= 0.05 THEN 240
190 IF A1 = 0.01 THEN 260
200 PRINT "INCORRECT INPUT ";"TRY AGAIN"
210 GO TO 160
220 LET Z4=1.282
230 GO TO 270
240 LET Z4=1.645
250 GO TO 270
260 LET Z4=2.326
270 IF A=2 THEN 450
280 LET U = INT(0.5 +N*P4+SQR(N*P4*(1-P4))*Z4)+1
285 PRINT "POWER FUNCTION"
287 PRINT "P","POWER AT P"
290 FOR P1 = P4 TO .99 STEP .05
300 LET Z = (U-0.5-N*P1)/SQR(N*P1*(1-P1))
310 GOSUB 900
320 LET P2 = 1-P
330 PRINT P1, P2
335 NEXT P1
340 PRINT "WHAT IS THE";" NUMBER OF SUCESSES X?"
350 INPUT X
360 PRINT "THE CRITICAL REGION FOR A ";A1;"LEVEL TEST IS"
370 PRINT "X>= ";U
380 IF X >= U THEN 410
390 PRINT "DECISION IS ";"ACCEPT P =";P4
400 GO TO 1030
410 PRINT "DECISION IS ";"REJECT P= ";P4
420 GO TO 1030
450 LET L=INT(N*P4-.5-SQR(N*P4*(1-P4))*Z4)
455 PRINT "POWER FUNCTION"
457 PRINT "P","POWER AT P"
460 FOR P1=P4 TO 0.01 STEP -0.05
470 LET Z =(L+0.5-N*P1)/SQR(N*P1*(1-P1))
480 GOSUB 900
490 PRINT P1, P
500 NEXT P1
*
```

```
510 PRINT "WHAT IS THE ";"NUMBER OS SUCCESSES X?"
520 INPUT X
530 PRINT "THE CRITICAL REGION FOR A ";A1;"LEVEL TEST IS"
540 PRINT "X<= ";L
550 IF X<= L THEN 580
560 PRINT "DECISION IS ";"ACCEPT P = ";P4
570 GO TO 1030
580 PRINT "DECISION IS ";"REJECT P = ";P4
590 GO TO 1030
600 PRINT "WHAT IS THE ";"SIGNIFICANCE LEVEL, ";"0.10,0.05, OR 0.01?"
610 INPUT A1
620 IF A1=0.10 THEN 650
630 IF A1= 0.05 THEN 670
640 IF A1=0.01 THEN 690
642 PRINT "INCORRECT INPUT, TRY AGAIN"
644 GO TO 610
650 LET Z4=1.645
660 GO TO 700
670 LET Z4=1.96
680 GO TO 700
690 LET Z4=2.576
700 LET L = INT(N*P4-.5-SQR(N*P4*(1-P4))*Z4)
710 LET U= INT (N*P4+.5+SQR(N*P4*(1-P4))*Z4)+1
720 PRINT "POWER FUNCTION"
725 PRINT "P","POWER AT P"
730 FOR P1=0.05 TO 0.95 STEP 0.05
740 LET Z=(L+0.5-N*P1)/SQR(N*P1*(1-P1))
750 GOSUB 900
760 LET P5=P
770 LET Z=(U-.5-N*P1)/SQR(N*P1*(1-P1))
780 GOSUB 900
790 LET P5=P5+(1-P)
800 PRINT P1,P5
805 LET P5=0
810 NEXT P1
820 PRINT "WHAT IS THE ";"NUMBER OF SUCCESSES X?"
830 INPUT X
840 PRINT "THE CRITICAL REGION FOR A ";A1;"LEVEL TEST IS"
850 PRINT "X<= ";L;"OR X>= ";U
860 IF X <= L THEN 890
870 IF X >= U THEN 890
880 PRINT "DECISION IS ACCEPT P = ";P4
885 GO TO 1030
890 PRINT "DECISION IS REJECT P = ";P4
895 GO TO 1030
900 LET Z1= ABS(Z)
910 LET C=1/SQR(2)
920 LET C1= .14112821
930 LET C2= .08864027
940 LET C3=.02743349
950 LET C4= -.00039446
960 LET C5=.00328975
970 DEF FNZ(X)=1-1/(1+C1*X+C2*X+2+C3*X+3+C4*X+4+C5*X+5)+8
980 LET P = .5+.5*FNZ(Z1*C)
990 LET P = 1E-4*(INT(1E4*P))
1000 IF Z > 0 THEN 1020
1010 LET P = 1-P
1020 RETURN
1030 END
*
```

```
 RUN
WHAT ARE THE VALUES OF N AND P0?
? 180, .6
WHICH ALTERNATIVE 1, 2 OR 3?
? 1
WHAT IS THE SIGNIFICANCE LEVEL, 0.10, 0.05, OR 0.01?
? 0.05
POWER FUNCTION
P                      POWER AT P
 0.6                    0.0401
 0.65                   0.3481
 0.7                    0.8548
 0.75                   0.9961
 0.8                    0.9999
 0.85                   1
 0.9                    1
 0.95                   1
WHAT IS THE NUMBER OF SUCESSES X?
? 130
THE CRITICAL REGION FOR A  0.05 LEVEL TEST IS
X>=  120
DECISION IS REJECT P=  0.6

*
```

be 0.05, the actual probability of a type I error is 0.0401. The
value 0.05 is referred to as the *nominal significance level* of the
test and the value $\alpha = 0.0401$ is referred to as the *actual significance level*. The difference is due to the fact that the critical
region is of form $X \geq j$, for integral j. The critical regions given
by PTEST are conservative in the sense that the actual significance
level will always be less than the nominal significance level. A
graph of the power function for this test is given in Fig. 7.2.1,
showing the properties of this test.

Exercises 7.2

Use the program PTEST as appropriate in the following:

1. In an article in the *Academy of Management Journal*, September
    1976, entitled "Informal Helping Relationships in Work Occupa-
    tions," by Ronald J. Burke, Tamara Weir, and Gordon Duncan, it
    was reported that of 53 young managers, 41 were married and 12
    unmarried. We wish to test the hypothesis

    $$H_0: \quad p \leq 0.75 \qquad \text{versus} \qquad H_1: \quad p > 0.75$$

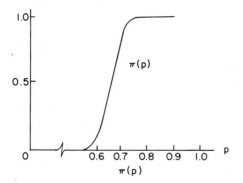

Fig. 7.2.1

where p is the true proportion of such managers who are married.
Find the critical region for the test using $\alpha = 0.05$. What
decision is made here?

2. In "Experiments on Plant Hybrids," reproduced in *The Origin of
   Genetics, A Mendel Source Book*, it is reported that of 8023 peas,
   2001 had green stems and 6022 had yellow. Mendel's theory predicted a 3:1 ratio of yellow to green using his genetic model.
   Test this hypothesis against a two-sided alternative using $\alpha =$
   0.05. Find the critical region for the test explicitly.

3. In 1973, Reggie Jackson, playing with the Oakland A's, had 158
   hits in 539 attempts. Find the critical region for testing
   H_0: $p \geq 0.300$ versus H_1: $p < 0.300$ using $\alpha = 0.01$. (Here p
   represents Jackson's lifetime batting average.) What decision
   is made in this case? Source: *The Baseball Encyclopedia*,
   Revised Edition, MacMillan, New York, 1974.

4. Assume that we wish to test at significance level $\alpha = 0.05$ the
   hypothesis that a computer program randomly prints a 0 or a 1
   (that is, $p = 0.5$) against the alternative that it does not.
   (a) Find the critical region for a test based on 60 such
        selections.
   (b) Find the critical region for a test based on 100 such
        selections.
   (c) Sketch the power curves for the tests in parts (a) and (b)
        and comment on the relationship between the two.

5. Suppose a holder of a political office wishes to determine whether
   a majority of his constituents are in favor of a particular bill
   on automobile speed limits. He wishes to test H_0: $p \leq 0.5$
   versus H_1: $p > 0.5$, where p is the proportion of his constituents favoring the bill. Suppose a random sample of 397 voters
   reveals 227 individuals in favor of the bill.
   (a) Find the critical region for the test for nominal signifi-
        cance levels 0.10, 0.05, and 0.01. Does the critical region
        become larger or smaller as α decreases?

(b) Plot the power curves for the tests with $\alpha = 0.01$ and $\alpha = 0.10$. Make a qualitative comparison of these power curves. Can you give a reason for the relationship?

7.3 TESTS CONCERNING μ IN NORMAL POPULATIONS

As explained in Sec. 4.4, many observations can be considered to come from normal distributions. Frequently we wish to test a hypothesis of the form $\mu \leq \mu_0$ versus $\mu > \mu_0$ for a fixed constant value μ_0, assuming a random sample from $X \sim N(\mu, \sigma^2)$ is available. For example, weights of newborn males may be considered to be normally distributed. In a study of the health of newborn males in an underdeveloped country we may want to test H_0: $\mu \leq 6.5$ lb versus H_1: $\mu > 6.5$ lb. Evidence for H_1 may indicate a satisfactory average birth weight. We shall consider testing hypotheses about μ in the normal case, first in the case that σ^2 is assumed known and then in the case that σ^2 is considered unknown.

7.3.1 Tests for μ for $X \sim N(\mu, \sigma^2)$, σ^2 Known

As in the case for testing the hypotheses about a population proportion p, the null hypothesis will be taken to be one of the forms listed in Table 7.3.1. We shall be interested in finding the critical region of values of a test statistic leading to the rejection of H_0.

Table 7.3.1   Tests for μ in the Normal Case, σ^2 Known

H_0	H_1	Critical region
$\mu \leq \mu_0$	$\mu > \mu_0$	$\bar{X} \geq \mu_0 + (\sigma/\sqrt{n})Z_\alpha$
$\mu \geq \mu_0$	$\mu < \mu_0$	$\bar{X} \leq \mu_0 - (\sigma/\sqrt{n})Z_\alpha$
$\mu = \mu_0$	$\mu \neq \mu_0$	$\bar{X} \geq \mu_0 + (\sigma/\sqrt{n})Z_{\alpha/2}$ or
		$\bar{X} \leq \mu_0 - (\sigma/\sqrt{n})Z_{\alpha/2}$

Again α will represent the maximum probability of rejecting H_0, under the assumption that H_0 is true. For a test of the hypothesis $\mu \leq \mu_0$, it is reasonable to base a test on the statistic $\bar{X}$ and to reject this hypothesis if $\bar{X}$ is large. An important theorem about independent observations from a normal distribution permits determination of the appropriate critical region for a test with significance level α. We state the theorem without proof.

THEOREM 7.3.1. If $(X_1, X_2, \ldots, X_n)$ are independent observations from $X \sim N(\mu, \sigma^2)$, then $(\bar{X} - \mu)/\sigma/\sqrt{n}$ is distributed as the standard normal variable Z.

To satisfy the requirements concerning the type I error for a test of H_0: $\mu \leq \mu_0$ versus H_1: $\mu > \mu_0$, we require

$$P((\bar{X} \geq c) | \mu \leq \mu_0) \leq \alpha$$

By determining c such that

$$P(\bar{X} \geq c | \mu = \mu_0) = \alpha$$

or

$$P\left((\frac{\bar{X} - \mu_0}{\sigma/\sqrt{n}}) \geq \frac{c - \mu_0}{\sigma/\sqrt{n}} \right) = \alpha$$

we obtain from Theorem 7.3.1

$$\frac{c - \mu_0}{\sigma/\sqrt{n}} = Z_\alpha \qquad \text{or} \qquad c = \mu_0 + \frac{\sigma}{\sqrt{n}} Z_\alpha$$

If $\mu < \mu_0$, we have

$$P(\bar{X} \geq c) = P(\bar{X} \geq \mu_0 + \frac{\sigma}{\sqrt{n}} Z_\alpha)$$

$$= P\left(\frac{\bar{X} - \mu}{\sigma/\sqrt{n}} \geq \frac{\mu_0 - \mu}{\sigma/\sqrt{n}} + Z_\alpha \right)$$

$$= P(Z \geq \frac{\mu_0 - \mu}{\sigma/\sqrt{n}} + Z_\alpha) \leq \alpha$$

as $\mu_0 - \mu > 0$.

Hence we obtain the critical region $\bar{X} > \mu_0 + (\sigma/\sqrt{n})Z_\alpha$. Critical regions for the other hypotheses are found similarly and are given in Table 7.3.1.

Suppose we consider the test mentioned above concerning male birth weights to decide between

$$H_0: \quad \mu \le 6.5 \quad \text{and} \quad H_1: \quad \mu > 6.5$$

Assume that $\sigma = 1$, the sample size $n = 25$, and $\alpha = 0.05$ and that we observe $\bar{x} = 6.95$. Using the first row of Table 7.3.1, we find $c = 6.5 + (1/5)1.645 = 6.829$. Hence the rejection region is given by $\bar{X} \ge 6.829$, and we decide to reject H_0. Theorem 7.3.1 also permits determination of the power function of the test. Recall that

$$\pi(\mu) = P(\text{rejecting } H_0|\mu) = P(\bar{X} \ge 6.829|\mu)$$

$$= P(\frac{\bar{X} - \mu}{0.2} \ge \frac{6.829 - \mu}{0.2})$$

$$= P(Z \ge \frac{6.829 - \mu}{0.2}) = 1 - F_Z(\frac{6.829 - \mu}{0.2})$$

For example, if $\mu = 7$, we find

$$\pi(7) = P(Z \ge \frac{6.829 - 7}{0.2}) = 1 - F_Z(-0.855)$$

$$= 1 - 0.1963 = 0.8037$$

A graph of the power function for this test is given in Fig. 7.3.1.

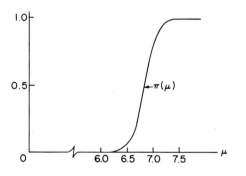

Fig. 7.3.1

For the moment suppose we consider only the hypothesis

H_0: $\mu = 6.5$ versus H_1: $\mu = 7$

with $\sigma = 1$ and $n = 25$ as before. The same critical region, namely
$\bar{X} \geq 6.829$, is appropriate for a test with significance level $\alpha = 0.05$.
As $\beta(\mu) = 1 - \pi(\mu)$, $\beta(7) = 1 - 0.8037 = 0.1963$. The areas representing
α and β are indicated in Fig. 7.3.2. Notice that the area to the *right*
of $c = 6.829$ under the normal density function centered at 6.5 is
$\alpha = 0.05$, while the area to the *left* of $c = 6.829$ under the normal
density function centered at 7 is $\beta = 0.1963$. If we attempt to make
α smaller by increasing c, β will become larger. If we attempt to
make β smaller by decreasing c, α will become larger. For a fixed
sample size both errors cannot simultaneously be made small. If we
allow n to vary, however, we can control α and β. Suppose we require
$\alpha = 0.05$ and $\beta = 0.05$. Then $P(\bar{X} \geq c | \mu = 6.5) = 0.05$ and $P(\bar{X} \leq c | \mu = 7)$
$= 0.05$. These two requirements lead to the equations

$$\frac{(c - 6.5)}{(1/\sqrt{n})} = 1.645 \quad \text{and} \quad \frac{(c - 7)}{(1/\sqrt{n})} = 1.645$$

Dividing the first of these equations by the second yields
$(c - 6.5)/(c - 7) = -1$ or $c = 6.75$. Substituting this value of c
into the first equation yields $\sqrt{n} = 6.58$ or $n = 43.3$. Hence using
$n = 44$ and $c = 6.75$ yields a test with (approximately) the required
values of α and β.

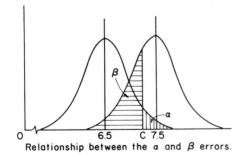

Relationship between the α and β errors.

Fig. 7.3.2.

Here we point out that there are five steps in testing a statistical hypothesis H_0 versus an alternative H_1. This reasoning process will apply quite generally in testing statistical hypotheses.

1. State H_0 and H_1.
2. Determine the values of α and n.
3. Select the appropriate test statistic.
4. Using steps 1 to 3, find the critical region of values of the test statistic which lead to the rejection of H_0.
5. Calculate the test statistic and make a decision between H_0 and H_1.

It is important to point out that steps 1 through 4 may be carried out before any data are collected. These four steps determine the properties of the statistical test. In this sense the decision procedure can be thought of as "objective." Either a decision for H_0 or H_1 may be incorrect, but we have knowledge of the probabilities of these errors. It is this knowledge which produces decision procedures which in a sense may be called "wise."

7.3.2   Tests for μ for X ~ N(μ,σ²), σ² unknown

In Sec. 6.2 we have seen that for a random sample $(X_1, X_2, \ldots, X_n)$ from $X \sim N(\mu,\sigma^2)$, the statistic

$$t = \frac{\bar{X} - \mu}{S/\sqrt{n}}$$

has a distribution known as Student's t distribution with $n - 1$ degrees of freedom. As this random variable depends only upon the unknown parameter μ, it is not surprising that it is used to test hypotheses about μ in the normal case if σ^2 is unknown. If we consider the first row of Table 7.3.1, we see that for the hypotheses

$$H_0: \ \mu \leq \mu_0 \quad \text{and} \quad H_1: \ \mu > \mu_0$$

that the critical region is of the form

$$\bar{X} \geq \mu_0 + (\frac{\sigma}{\sqrt{n}})Z_\alpha \quad \text{or} \quad \frac{(\bar{X} - \mu_0)}{\sigma/\sqrt{n}} \geq Z_\alpha$$

Table 7.3.2   Tests for μ in the Normal Case, σ^2 Unknown

H_0	H_1	Critical Region
$\mu \leq \mu_0$	$\mu > \mu_0$	$t \geq t_{\alpha, n-1}$
$\mu \geq \mu_0$	$\mu < \mu_0$	$t \leq -t_{\alpha, n-1}$
$\mu = \mu$	$\mu \neq \mu_0$	$\lvert t \rvert \geq t_{\alpha/2, n-1}$

For the same two hypotheses, in the case that σ^2 is *unknown*, the rejection region is of the form

$$t = \frac{(\bar{X} - \mu_0)}{S/\sqrt{n}} \geq t_{\alpha, n-1}$$

It is apparent that if H_1 is true, this statistic will tend to be positive. The larger the value of t, the more evidence exists in favor of H_1. Additionally, this rejection region satisfies the type I error requirement that

$$P(\frac{(\bar{X} - \mu_0)}{(S/\sqrt{n})} \geq t_{\alpha, n-1} \mid \mu \leq \mu_0) \leq P(t_{n-1} \geq t_{\alpha, n-1}) = \alpha$$

providing a critical region with a maximum type I error of α. In Table 7.3.2 we present the appropriate critical regions for the indicated hypotheses for a test with significance level α.

The program MTEST has been written to calculate the value of $t = (\bar{x} - \mu_0)/s/\sqrt{n}$. The values of μ_0 and n are entered in input statements beginning at line 600. In the sample run we have calculated the value of t for the heights of 25 kindergarten boys given in Sec. 6.2 to test the hypothesis

$$H_0: \ \mu = 42.5 \quad \text{versus} \quad H_1: \ \mu \neq 42.5$$

with $\alpha = 0.05$. The calculated value of t is found to be t = 0.879, and referring to Table B.II, we see that $t_{0.025, 24} = 2.064$. As $\lvert t \rvert < 2.064$, we retain H_0. Critical regions for two-sided alternatives are computed for $\alpha = 0.10$, 0.05, 0.01 or for one-sided alternatives with $\alpha = 0.05$, 0.025, or 0.005. The maximum value of n is 50.

MTEST

```
10 REM THIS PROGRAM TESTS THE HYPOTHESIS: MU=MUO AT SIGNIFICANCE
20 REM LEVEL ALPHA FOR A TWO-SIDED TEST, ALPHA/2 FOR A ONE SIDED
30 REM TEST AGAINST ONE OF THE ALTERNATIVES
40 REM 1) MU > MUO
50 REM 2) MU < MUO
60 REM 3) MU <>MUO
70 REM THE POPULATION SAMPLED IS ASSUMED TO BE NORMALLY
80 REM DISTRIBUTED WITH UNKNOWN VARIANCE. THE SAMPLE SIZE
90 REM IS N.
100 PRINT "WHAT ARE THE VALUES OF N AND MUO?"
110 INPUT N, UO
120 PRINT "WHICH ALTERNATIVE ";"1,2 OR 3?"
130 INPUT A
140 IF A=1 THEN 190
150 IF A =2 THEN 190
160 IF A = 3 THEN 190
170 PRINT "INCORRECT FORMAT ";"TRY AGAIN"
180 GO TO 110
190 GOSUB 380
195 DIM X(50), T(50), U(50), V(50)
200 FOR I = 1 TO N
205 READ X(I)
210 LET S = S+X(I)
215 LET S1= S1+X(I)↑2
220 NEXT I
225 LET X1=S/N
230 LET S2=SQR((S1-N*X1↑2)/(N-1))
235 PRINT "WHAT IS THE VALUE OF ALPHA, 0.10,0.05,0.01?"
240 INPUT A1
245 LET T=(X1-UO)/(S2/SQR(N))
250 IF A = 1 THEN 282
255 IF A = 2 THEN 302
260 PRINT "THE CRITICAL REGION ";"FOR A TWO-SIDED TEST"
262 PRINT "AT SIGNIFICANCE LEVEL= ";A1
264 IF A1=0.10 THEN 278
266 IF A1=0.05 THEN 274
270 PRINT "IS ABSOLUTE T >= ";V(N-1)
272 GO TO 320
274 PRINT "IS ABSOLUTE T>= ";U(N-1)
276 GO TO 320
278 PRINT "IS ABSOLUTE T >= ";T(N-1)
280 GO TO 320
282 PRINT "THE CRITICAL REGION ";"FOR A ONE=SIDED TEST"
284 PRINT "AT SIGNIFICANCE ";"LEVEL = ";A1/2
286 IF A1=0.10 THEN 298
288 IF A1=0.05 THEN 294
290 PRINT "IS T >=";V(N-1)
292 GO TO 320
294 PRINT "IS T >= ";U(N-1)
296 GO TO 320
298 PRINT "IS T >= ";T(N-1)
300 GO TO 320
*
```

```
302 PRINT "THE CRITICAL REGION ";"FOR A ONE-SIDED TEST"
304 PRINT "AT SIGNIFICANCE LEVEL = ";A1/2
306 IF A1=0.10 THEN 318
308 IF A1= 0.05 THEN 314
310 PRINT "IS T <= ";-V(N-1)
312 GO TO 320
314 PRINT "IS T <= ";-U(N-1)
316 GO TO 320
318 PRINT "IS T <= ";-T(N-1)
320 PRINT
325 PRINT "THE COMPUTED VALUE OF T =";T
340 GO TO 1000
380 FOR I = 1 TO 50
382 READ T(I)
384 NEXT I
386 FOR I = 1 TO 50
388 READ U(I)
390 NEXT I
392 FOR I = 1 TO 50
394 READ V(I)
396 NEXT I
398 RETURN
*
 EDIT LIST 600-1000

600 DATA 42.58, 41.82, 41.94, 40.25, 42.37, 44.6, 42.74, 43.14, 41.17, 41.9, 46.2
610 DATA 38.89, 45.02, 38.1, 48.59, 34.54, 48.07, 42.45, 47.97, 41.5, 43.9, 43.2
620 DATA 42.02, 48.3, 45.9
1000 END
*
 RUN
WHAT ARE THE VALUES OF N AND MUO?
? 25, 42.5
WHICH ALTERNATIVE 1, 2 OR 3?
? 3
WHAT IS THE VALUE OF ALPHA, 0.10, 0.05, 0.01?
? 0.05
THE CRITICAL REGION FOR A TWO-SIDED TEST
AT SIGNIFICANCE LEVEL=  0.05
IS ABSOLUTE T>=  2.0639

THE COMPUTED VALUE OF T = 0.879079

*
```

The example in the preceding paragraph shows an interesting relationship between a confidence interval for μ and a test of hypothesis about μ. A confidence interval with confidence coefficient $1 - \alpha$ contains those points μ_0 for which

$$\bar{x} - (\frac{s}{\sqrt{n}})t_{\alpha/2,n-1} < \mu_0 < \bar{x} + (\frac{s}{\sqrt{n}})t_{\alpha/2,n-1}$$

or equivalently by

$$\left|\frac{(\bar{x} - \mu_0)}{s/\sqrt{n}}\right| < t_{\alpha/2,n-1} \qquad\qquad (7.3.1)$$

However relation (7.3.1) describes exactly all those values of μ_0 which would be accepted in a test of $H_0: \mu = \mu_0$ versus $H_1: \mu \neq \mu_0$ at level of significance level α. Conversely, all those points satisfying relation (7.3.1) will lie in the $(1 - \alpha)100\%$ confidence interval for μ. Hence as we found a 95% confidence interval to be (41.74, 44.46) in Sec. 6.2, we know that $H_0: \mu = 42.5$ will be accepted at significance level $\alpha = 0.05$. However $H_0: \mu = 41$ will be rejected at the same significance level, which is verified by calculating t = (43.08 - 41)/3.34/5 = 3.128 > 2.064. Thus, the confidence interval gives all the information required for a two-sided test.

Problems 7.3

1. Assume that the weights of "5-lb" bags of sugar from a production line can be considered to be distributed as $X \sim N(\mu,0.01)$. We want to decide, based on a sample of size 16, if the average weight is acceptable. We test

   $$H_0: \mu \leq 5 \qquad \text{versus} \qquad H_1: \mu > 5$$

   where H_1 can be thought of as indicating acceptable weights.
   (a) Find the critical region for this test if $\alpha = 0.05$.

```
  RUN
WHAT ARE THE VALUES OF N AND MU0?
? 25, 41
WHICH ALTERNATIVE 1, 2 OR 3?
? 3
WHAT IS THE VALUE OF ALPHA, 0.10, 0.05, 0.01?
? 0.05
THE CRITICAL REGION FOR A TWO-SIDED TEST
AT SIGNIFICANCE LEVEL=   0.05
IS ABSOLUTE T>=   2.0639

THE COMPUTED VALUE OF T = 3.12775

*
```

(b) Find an expression for the power function of this test and evaluate the expression for $\mu = 5(0.02)(5.1)$. The expression will be a function of μ. Sketch a graph of this power function.

2. Assume that SAT scores can be taken to be normally distributed with $\sigma = 100$. A sampling of 25 members of an entering freshman class yields $\bar{x} = 580$. We wish to test the hypotheses

$$H_0: \quad \mu \geq 600 \qquad versus \qquad H_1: \quad \mu < 600$$

at significance level 0.05. Evidence for H_1 will suggest a decline in the test scores for this entering class.

(a) Find the critical region for this test.

(b) What decision is made between H_0 and H_1?

3. The EPA is suspicious that a manufacturing company producing batteries is discharging mercury into a river. On 8 days a sample of the mercury content of the water downstream from the factory is made. The following data (in micrograms/liter) are obtained.

Day	1	2	3	4	5	6	7	8
Observation	2.2	1.8	1.7	1.9	1.6	1.3	1.9	2.0

(a) An acceptable level of mercury content is less than *2* μg/liter. Test $H_0: \quad \mu \geq 2$ versus $H_1: \quad \mu < 2$ assuming normality of the observations. Let $\alpha = 0.05$.

(b) Find a 95% confidence interval for μ. What can the EPA state statistically concerning the water quality?

4. Consider again testing $H_0: \quad \mu \leq \mu_0$ against $H_1: \quad \mu > \mu_0$, where $X \sim N(\mu, \sigma^2)$ with σ^2 known.

(a) Show that the power function of this test can be written as

$$\pi(\mu) = 1 - F_Z\left(Z_\alpha + \frac{\sqrt{n}(\mu_0 - \mu)}{\sigma}\right)$$

(b) Prove that $\pi(\mu)$ is an increasing function of μ.

5. In an article by S. Josephina Concannon in *The Journal of Educa-tional Research*, November 1975, entitled "Comparison of the Stanford-Binet Scale with the Peabody Picture Vocabulary Test," a report is given about a sample of 32 five-year-old children of above average ability. The average IQ score on the Standord-Binet test is given to be 134.4 for these children with s = 6.53.

    (a) Assuming that IQ scores are $N(\mu, 225)$, test the hypothesis

    $$H_0: \ \mu \leq 130 \qquad \text{versus} \qquad H_1: \ \mu > 130$$

    for these children. Use $\alpha = 0.05$.

    (b) Assuming normality, but that σ^2 is not known, test the hypotheses ($\alpha = 0.05$) given in part (a).

6. A newspaper states that the average income of high school teachers is \$11,000. Suppose a random sample of 16 teachers' salaries yields $\bar{x} = 10,500$ and s = 500. Assuming that such salaries are normally distributed, we wish to test

    $$H_0: \ \mu = 11,000 \qquad \text{versus} \qquad H_1: \ \mu < 11,000$$

    (a) Find the rejection region for this test using $\alpha = 0.05$.

    (b) Find the computed value of the appropriate statistic for these data.

    (c) Which of the two hypotheses would be chosen?

7. Suppose that the time X that an experimental engine will operate with 1 gallon of a certain kind of fuel has a normal distribution, that is, $X \sim N(\mu, \sigma^2)$. Test runs with five models of this engine showed that they operated, respectively, for 21, 19, 23, 18, and 19 min with this kind of fuel.

    (a) What is the BLUE estimate of μ?

    (b) What is the value of the sample standard deviation s?

    (c) Find a 95% confidence interval for μ.

    (d) Based on the data given would one accept the null hypothesis $H_0: \ \mu = 23$ (versus $H_1: \ \mu \neq 23$) at significance level $\alpha = 0.05$?

8. Suppose that for many years a certain kind of power unit produced by the E. G. Company has, on the average, produced 1000 V of power/hr. The unit has recently been redesigned and E. G. now claims it produces more power, on the average, than the old design. As a user of such a power unit, you must decide to accept the claim or not, so you sample four units observing $\bar{v}$ = 1250 V and s = 100 V.

   (a) How should the null and alternative hypotheses be stated by you, a user?

   (b) Find the critical region for α = 0.05 and α = 0.01 level tests.

   (c) What decisions are made in part (b)?

Exercises 7.4

1. Use the program MTEST to find a critical region for the test of H_0: μ = 72 versus H_1: μ $\neq$ 72 at significance levels α = 0.05 and α = 0.01, for the observations of 50 heights given in Exercise 5.2.2.

2. (a) Use the data of Sec. 6.3 on the weights of 20 "1-lb" bags of flour to test the hypotheses

   $$H_0: \; μ \leq 16 \quad \text{versus} \quad H_0: \; μ > 16$$

   at the 5% level of significance.

   (b) Use these same data to test

   $$H_0: \; μ = 16 \quad \text{versus} \quad H_1: \; μ \neq 16$$

   at the 5% level of significance.

   Use the program MTEST.

3. The following observations are the lengths of time that driers set for 20 min actually operated: 19.69, 19.06, 19.15, 17.80, 19.49, 21.38, 19.79, 20.11, 18.54, 19.12, 22.61, 16.71, 21.61, 16.08, 24.47, 13.23, 24.06, 19.56, 23.97, 18.80. Using MTEST,

(a) Test the hypothesis $\mu = 20$ versus $\mu \neq 20$ at level of significance $\alpha = 0.05$.

(b) Carry out tests as in part (a) of $\mu = 21$ and $\mu = 22$ versus a two-sided alternative at level of significance $\alpha = 0.05$.

7.4   TESTS FOR μ IN LARGE SAMPLES

In Chap. 5 we have seen that the central limit theorem could be employed to find a confidence interval for μ, assuming a large sample from a continuous population is available. Due to the close relationship between confidence intervals and tests of hypotheses just mentioned, it is not surprising that tests of hypotheses concerning μ, in the large sample case can be based upon the normal distribution. Consider a test of

$$H_0: \ \mu = \mu_0 \qquad \text{versus} \qquad H_1: \ \mu \neq \mu_0 \qquad\qquad (7.4.1)$$

For large n, if H_0 is true, $Z = (\bar{X} - \mu_0)/\sigma/\sqrt{n}$ is approximately distributed as the standard normal variable. Hence

$$P(\frac{|\bar{X} - \mu_0|}{\sigma/\sqrt{n}} \geq Z_{\alpha/2}) = \alpha$$

or estimating σ by the sample standard deviation s,

$$P(\frac{|\bar{X} - \mu_0|}{s/\sqrt{n}} \geq Z_{\alpha/2}) = \alpha$$

If H_1 were true, then $(\bar{X} - \mu_0)/s/\sqrt{n}$ would tend to be large in magnitude, and the larger the magnitude the more evidence for H_1. Hence an approximate α level test for the hypotheses given in Eq. (7.4.1) is defined by the critical region $|\bar{X} - \mu_0|/s/\sqrt{n} \geq Z_{\alpha/2}$ if n is large. Table 7.4.1 gives the appropriate critical regions for tests in the case of one-sided alternatives.

Another concept used in statistical testing is the "likelihood" that an observed mean $\bar{x}$ comes from a population with theoretical expectation μ_0. Assuming a test against a two-sided alternative, we consider the probability

Table 7.4.1 Rejection Regions for Large Sample Tests

H_0	H_1	Critical region
$\mu \leq \mu_0$	$\mu > \mu_0$	$(\bar{X} - \mu_0)/s/\sqrt{n} \geq Z_\alpha$
$\mu \geq \mu_0$	$\mu < \mu_0$	$(\bar{X} - \mu_0)/s/\sqrt{n} \leq -Z_\alpha$
$\mu = \mu_0$	$\mu \neq \mu_0$	$|\bar{X} - \mu_0|/s/\sqrt{n} \geq Z_{\alpha/2}$

$$P(|Z| \geq \frac{|\bar{x} - \mu_0|}{s/\sqrt{n}})$$

where Z represents the standard normal variable. This gives the
probability of observing a value of $\bar{X}$ as far from μ_0 or farther
(in magnitude) than the value $\bar{x}$ observed. Such a probability is
referred to as a P-VALUE. If this probability is small, doubt is
cast on the assumption that $\mu = \mu_0$. Rather than relying on arbitrary
values of the significance level α, many statisticians prefer to use
the P-VALUE as an indication of the truth of the assertion $\mu = \mu_0$.
If the alternative to H_0: $\mu = \mu_0$ is considered to be one-sided,
the P-VALUE (in the upper tailed case) is

$$P(Z \geq \frac{\bar{x} - \mu_0}{s/\sqrt{n}})$$

i.e., one-half of the probability in the two-sided case.

The program MUTEST considers the hypothesis H_0: $\mu = \mu_0$ in the
large sample case. The value of n and μ_0 are provided in INPUT
statements. The program requests the alternative (1) $\mu > \mu_0$, (2)
$\mu < \mu_0$, or (3) $\mu \neq \mu_0$. The observations are in DATA statements
beginning with line 400. The value of $z = (\bar{x} - \mu_0)/s/\sqrt{n}$ is computed
and a test may be carried out using Table B.I for any desired significance level. Additionally the P-VALUE defined in the previous
paragraph is calculated. In the example, run the first 50 "ages"
of legislators from Exercise 5.1.3 have been used as data to test

```
MUTEST
10 REM THIS PROGRAM TESTS THE HYPOTHESIS: MU = MUO
20 REM AGAINST ONE OF THE ALTERNATIVES
30 REM 1) MU > MUO
40 REM 2) MU < MUO
50 REM 3) MU <> MUO
60 REM THE POPULATION SAMPLED IS CONSIDERED TO HAVE AN
70 REM EXPECTATION AND VARIANCE
80 REM THE SAMPLE SIZE N IS CONSIDERED TO BE "LARGE"
90 REM THE PROGRAM CALCULATES Z AND THE CORRESPONDING P-VALUE
100 PRINT "WHAT ARE THE VALUES OF N AND MUO?"
110 INPUT N, UO
120 PRINT "WHICH ALTERNATIVE, ";"1,2,OR 3?"
130 INPUT A
140 IF A = 1 THEN 190
150 IF A = 2 THEN 190
160 IF A = 3 THEN 190
170 PRINT "INCORRECT FORMAT ";"TRY AGAIN"
180 GO TO 110
190 FOR I = 1 TO N
200 READ X
210 LET S = S + X
220 LET S1 = S1 + X↑2
230 NEXT I
240 LET X1 = S/N
250 LET S2=SQR((S1-N*X1↑2)/(N-1))
260 LET Z = (X1-UO)/(S2/SQR(N))
270 GOSUB 900
290 PRINT "THE CALCULATE Z VALUE = ";Z
300 IF A = 3 THEN 330
310 PRINT
320 PRINT "THE P-VALUE FOR ";"ALTERNATIVE ";A;" = ";P
325 GO TO 1030
330 PRINT
340 PRINT "THE P-VALUE FOR ";"ALTERNATIVE ";A;" = ";P*2
350 GO TO 1030
400 DATA 46, 44, 45, 40, 46, 52, 47, 48, 42, 44
410 DATA 57, 36, 53, 34, 63, 24, 62, 46, 61, 43
420 DATA 50, 48, 45, 62, 56, 46, 46, 40, 44, 52
430 DATA 43, 39, 49, 47, 42, 53, 53, 43, 41, 47
440 DATA 46, 34, 38, 44, 43, 48, 48, 45, 41, 61
900 LET Z1= ABS(Z)
910 LET C=1/SQR(2)
920 LET C1= .14112821
930 LET C2= .08864027
940 LET C3=.02743349
950 LET C4= -.00039446
960 LET C5=.00328975
970 DEF FNZ(X)=1-1/(1+C1*X+C2*X↑2+C3*X↑3+C4*X↑4+C5*X↑5)↑8
980 LET P = .5+.5*FNZ(Z1*C)
990 LET P = 1E-4*(INT(1E4*P))
1010 LET P = 1-P
1020 RETURN
1030 END
*
```

```
 RUN
WHAT ARE THE VALUES OF N AND MUO?
? 50, 50
WHICH ALTERNATIVE, 1, 2, OR 3?
? 3
THE CALCULATE Z VALUE = -3.17236

THE P-VALUE FOR ALTERNATIVE  3  =  0.0016

*RUN
WHAT ARE THE VALUES OF N AND MUO?
? 6-50, 48
WHICH ALTERNATIVE, 1, 2, OR 3?
? 3
THE CALCULATE Z VALUE = -1.33863

THE P-VALUE FOR ALTERNATIVE  3  =  0.1808

*
```

the hypothesis μ = 50. The computed value of z = -3.17, and since |z| ≥ 1.96, we would reject this hypothesis at the 5% level of significance against a two-sided alternative. The P-VALUE of 0.0016 indicates that μ = 50 is an "unlikely" value of μ. The same data used to test μ = 48 yield z = -1.34 a P-VALUE = 0.181, which leads to acceptance of H_0 at any significance level less than 0.181. Hence the P-VALUE is the smallest significance level at which the null hypothesis can be rejected. If a predetermined α is less than the P-VALUE, the null hypothesis is retained.

Exercises 7.4

1. Use the height data of Exercise 5.2.2 with MUTEST to test the hypotheses μ = 72 versus μ ≠ 72 at significance level α = 0.05. What is the P-VALUE for this test?

2. The list below gives the age at first inauguration of 38 presidents*:

George Washington	57	James Buchanan	65
John Adams	61	Abraham Lincoln	52
Thomas Jefferson	57	Andrew Johnson	56
James Madison	57	U. S. Grant	46
James Monroe	58	Rutherford B. Hayes	54
John Q. Adams	57	James A. Garfield	49
Andrew Jackson	61	Chester A. Arthur	50
Martin Van Buren	54	Grover Cleveland	47
William Henry Harrison	68	Benjamin Harrison	55
John Tyler	51	William McKinley	54
James K. Polk	49	Theodore Roosevelt	42
Zachary Taylor	64	William H. Taft	51
Millard Filmore	50	Woodrow Wilson	56
Franklin Pierce	48	Warren G. Harding	55

---

*Source: *The World Almanac and Book of Facts,* 1975.

Calvin Coolidge	51	John F. Kennedy	43
Herbert Hoover	54	Lyndon B. Johnson	55
F. D. Roosevelt	51	Richard Nixon	56
Harry S. Truman	60	Gerald R. Ford	61
Dwight D. Eisenhower	62	James E. Carter	52

Use MUTEST to test the hypothesis that the true average at inauguration is 55 versus the alternative that μ is greater than 55 at level of significance $\alpha = 0.02$. What is the P-VALUE for this test? What is the P-VALUE if the alternative is written as $\mu \neq 55$?

3. Use the 50 numbers generated by the program RANDSAM in Sec. 5.1 to test the hypothesis $\mu = 500.5$ versus $\mu \neq 500.5$ using the P-VALUE to make a decision.

7.5   DISTRIBUTION-FREE TESTS OF LOCATION

In the preceding two sections of this chapter we have considered tests concerning the expectation μ of a population assumed to be normally distributed, or one for which the sample size was sufficiently large that the sample mean could be assumed to be normally distributed. In this section we consider tests of location in the case that the underlying population is considered only to be continuous. Specifically, the cumulative distribution function of the random variable X is considered to be continuous. Such tests are referred to as distribution-free tests because the observations are not assumed to come from a specific family of distributions such as the normal or exponential. The term *nonparametric* is also used to describe these tests. This term is unfortunate, however, because the tests to be discussed will concern parameters of a distribution.

7.5.1   *The Sign Test*

We assume that we have a random sample $(X_1, X_2, \ldots, X_n)$ from a continuous population with a unique median η. In Sec. 6.2 we found confidence intervals for η based on the binomial distribution. As there is a close relationship between confidence intervals and tests of hypotheses, it should not be surprising that a test of

$$H_0: \eta = \eta_0 \qquad \text{versus} \qquad H_1: \eta \neq \eta_0$$

can be based on the binomial distribution. This test is referred to as the *sign test*. We use the statistic

$$S = \sum_{X_i > \eta_0} \text{sgn}(X_i - \eta_0)$$

i.e., the number of observations which exceed η_0 as the test statistic. If the true median is η_0, the observations $X_i - \eta_0$ will have a positive sign with probability 1/2 and a negative sign with probability 1/2 [the assumption that X, the parent distribution is continuous, implies that $P(X_i = \eta_0) = 0$]. Thus the statistic S has the binomial distribution $B(n,1/2)$ under H_0. Against a two-sided alternative we would reject H_0 if either S is very large or small. We reject H_0 if either $S \leq r$ or $S \geq n - r$, yielding a test with type I error $\alpha = P(S \leq r \text{ or } S \geq n - r | p = 1/2) = 2P(B(n,1/2) \leq r)$, as the probability in the two tails is the same.

Let us consider the baseball data in Table 6.2.1. The 20 ordered league leading numbers of hits (1952-1971) are 190, 192, 194, 200, 200, 204, 205, 205, 208, 209, 209, 210, 211, 212, 215, 218, 223, 230, 230, 231. Assume we wish to test

$$H_0: \eta = 201 \qquad \text{versus} \qquad H_1: \eta \neq 201$$

at approximately the 0.05 level of significance. Running the program BINOM for n = 20 and p = 1/2, we see that for r = 6, $2P(B(20,1/2) \leq 6)$ = 0.0414. Hence we would reject H_0 if $S \leq 6$ or $S \geq 14$. In this case S = 15 because 15 of the 20 observations exceed $\eta_0 = 201$, and thus the hypothesis H_0 would be rejected. Tests against one-sided alternatives are obtained similarly and are given in Table 7.5.1.

For large samples one can use the normal approximation to the binomial distribution to find the critical region for the tests. Suppose we wish to test

$$H_0: \eta \leq \eta_0 \qquad \text{versus} \qquad H_1: \eta > \eta_0$$

Table 7.5.1   Critical Regions for the Sign Test

H_0	H_1	Critical region
$\eta \le \eta_0$	$\eta > \eta_0$	$S \ge n - r_1$; $\alpha = P(B(n,1/2) \le r_1)$
$\eta \ge \eta_0$	$\eta < \eta_0$	$S \le r_2$; $\alpha = P(B(n,1/2) \le r_2)$
$\eta \quad \eta_0$	$\eta \ne \eta_0$	$S \le r$ or $S \ge n - r$; $\alpha = 2P(B(n,1/2) \le r)$

using the sign test. One calculates the statistic

$$S = \sum_{X_i > \eta_0} sgn(X_i - \eta_0)$$

If the true population median η exceeds η_0, then there will tend to be a large number of positive signs among the comparisons $X_i - \eta_0$, $i = 1, 2, \ldots, n$. To achieve a test of significance level α we require $P(S \ge c|H_0) = \alpha$. Using the normal approximation to the binomial distribution we obtain

$$P(\frac{S - n/2}{\sqrt{n}/2} \ge Z_\alpha) = \alpha$$

Hence we can base the test on the statistic $Z = (S - n/2)/\sqrt{n}/2$. If $Z \ge Z_\alpha$, we reject H_0 in favor of H_1. Similarly for the other two alternatives of Table 7.5.1 we obtain the rejection regions $Z \le -Z_\alpha$ $|Z| \ge Z_{\alpha/2}$. The P-VALUE for the null hypothesis can be obtained as for tests of μ in the large sample case.

The program MEDTEST uses the sign test statistic S to test the hypothesis H_0: $\eta = \eta_0$ in moderate or large samples. The values of n, η_0, and the alternative (1) $\eta > \eta_0$, (2) $\eta < \eta_0$ or (3) $\eta \ne \eta_0$ are input in response to the program. The observations should appear in DATA statements beginning at line 400. The program computes S, Z, and the P-VALUE for the indicated alternative. In the sample run, the height data of Sec. 6.2 have been used to test $\eta = 42.5$ and $\eta = 41$ against alternative (3). The respective P-VALUEs are 0.8416 and 0.0008 indicating acceptance of $\eta = 42.5$ and rejection of $\eta = 41$, agreeing with the parametric test results of Sec. 7.3.

MEDTEST

```
10 REM THIS PROGRAM USES THE SIGN TEST STATISTIC TO TEST THE
20 REM NULL HYPOTHESIS THE MEDIAN OF A CONTINUOUS POPULATION
30 REM IS M0. THE PROGRAM ASSUMES A MODERATE OR LARGE SAMPLE SIZE N
40 REM THE ALTERNATIVES ARE OF FORM
50 REM 1) MEDIAN > M0
60 REM 2) MEDIAN < M0
70 REM 3) MEDIAN <> M0
80 REM THE PROGRAM COMPUTES THE VALUE OF S,Z AND THE
90 REM CORRESPONDING P-VALUE
100 PRINT "WHAT ARE THE VALUES OF N AND MO?"
110 INPUT N,M0
120 PRINT "WHICH ALTERNATIVE, ";"1,2,OR 3?"
130 INPUT A
140 IF A = 1 THEN 190
150 IF A = 2 THEN 190
160 IF A = 3 THEN 190
170 PRINT "INCORRECT FORMAT ";"TRY AGAIN"
180 GO TO 110
190 FOR I = 1 TO N
200 READ X
205 IF X = M0 THEN 225
210 IF X < M0 THEN 230
215 LET S = S+1
220 GO TO 230
225 LET T = T+1
230 NEXT I
240 PRINT "THE NUMBER OF X = M0 IS ";T
250 PRINT "THE VALUE OF S =";S
260 LET Z =(2*S-(N-T))/SQR(N-T)
270 GOSUB 900
290 PRINT "THE CALCULATE Z VALUE = ";Z
300 IF A = 3 THEN 330
310 PRINT
320 PRINT "THE P-VALUE FOR ";"ALTERNATIVE ";A;" = ";P
325 GO TO 1030
330 PRINT
340 PRINT "THE P-VALUE FOR ";"ALTERNATIVE ";A;" = ";P*2
350 GO TO 1030
400 DATA 42.58,41.82,41.94,40.25,42.37,44.6,42.74,43.14,41.17,41.9,46.2
410 DATA 38.89,45.02,38.1,48.59,34.54,48.07,42.45,47.97,41.5,43.9,43.2
420 DATA 42.02,48.3,45.9
900 LET Z1= ABS(Z)
910 LET C=1/SQR(2)
920 LET C1= .14112821
930 LET C2= .08864027
940 LET C3=.02743349
950 LET C4= -.00039446
960 LET C5=.00328975
970 DEF FNZ(X)=1-1/(1+C1*X+C2*X↑2+C3*X↑3+C4*X↑4+C5*X↑5)↑8
980 LET P = .5+.5*FNZ(Z1*C)
990 LET P = 1E-4*(INT(1E4*P))
1010 LET P = 1-P
1020 RETURN
1030 END
```

```
 RUN
WHAT ARE THE VALUES OF N AND M0?
? 25,42.5
WHICH ALTERNATIVE 1,2,OR 3?
? 3
THE NUMBER OF X = M0 IS 0
THE VALUE OF S = 13
THE CALCULATE Z VALUE=  0.2

THE P-VALUE FOR ALTERNATIVE  3  =  0.8416

*RUN
WHAT ARE THE VALUES OF N AND M0?
? 25,41
WHICH ALTERNATIVE 1,2,OR 3?
? 3
THE NUMBER OF X = M0 IS 0
THE VALUE OF  S = 21
THE CALCULATE Z VALUE =  3.4

THE P-VALUE FOR ALTERNATIVE  3  =  7.99999 E-4


*
```

7.5.2   *The Wilcoxon One-Sample Test*

We have seen that the sign test provides a simple test of hypothesis
for the population median of a continuous distribution. However, as
the statistic S depends only upon the signs of $X_i - \eta_0$ information
about the magnitude of the original observations is lost. We consider here a more sensitive test of location generally referred to
as the Wilcoxon one-sample test or the Wilcoxon signed-rank test.
We consider that we are sampling from a continuous population. We
assume that the random variable in question has a continuous probability density function $f(x;\theta)$ dependent on a location parameter θ.
The density function is assumed to be symmetric about θ, that is,
$f(\theta - x) = f(\theta + x)$ for all x. It is straightforward to show that
the median of the distribution is θ and that $E(X) = \theta$, if the expectation exists. A typical probability density function of this kind
appears in Fig. 7.5.1.

We wish to test the hypotheses H_0: $\theta = \theta_0$ versus H_1: $\theta \neq \theta_0$,
as well as one-sided tests. The test statistic is again based on
the differences $X_i - \theta_0$. The statistic is a function of the ranks
(1, 2, ..., n) of the magnitude of these differences. Following
common notation, we define R_i to be the rank of the magnitude of the
ith difference. We further define $T_+ = \Sigma_{X_i > \theta_0} R_i$ and $T_- = \Sigma_{X_i < \theta_0} R_i$.
It is clear that $T_+ + T_- = n(n + 1)/2$ as this is just the sum of the

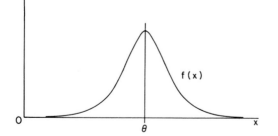

Fig. 7.5.1

first n integers. Hence T_+ can be computed as $n(n + 1)/2 - T_-$. The test is based on the statistic $T = \min(T_+, T_-)$. We indicate the computation of T for the baseball data of the preceding section using

X_i	$\left|X_i - 201\right|$	R_i	X_i	$\left|X_i - 201\right|$	R_i
190	11	13.5	209	8	8.5
192	9	10.5	210	9	10.5
194	7	6.5	211	10	12
200	1	1.5	212	11	13.5
200	1	1.5	215	14	15
204	3	3	218	17	16
205	4	4.5	223	22	17
205	4	4.5	230	29	18.5
208	7	6.5	230	29	18.5
209	8	8.5	231	30	20

$$T_- = 33.5, \; T_+ = 210 - 33.5 = 176.5, \; T = 33.5$$

Note that for tied observations the average of the corresponding ranks is assigned to each such observation. It is clear that small values of T indicate that the ranks assigned to the positive or negative differences are inconsistent with $\theta = \theta_0$.

In order to answer the question of whether T is sufficiently small to reject H_0: $\theta = \theta_0$, we must know the distribution of T under the null hypothesis. In Table B.IV we have values of T satisfying $\alpha'' = 2P(T \leq d)$, which are appropriate for two-sided tests with significance level α''. Values of d are given for values of α'' approximately equal to 0.01, 0.05, and 0.10. These values of d are appropriate for one-sided alternatives with a significance level of $\alpha' = \alpha''/2$. We see that for n = 20, $2P(T \leq 38) = 0.009$, so that the decision to reject $\theta = 201$ at significance level $\alpha = 0.01$ would be correct.

With a little reflection we see that the distributions of T_+ and T_-, excluding ties, are the same under H_0. This is true because each *combination* of the n integers summing to a particular value d can be signed either plus or minus. Hence, as $T_+ + T_- = n(n + 1)/2$, we have $E(T_+) + E(T_-) = n(n + 1)/2 = 2E(T_+)$, and thus $E(T_+) = n(n + 1)/4$ under H_0. Additionally the distribution of T_+ is symmetric about $n(n + 1)/4$ as

$$P(T_+ \leq \frac{n(n + 1)}{4} - d) = P(\frac{n(n + 1)}{2} - T_- \leq \frac{n(n + 1)}{4} - d)$$

$$= P(\frac{n(n + 1)}{4} + d \leq T_-)$$

$$= P(\frac{n(n + 1)}{4} + d \leq T_+)$$

The last equality follows from the fact that T_- and T_+ have the same distribution. It can additionally be shown that $Var(T_+) = Var(T_-) = n(n + 1)(2n + 1)/24$ under H_0. These facts about the null distribution of T_+ permit large sample tests of H_0: $\theta = \theta_0$ to be carried out using the normal distribution, as in the case of the sign test.

The statistic T_+ will be used in the large sample case. It may be proved that the standardized statistic $(T_+ - n(n + 1)/4)/\sqrt{n(n + 1)(2n + 1)/24}$ is approximately distributed as the standard normal variable Z under H_0 if n is large, say $n \geq 25$. Let us consider testing

$$H_0: \quad \theta \leq \theta_0 \qquad \text{versus} \qquad H_1: \quad \theta > \theta_0$$

using the statistic T_+. If the true value of θ exceeds θ_0, then the larger ranks of $|X_i - \theta_0|$ will tend to have positive signs. Hence we will reject H_0 if T_+ is large or analogously if

$$Z = \frac{(T_+ - n(n + 1)/4)}{\sqrt{n(n + 1)(2n + 1)/24}} \tag{7.5.1}$$

is large. A test with approximate significance level α for large n is defined by the critical region $Z \geq Z_\alpha$. The critical regions for commonly used alternatives are presented in Table 7.5.2, where Z is defined by Eq. (7.5.1). For the height data of Sec. 6.2 we have tested H_0: $\mu = 42.5$ in Sec. 7.3 using the t distribution. If we carry out the same test against a two-sided alternative, we calculate

Table 7.5.2   Critical Regions for the Signed-Rank Test

H_0	H_1	Critical Region
$\theta \leq \theta_0$	$\theta > \theta_0$	$Z \geq Z_\alpha$
$\theta \geq \theta_0$	$\theta < \theta_0$	$Z \leq -Z_\alpha$
$\theta = \theta_0$	$\theta \neq \theta_0$	$|Z| \geq Z_{\alpha/2}$

$T_+ = 193$ and hence $Z = (193 - 162.5)/\sqrt{25(26)(51)/24} = 0.821$. Hence again the decision is made to retain H_0. The signed-rank test is appropriate here as the normal probability density function is symmetric about μ.

    The program WILTEST is similar to other programs of this chapter for testing the value of a location parameter. The program calls for the sample size n, θ_0, and the appropriate alternative. A test of the hypotheses described in Table 7.5.2 is carried out based on a large sample approximation to the distribution of T_+. The program calculates T_+, Z, and the P-VALUE for the alternative selected. For the height data of Sec. 6.2 tests of $\theta_0 = 42.5$ and $\theta_0 = 41$ have been carried out against alternative (3). The P-VALUEs are 0.412 and 0.0018 indicating the same decisions as before. The Z values of 0.821 and 3.135 are quite close to the t values of 0.879 and 3.128 found using the parametric test. The strength of the rank test lies in the fact that the class of distributions for which it is appropriate is larger than for the corresponding aprametric tests.

Problems 7.5

1. (a) Prove that if $X \sim B(n, 1/2)$, then $P(X \leq r) = P(X \geq n - r)$ for any integer $0 \leq r \leq [n/2]$.

    (b) Illustrate this result directly for n = 6 and r = 2 by calculating the required probabilities.

WILTEST

```
10 REM THIS PROGRAM USES THE SIGNED RANK STATISTIC TO TEST THE NULL
20 REM HYPOTHESIS THAT THE MEDIAN OF A SYMMETRIC CONTINUOUS POPULATION
30 REM IS M0. THE PROGRAM ASSUMES A MODERATE OR LARGE SAMPLE SIZE N
40 REM THE ALTERNATIVES ARE OF FORM
50 REM 1) MEDIAN > M0
60 REM 2) MEDIAN < M0
70 REM 3) MEDIAN <> M0
80 REM THE PROGRAM COMPUTES THE VALUE OF S,Z AND THE
90 REM CORRESPONDING P-VALUE
100 PRINT "WHAT ARE THE VALUES OF N AND M0?"
110 INPUT N,M0
120 PRINT "WHICH ALTERNATIVE, ";"1,2,OR 3?"
130 INPUT A
140 IF A = 1 THEN 190
150 IF A = 2 THEN 190
160 IF A = 3 THEN 190
170 PRINT "INCORRECT FORMAT ";"TRY AGAIN"
180 GO TO 110
185 DIM X(100),T(100),Z(100)
190 FOR I = 1 TO N
195 READ X(I)
200 LET X(I)=X (I)-M0
205 IF X(I)<=0 THEN 215
210 LET Z(I)=1
215 NEXT I
220 FOR I = 1 TO N
225 LET T(I)=I
230 LET X(I)=ABS(X(I))
232 NEXT I
235 FOR J = 1 TO N-1
240 FOR I = 1 TO N-J
245 IF X(I)<=X(I+1) THEN 275
250 LET T=X(I+1)
255 LET T1=T(I+1)
260 LET X(I+1)=X(I)
262 LET T(I+1)=T(I)
265 LET X(I)=T
270 LET T(I)=T1
275 NEXT I
280 NEXT J
285 FOR I = 1 TO N
290 LET X(I)=I
292 NEXT I
295 FOR J= 1 TO N-1
300 FOR I = 1 TO N-J
305 IF T(I)<T(I+1) THEN 340
310 LET T=T(I+1)
315 LET T1=X(I+1)
320 LET T(I+1)=T(I)
325 LET X(I+1)=X(I)
330 LET T(I)=T
335 LET X(I)=T1
340 NEXT I
345 NEXT J
350 FOR I = 1 TO N
355 LET T2=T2+X(I)*Z(I)
360 NEXT I
```

```
365 PRINT "THE VALUE OF T+ =";T2
370 LET Z =(T2-N*(N+1)/4)/SQR(N*(N+1)*(2*N+1)/24)
380 GOSUB 900
390 PRINT "THE CALCULATED VALUE OF Z = ";Z
400 IF A = 3 THEN 430
410 PRINT
420 PRINT "THE P-VALUE FOR ";"ALTERNATIVE ";A;" = ";P
425 GO TO 1030
430 PRINT
440 PRINT "THE P-VALUE FOR ";"ALTERNATIVE ";A;" = ";P*2
450 GO TO 1030
500 DATA 42.58,41.82,41.94,40.25,42.37,44.6,42.74,43.14,41.17,41.9,46.2
510 DATA 38.89,45.02,38.1,48.59,34.54,48.07,42.45,47.97,41.5,43.9,43.2
520 DATA 42.02,48.3,45.9
900 LET Z1= ABS(Z)
910 LET C=1/SQR(2)
920 LET C1= .14112821
930 LET C2= .08864027
940 LET C3=.02743349
950 LET C4= -.00039446
960 LET C5=.00328975
970 DEF FNZ(X)=1-1/(1+C1*X+C2*X†2+C3*X†3+C4*X†4+C5*X†5)†8
980 LET P = .5+.5*FNZ(Z1*C)
990 LET P = 1E-4*(INT(1E4*P))
1010 LET P = 1-P
1020 RETURN
1030 END
*
 RUN
WHAT ARE THE VALUES OF N AND MO?
? 25,42.5
WHICH ALTERNATIVE  1,2,OR 3?
? 3
THE VALUE OF T+ = 193
THE CALCULATED VALUE OF Z =   0.820661

THE P-VALUE FOR ALTERNATIVE  3  =  0.412

*RUN
WHAT ARE THE VALUES OF N AND MO?
? 25,41
WHICH ALTERNATIVE  1,2,OR 3?
? 3
THE VALUE OF T+ = 274
THE CALCULATED VALUE OF Z =   3.00012

THE P-VALUE FOR ALTERNATIVE  3  = 0.0028


*
```

2. Consider the sample of 20 observations from the Cauchy distribution given in the output of CAUCHY in Sec. 4.6. The pdf for this population is

$$f(x) = \frac{1}{\pi} \frac{1}{1 + x^2} \qquad -\infty < x < \infty$$

so that the variance does not exist.

(a) Test the hypothesis $\eta = 0$ using the sign test. Consider the alternative to be two-sided and the significance level to be 0.0414.

(b) Test the hypothesis $\eta = 0$ as in part (a) using the Wilcoxon signed-rank test at significance level approximately equal to 0.05.

3. (a) Use the data of Exercise 7.3.3 on the running time of 20 driers set for 20 min to test the hypothesis $\mu = 20$ versus $\mu \neq 20$. Calculate the value of the test statistic T and carry out the test at significance level 0.05 (approximately) against a two-sided alternative.

(b) Carry out tests as in part (a) of $\mu = 21$ and $\mu = 22$ versus two-sided alternatives at level of significance 0.05 approximately. Do your decisions agree with those in Exercise 7.3.3?

4. The following are the values of the weights of 20 packages of butter produced in a processing plant: 4.01, 4.02, 4.00, 4.09, 4.15, 4.07, 4.01, 3.95, 3.90, 3.99, 3.95, 4.01, 3.99, 4.08, 4.12, 4.20, 4.25, 4.21, 4.16, 4.04.

(a) Test the hypothesis $\eta \leq 4.0$ versus $\eta > 4.0$ using the sign test. Discard any value equal to 4.0. Carry out the test at level of significance $\alpha = 0.0318$.

(b) Test the hypotheses in part (a) using the Wilcoxon signed-rank test at level of significance 0.05 (approximately).

5. Suppose that a test is carried out on a drug which is supposed to reduce blood pressure. Assume that of 225 individuals tested,

140 show definite reduction in blood pressure. Does the sign test indicate the effectiveness ($p > 0.5$) of the drug at significance level 0.05? At significance level 0.01?

6. Show that if a random variable X has a continuous pdf symmetric about θ, that is, $f(\theta - x) = f(\theta + x)$ for all x, then the corresponding random variable X has median θ.

7. Assume that each of the 32 possible assignments of signs (e.g., + - + - -) to the ranks (1, 2, 3, 4, 5) is equally likely.
   (a) What are the smallest and largest values which T_+ may attain.
   (b) Find explicitly the probability distribution of T_+ under the assumption made.
   (c) Verify directly that $E(T_+) = 7.5$ and $Var(T_+) = 55/4$ from the result in part (b).

8. Assume that we are testing

$$H_0: \ \theta = \theta_0 \quad \text{versus} \ H \ H_1: \ \theta > \theta_0$$

for θ the median of a continuous distribution. If we carry out the test using the Wilcoxon statistic T and sample size n = 17, find the critical region for the test with $\alpha = 0.05$ (approximately).

Exercises 7.5

1. Use the program WILTEST to test the hypothesis H_0: $\mu = 72$ versus H_1: $\mu = 72$ at significance levels $\alpha = 0.05$ and $\alpha = 0.01$ using the height data in Exercise 5.2.2. Compare the Z values with those found in Exercise 7.4.1.

2. In 1971, a draft lottery was held. The following are the priority dates assigned to those with birthdays in November and December*:

---

*Source: *U. S. News and World Report*, July 13, 1970.

November (priority)			*December (priority)*		
1 (243)	11 (123)	21 (35)	1 (347)	11 (73)	21 (181)
2 (205)	12 (255)	22 (253)	2 (321)	12 (19)	22 (194)
3 (294)	13 (272)	23 (193)	3 (110)	13 (151)	23 (219)
4 (39)	14 (11)	24 (81)	4 (305)	14 (348)	24 (2)
5 (286)	15 (362)	25 (23)	5 (27)	15 (87)	25 (361)
6 (245)	16 (197)	26 (52)	6 (198)	16 (41)	26 (80)
7 (72)	17 (6)	27 (168)	7 (162)	17 (315)	27 (239)
8 (119)	18 (280)	28 (324)	8 (323)	18 (208)	28 (128)
9 (176)	19 (252)	29 (100)	9 (114)	19 (249)	29 (145)
10 (63)	20 (98)	30 (67)	10 (204)	20 (218)	30 (192)

Use the sample of 61 priority numbers to test $\mu = 183$ versus $\mu \neq 183$ at significance level $\alpha = 0.10$ using the program WILTEST. What is the P-VALUE for this data? If one concluded $\mu > 183$, what does this mean about the lottery procedure?

3. Use the data of Sec. 5.1 giving the number of days required by 50 students to complete a self-paced course to test the hypothesis $\mu = 51$ versus $\mu \neq 51$ at significance level $\alpha = 0.10$, using WILTEST. What is the P-VALUE for these data? Carry out the same analysis using $\mu = 55$.

7.6   TESTS CONCERNING σ^2 AND σ IN NORMAL POPULATIONS

In Chap. 6 we have seen that a confidence interval for σ^2 can be based on the χ^2_{n-1} distribution. Due to the relationship between confidence intervals and tests of hypotheses it is apparent that tests concerning σ^2 in the normal case will be based on the chi-square variable. In fact the $(1 - \alpha)100\%$ confidence interval for σ^2 given in Sec. 6.3.

$$[\frac{(n - 1)s^2}{\chi^2_{\alpha/2,n-1}} , \frac{(n - 1)s^2}{\chi^2_{1-\alpha/2,n-1}}]$$

provides a test of H_0: $\sigma^2 = \sigma_0^2$ against a two-sided alternative. If the hypothesized value σ_0^2 lies in the interval, we retain H_0 and otherwise, we reject it. For the example data giving the weights of 20 "1-lb" bags of flour in that section we have found the 95% confidence interval [0.118, 0.437] for σ^2. Hence any hypothesized value in this interval would be accepted for a 0.05 level test against a two-sided alternative. Values outside this interval would be rejected.

As control of the variation of a production process is important we often wish to have critical measurements satisfy $\sigma^2 \le \sigma_0^2$, where σ_0^2 is some maximum tolerable variance. We state the null and alternative hypotheses as

$$H_0:\ \sigma^2 \le \sigma_0^2 \qquad \text{versus} \qquad H_1:\ \sigma^2 > \sigma_0^2$$

We calculate the variance of a random sample of size n and clearly will reject H_0 if the observed value of $S^2 \ge k$ for some positive constant k. The requirement for an α level test,

$$P(S^2 \ge k \mid H_0) \le \alpha$$

will determine the value of k. For $\sigma^2 \le \sigma_0^2$, we have

$$P(\frac{(n - 1)S^2}{\sigma^2} \ge \frac{(n - 1)k}{\sigma_0^2}) \le P(\frac{(n - 1)S^2}{\sigma_0^2} \ge \frac{(n - 1)k}{\sigma_0^2})$$

Using Theorem 6.3.1, the last probability will equal α if

$$\frac{k(n - 1)}{\sigma_0^2} = \chi^2_{\alpha,n-1}$$

or

$$k = \frac{\sigma_0^2 \chi^2_{\alpha,n-1}}{n - 1}$$

Generally the test statistic used is denoted $\chi^2 = (n - 1)S^2/\sigma_0^2$, with the critical region in this case defined to be $\chi^2 \ge \chi^2_{\alpha,n-1}$. The analogous critical regions for α level tests in the normal case appear in Table 7.6.1.

Table 7.6.1   Critical Regions for Tests of σ^2 in the Normal Case

H_0	H_1	Critical region
$\sigma^2 \leq \sigma_0^2$	$\sigma^2 > \sigma_0^2$	$\chi^2 \geq \chi_{\alpha,n-1}^2$
$\sigma^2 \geq \sigma_0^2$	$\sigma^2 < \sigma_0^2$	$\chi^2 \leq \chi_{1-\alpha,n-1}^2$
$\sigma^2 = \sigma_0^2$	$\sigma^2 \neq \sigma_0^2$	$\chi^2 \geq \chi_{\alpha/2,n-1}^2$ or $\chi^2 \leq \chi_{1-\alpha/2,n-1}^2$

In Sec. 6.6 we introduced the Downton unbiased estimator of σ. Using the notation employed there we have seen that for $n \geq 20$,

$$Z = \frac{\hat{\sigma} - \sigma}{\sqrt{A(n)}\sigma}$$

is approximately normally distributed.   Hence tests of hypothesis about σ in the normal case may be based on the statistic $Z = ((\hat{\sigma} - \sigma_0)/\sqrt{A(n)}\sigma_0)$.   These tests are summarized in Table 7.6.2.

Table 7.6.2   Critical Regions for Tests of σ in the Normal Case

H_0	H_1	Critical region
$\sigma \leq \sigma_0$	$\sigma > \sigma_0$	$Z \geq Z_\alpha$
$\sigma \geq \sigma_0$	$\sigma < \sigma_0$	$Z \leq -Z_\alpha$
$\sigma = \sigma_0$	$\sigma \neq \sigma_0$	$|Z| \geq Z_{\alpha/2}$

Problems 7.6

1. The heights of 25 males, assumed to come from a normally distri-
    buted population, yield $\bar{x} = 43.086$ and $s = 3.335$.   Test the
    hypotheses H_0: $\sigma^2 \leq 4$ versus H_1: $\sigma^2 > 4$ using the chi-square
    test.   Let $\alpha = 0.05$.

2. A random sample of 17 bass yields $\bar{x}$ = 14.1 in. and s = 1.20 in. Assuming normality, find a 95% confidence interval for σ^2 to decide between H_0: σ^2 = 2 and H_1: $\sigma^2 \neq 2$ at significance level α = 0.05.

3. (a) Show that for a test of H_0: $\sigma^2 \leq \sigma_0^2$ against H_1: $\sigma^2 > \sigma_0^2$ in the normal case, the power function may be expressed as

$$\pi(\sigma^2) = 1 - F_{\chi^2}\left(\frac{\chi_{\alpha,n-1}^2 \sigma_0^2}{\sigma^2}\right)$$

   Here $F_{\chi^2}(\ )$ represents the cdf of a chi-square random variable with n - 1 degrees of freedom.

   (b) Show $\lim_{\sigma^2 \to \infty} \pi(\sigma^2) = 1$.

   (c) Show that $\pi(\sigma^2)$ is an increasing function of σ^2.

Exercises 7.6

1. Write a program similar to MUTEST based on the large sample approximation of the Downton estimator to test H_0: $\sigma = \sigma_0$ against any of the usual alternatives. The program should output the calculated value of Z and the appropriate P-VALUE.

2. Use the program written in Exercise 7.6.1 together with the data of Exercise 5.2.2 to test the hypothesis H_0: σ = 2 versus H_1 : $\sigma \neq 2$ for these height data. What are the values of Z and P-VALUE for this test. What is the correct decision at α = 0.01?

TWO-SAMPLE TESTS

8.1 INTRODUCTION

In this chapter we consider comparisons of two statistical populations.
One is often interested in testing whether there are differences in
two methods or ways of accomplishing an action and in estimating the
magnitude of the differences. In statistics we say that we are testing whether a difference exists between two effects. As examples we
cite the following:

a. Are there differences in the results achieved by two different
    teaching methods?
b. Are there differences in the weight gains of similar animals on
    two different diets over a fixed period of time?
c. Do differences exist in the density of cakes prepared by two
    different cake mixes?
d. Do students in different colleges of a university, say engineer-
    ing and liberal arts, spend different amounts of time studying
    on the average?

Such questions fall under the general heading of two-sample problems
in statistics. The basic idea is that we obtain random samples from
each of two populations. We first wish to answer the question of
whether these populations are statistically identical. If not, we
want to have some measure of the magnitude of the differences between
the populations.

8.2   DIFFERENCES OF POPULATION EXPECTATIONS IN THE LARGE SAMPLE
      CASE

In this section we assume that "large" independent samples are available from each of two populations described by the random variables
X_1 and X_2. We use the following notation:

$$E(X_1) = \mu_1 \qquad E(X_2) = \mu_2$$

$$Var(X_1) = \sigma_1^2 \qquad Var(X_2) = \sigma_2^2$$

The null hypothesis which we consider is

$$H_0: \quad \mu_1 = \mu_2$$

against alternatives of three general types: (1) $\mu_1 > \mu_2$, (2)
$\mu_1 < \mu_2$, or (3) $\mu_1 \neq \mu_2$. It seems reasonable that a test of this
null hypothesis, that the two populations have the same expectation,
should be tested using a statistic based on the difference of the
sample means, $\bar{X}_1$ and $\bar{X}_2$.

To be precise, it is assumed that we have two independent
random samples:

$$X_{11}, X_{12}, \ldots, X_{1n_1} \qquad \text{and} \qquad X_{21}, X_{22}, \ldots, X_{2n_2}$$

The samples of sizes n_1 and n_2 are from populations one and two,
respectively. Of course,

$$\bar{X}_1 = \sum_{i=1}^{n_1} \frac{X_{1i}}{n_1} \qquad \text{and} \qquad \bar{X}_2 = \sum_{i=1}^{n_2} \frac{X_{2i}}{n_2}$$

Due to the properties of sample means and the variance of a difference of two independent random variables, we have

$$E(\bar{X}_1 - \bar{X}_2) = E(\bar{X}_1) - E(\bar{X}_2) = \mu_1 - \mu_2$$

$$Var(\bar{X}_1 - \bar{X}_2) = Var(\bar{X}_1) + Var(\bar{X}_2) = \frac{\sigma_1^2}{n_1} + \frac{\sigma_2^2}{n_2}$$

For "large" n_1 and n_2, one can assume that the random variable

$$\frac{(\bar{X}_1 - \bar{X}_2) - (\mu_1 - \mu_2)}{\sqrt{\sigma_1^2/n_1 + \sigma_2^2/n_2}}$$

is approximately distributed as the standard normal variable Z.
Under the hypothesis $\mu_1 = \mu_2$, using the sample variances S_1^2 and
S_2^2 as estimates of the corresponding population variances, we have
the statistic

$$Z = \frac{(\bar{X}_1 - \bar{X}_2)}{\sqrt{S_1^2/n_1 + S_2^2/n_2}}$$

Clearly the appropriate critical regions for the alternatives (1),
(2), and (3) are $Z \geq Z_\alpha$, $Z \leq -Z_\alpha$, and $|Z| \geq Z_{\alpha/2}$ for tests at signif-
icance level α.

The program 2SAMP has been written to test the hypothesis
$\mu_1 = \mu_2$ in the large sample case. The program requests the values
of n_1 and n_2. The observations in sample 1 should appear in DATA
statements in lines 400-499 and those in sample 2 in DATA statements
in lines 500-599. The program computes $\bar{x}_1$, $\bar{x}_2$, s_1, s_2,
$\sqrt{s_1^2/n_1 + s_2^2/n_2}$, as well as the computed Z value with the corresponding P-VALUE. In the sample run the first 50 "ages" in Exercise
5.1.3 has been used as sample 1 and the last 50 "ages" as sample 2.
The value Z = -0.401 and the P-VALUE of 0.6886 suggest that one
should retain $\mu_1 = \mu_2$ at any reasonable level of significance.

To find a confidence interval for $\mu_1 - \mu_2$, let us consider a
general statistic $\hat{\theta}_n$ with $E(\hat{\theta}_n) = \theta$ which is distributed as the
standard normal variable Z for large n. The statement

$$P\left(-Z_{\alpha/2} \leq \frac{\hat{\theta}_n - \theta}{\sqrt{\mathrm{Var}(\hat{\theta}_n)}} \leq Z_{\alpha/2}\right) = 1 - \alpha$$

may be written as

$$P\left(\hat{\theta}_n - \sqrt{\mathrm{Var}(\hat{\theta}_n)}Z_{\alpha/2} \leq \theta \leq \hat{\theta}_n + \sqrt{\mathrm{Var}(\hat{\theta}_n)}Z_{\alpha/2}\right) = 1 - \alpha$$

2SAMP

```
10 REM THIS PROGRAM TESTS THE HYPOTHESIS: MU1 = MU2
20 REM AGAINST ONE OF THE ALTERNATIVES
30 REM 1) MU1 > MU2
40 REM 2) MU1 < MU2
50 REM 3) MU1 <> MU2
60 REM ASSUMING INDEPENDENT SAMPLES OF SIZE N1 AND N2
70 REM FROM TWO POPULATIONS WITH FINITE EXPECTATIONS AND VARIANCES
80 REM THE SAMPLE SIZES N1 AND N2 ARE CONSIDERED TO BE LARGE
90 REM THE PROGRAM COMPUTES Z, THE CORRESPONDING P-VALUE AND
95 REM OTHER STATISTICS OF INTEREST
100 PRINT "WHAT ARE THE VALUE OF N1 AND N2?"
110 INPUT N1, N2
120 PRINT "WHICH ALTERNATIVE, ";"1, 2 OR 3?"
130 INPUT A
140 IF A = 1 THEN 190
150 IF A = 2 THEN 190
160 IF A = 3 THEN 190
170 PRINT "INCORRECT FORMAT ";"TRY AGAIN"
180 GO TO 110
190 FOR I = 1 TO N1
195 READ X
200 LET S = X + S
205 LET S1 = S1 +X↑2
210 NEXT I
215 LET X1=S/N1
220 LET S1= SQR((S1-N1*X1↑2)/(N1-1))
225 FOR I = 1 TO N2
230 READ X
235 LET T = T +X
240 LET T1 = T1 + X↑2
245 NEXT I
250 LET X2=T/N2
255 LET S2=SQR((T1-N2*X2↑2)/(N2-1))
260 LET Z = (X1-X2)/SQR(S1↑2/N1+S2↑2/N2)
265 GOSUB 900
270 PRINT "SAMPLE","MEAN","STD DEVIATION"
275 PRINT "  1    ",X1, S1
280 PRINT "  2    ",X2, S2
285 PRINT "STD DEV OF DIFF = "; SQR(S1↑2/N1+S2↑2/N2)
290 PRINT "THE CALCULATE Z VALUE = ";Z
300 IF A = 3 THEN 330
310 PRINT
320 PRINT "THE P-VALUE FOR ";"ALTERNATIVE ";A;" = ";P
325 GO TO 1030
330 PRINT
340 PRINT "THE P-VALUE FOR ";"ALTERNATIVE ";A;" = ";P*2
350 GO TO 1030
400 DATA 46, 44, 45, 40, 46, 52, 47, 48, 42, 44
410 DATA 57, 36, 53, 34, 63, 24, 62, 46, 61, 43
420 DATA 50, 48, 45, 62, 56, 46, 46, 40, 44, 52
430 DATA 43, 39, 49, 47, 42, 53, 53, 43, 41, 47
440 DATA 46, 34, 38, 44, 43, 48, 48, 45, 41, 61
500 DATA 48, 50, 41, 41, 30, 51, 51, 43, 54, 55
510 DATA 41, 45, 48, 54, 54, 49, 46, 52, 56, 45
520 DATA 45, 50, 35, 52, 49, 44, 45, 50, 52, 52
530 DATA 45, 47, 45, 60, 48, 49, 32, 50, 52, 47
540 DATA 57, 42, 48, 41, 40, 48, 39, 39, 53, 45
*
```

```
900 LET Z1= ABS(Z)
910 LET C=1/SQR(2)
920 LET C1= .14112821
930 LET C2= .08864027
940 LET C3=.02743349
950 LET C4= -.00039446
960 LET C5=.00328975
970 DEF FNZ(X)=1-1/(1+C1*X+C2*X↑2+C3*X↑3+C4*X↑4+C5*X↑5)↑8
980 LET P = .5+.5*FNZ(Z1*C)
990 LET P = 1E-4*(INT(1E4*P))
1010 LET P = 1-P
1020 RETURN
1030 END
*

  RUN
WHAT ARE THE VALUE OF N1 AND N2?
?  50, 50
WHICH ALTERNATIVE, 1,2 OR 3?
?  3
SAMPLE           MEAN            STD DEVIATION
   1             46.54            7.7122
   2             47.1             6.16855
STD DEV OF DIFF =   1.39663
THE CALCULATE Z VALUE = -0.400965

THE P-VALUE FOR ALTERNATIVE  3  =   0.6886

   *
```

Hence the endpoints of the confidence interval are of the form $\hat{\theta}_n \pm \sqrt{Var(\hat{\theta}_n)}Z_{\alpha/2}$. We have seen this before in Sec. 5.5 where $\bar{X}_n$ is a statistic used to find a confidence interval for μ. If $Var(\hat{\theta}_n)$ depends on an unknown parameter, then $Var(\hat{\theta}_n)$ is replaced by an estimator, say $\widehat{Var}(\hat{\theta}_n)$, and the confidence interval becomes

$$\hat{\theta}_n \pm \sqrt{\widehat{Var}(\hat{\theta}_n)}Z_{\alpha/2} \tag{8.2.1}$$

For the difference $\mu_1 - \mu_2$ using statement (8.2.1), we find

$$\bar{X}_1 - \bar{X}_2 \pm \sqrt{\frac{S_1^2}{n_1} + \frac{S_2^2}{n_2}} \; Z_{\alpha/2}$$

to be an appropriate $(1 - \alpha)100\%$ confidence interval. For the data used in 2SAMP we find -0.560 ± 1.397 (1.96), that is, $[-3.297, 2.177]$ to be a 95% confidence interval for $\mu_1 - \mu_2$.

Problems 8.2

1. Assume that 100 boys and 100 girls are given a test with the following results: $\bar{x}_1 = 565$, $\bar{x}_2 = 540$, $s_1 = 30$, and $s_2 = 40$.
    (a) Test the hypothesis that the average performance is the same for boys and girls.
    (b) Find a 95% confidence interval for $\mu_1 - \mu_2$.

2. Assume that it is of interest to know whether the CPU time of faculty members program is greater than that of student programs. A sample of 100 programs of each group yields $\bar{x}_1 = 10$ and $s_1^2 = 10$ for the faculty programs, while $\bar{x}_2 = 8$ and $s_2^2 = 15$ for the student programs. Time is measured in seconds.
    (a) Test the hypothesis that the average CPU time is the same against the alternative that the CPU time for faculty programs exceeds that of student programs. $(\alpha = 0.01)$
    (b) Find a 98% confidence interval for $\mu_1 - \mu_2$.

3. In an article in the *American Psychologist* by Sandra Scarr and Richard A. Weinberg (August, 1976) entitled "IQ Test Performance

of Black Children Adopted by White Families," the IQ scores of
natural and adopted children were reported. The results showed
the average Stanford-Binet IQ of the 48 natural children was
$\bar{x}_1$ = 113.8 with s_1 = 16.7 and the average IQ for 122 adopted
children was $\bar{x}_2$ = 106.5 with s_2 = 13.9.

(a) Do the data indicate a significant difference between these
two groups (α = 0.05)?

(b) Find a 95% confidence interval for $\mu_1 - \mu_2$.

4. Vidhu Moran and Dalip Kumar report in the *British Journal of
Psychology* (August, 1976) the results of a study on the relationship between introvertism and extrovertism and intelligence in
an article entitled "Qualitative Analysis of the Performance of
Introverts and Extroverts on Standard Progressive Matrices."
The number of correct responses (of 12 possible) on the most
difficult tests are summarized below:

	Introverts	Extroverts
$\bar{x}_i$	7.88	·6.08
s_i	1.65	2.13
n_i	50	50

Test the hypothesis that there is no difference between the two
groups using α = 0.05.

5. Roberta M. Milgram and Norman A. Milgram report on the relationship between anxiety and warfare in an article, "The Effect of
the Yom Kippur War on Anxiety Level in Israeli Children," in
The Journal of Psychology (September, 1976). The following
data are presented where the measurements are made on the Sarason
Scale for anxiety (20 maximum).

	Israeli peacetime		Israeli wartime	
	$\bar{x}$	s	$\bar{x}$	s
Boys (n = 42)	5.83	4.51	14.94	3.79
Girls (n = 43)	8.10	5.11	11.91	4.59

(a) Compare anxiety levels for boys and girls in peacetime by testing the null hypothesis of no difference between the two groups ($\alpha = 0.10$).

(b) Test the hypothesis that anxiety increases during war in children using the data for the girls ($\alpha = 0.05$). What assumptions of the two-sample test situation are violated in comparison?

6. Assume that we have independent samples from each of two populations of sizes n_1 and n_2, where the sample sizes are considered to be large. Find a $(1 - \alpha)100\%$ confidence interval for $(\mu_1 + \mu_2)/2$. Assume that σ_1 and σ_2 are unknown and may be unequal. Use expression (8.2.1)

Exercises 8.2

1. (a) Use the second 50 "ages" and the next to last 50 "ages" from the data of Exercise 5.1.3 to test the hypothesis $\mu_1 = \mu_2$ versus $\mu_1 \neq \mu_2$ with $\alpha = 0.05$. Use the program 2SAMP.

(b) Find a 95% confidence interval for the difference $\mu_1 - \mu_2$.

2. (a) The data given below represent the weights of 40 men and 60 women to the nearest pound. Of course we expect $\mu_1 > \mu_2$. Test $\mu_1 = \mu_2$ versus $\mu_1 > \mu_2$ using 2SAMP with $\alpha = 0.01$.

(b) Estimate the difference $\mu_1 - \mu_2$ and find a 95% confidence interval for this difference. How does your estimate compare with the true difference of 10?

Sample 1: Weights of 40 Men

148	156	163	116	153	148	143	157	145	148
133	170	151	147	137	158	145	150	158	147
146	139	151	143	139	142	130	169	171	145
149	140	147	145	172	144	161	156	142	149

Sample 2: Weights of 60 women

137	147	155	99	144	137	131	149	138	141
131	134	134	138	120	164	141	137	125	150
117	141	154	143	134	140	149	137	135	127
141	132	124	135	127	140	126	131	116	163
165	134	139	128	133	129	142	136	136	134
166	132	153	147	130	139	132	163	147	163

3. Below we present the scores of 31 females and 44 males on a
    final examination in an elementary statistics course. Use 2SAMP
    to test the hypothesis of no difference in average performance
    with $\alpha = 0.10$. (Note: Maximum possible score is 35.)

Females

19	16	12	16	26	27	24	31	20	17	11	33	13	24	12	27
25	27	21	13	27	20	23	20	22	21	20	32	22	21	29	

Males

29	29	31	27	29	25	24	26	16	17	22	25	28	31	30
20	28	28	35	25	24	22	21	33	22	22	34	25	24	34
14	15	16	23	34	28	28	33	28	22	35	21	28	16	

8.3   DIFFERENCES OF POPULATION EXPECTATIONS IN THE NORMAL CASE

8.3.1   *The Case* $\sigma_1^2 = \sigma_2^2$

Here we assume the same situation as in Sec. 8.2 with the additional
information that $X_1 \sim N(\mu_1, \sigma_1^2)$ and $X_2 \sim N(\mu_2, \sigma_2^2)$. Again we are
interested in inference concerning $\mu_1 - \mu_2$, based on independent
samples of size n_1 and n_2, not necessarily "large." We consider
first the case of testing H_0: $\mu_1 = \mu_2$ assuming that $\sigma_1^2 = \sigma_2^2 = \sigma^2$,
where σ^2 is considered to be unknown. Here the variance of $\bar{X}_1 - \bar{X}_2$
is $\sigma^2(1/n_1 + 1/n_2)$ where σ^2 is the assumed common variance of the
two populations. To estimate σ^2 we use information from both samples
to obtain the *pooled estimate of the variance*

$$\hat{\sigma}^2 = \frac{\sum_{i=1}^{n_1} (X_{1i} - \bar{X}_1)^2 + \sum_{i=1}^{n_2} (X_{2i} - \bar{X}_2)^2}{n_1 + n_2 - 2}$$

It can be shown that $E(\hat{\sigma}^2) = \sigma^2$. The test statistic

$$t = \frac{\bar{X}_1 - \bar{X}_2}{\hat{\sigma}\sqrt{1/n_1 + 1/n_2}}$$

can be shown to have Student's distribution with $n_1 + n_2 - 2$ degrees of freedom under the assumptions of this paragraph. For the alternatives (1) $\mu_1 > \mu_2$, (2) $\mu_1 < \mu_2$, and (3) $\mu_1 \neq \mu_2$, the critical regions for level tests are $t > t_{\alpha, n_1 + n_2 - 2}$, $t < -t_{\alpha, n_1 + n_2 - 2}$, and $|t| > t_{\alpha/2, n_1 + n_2 - 2}$, respectively.

The assumptions that $\sigma_1^2 = \sigma_2^2$ is, of course, not necessarily true. Good statistical practice requires carrying out a test of the hypothesis H_0: $\sigma_1^2 = \sigma_2^2$ versus H_1: $\sigma_1^2 \neq \sigma_2^2$ and conducting the t test described above if H_0 is retained. In the following theorem, stated without proof, we introduce a new distribution, related to the chi-square distribution, which is used in determining a critical region for this test.

THEOREM 8.3.1. If χ_1^2 and χ_2^2 are two independent chi-square variables with ν_1 and ν_2 degrees of freedom, then the ratio $F = (\chi_1^2/\nu_1)/(\chi_2^2/\nu_2)$ has the F distribution with ν_1 and ν_2 degrees of freedom.

The F distribution has two parameters, ν_1 and ν_2, referred to as the numerator and denominator "degrees of freedom." We write $F(\nu_1, \nu_2)$ to indicate the value of the F statistic with ν_1 and ν_2 degrees of freedom exceeded with probability α. Table B.V (Appendix B) gives values of $F_{0.05}$ and $F_{0.01}$ for various combinations of the degrees of freedom. Notice that $F(\nu_1, \nu_2) = 1/F(\nu_2, \nu_1)$ according to the definition in Theorem 8.3.1. Hence $P(F(\nu_1, \nu_2) \le 1/F(\nu_2, \nu_1)_\alpha) = P(F(\nu_2, \nu_1)_\alpha \le 1/F(\nu_1, \nu_2)) = F(F(\nu_2, \nu_1)_\alpha \le F(\nu_2, \nu_1)) = \alpha$. Thus Table B.V can be used to find $F_{0.95}$ and $F_{0.99}$ points by using the reciprocals of the $F_{0.05}$ and $F_{0.01}$ points with the reversed degrees of freedom.

In order to test H_0: $\sigma_1^2 = \sigma_2^2$ we recall that in sampling from a normal population, $(n-1)S^2/\sigma^2$ has the chi-square distribution with $n-1$ degrees of freedom. Using Theorem 8.3.1, we see that if S_1^2 and S_2^2 represent the sample variances from two independent normal populations, then

$$F = \frac{S_1^2/\sigma_1^2}{S_2^2/\sigma_2^2}$$

has the $F(n_1 - 1, n_2 - 1)$ distribution. If $\sigma_1^2 = \sigma_2^2$, then $F = S_1^2/S_2^2$, which is simply the ratio of the sample variances. It makes sense to reject H_0 if this ratio is either very large or very small. For an α level test the appropriate critical region would be $F < F(n_1 - 1, n_2 - 1)_{1-\alpha/2}$ or $F > F(n_1 - 1, n_2 - 1)_{\alpha/2}$. For example, suppose two independent samples from normal populations yield the following: $n_1 = 16$, $s_1^2 = 16$, $n_2 = 9$, and $s_2^2 = 8$. At significance level 0.10 we find from Table B.V that we reject H_0 if $F > F(15,8)_{0.05} = 3.22$ or if $F < 1/F(8,15)_{0.05} = 1/2.64 = 0.379$. As $F = 2$, we retain the hypothesis $\sigma_1^2 = \sigma_2^2$.

The program N2SAMP has been written to test the hypothesis H_0: $\mu_1 = \mu_2$ in the normal case. The program accomplishes two analyses, the first under the assumption $\sigma_1 = \sigma_2 = \sigma$, where σ is unknown. The program requests the values of n_1 and n_2. The sample data for the first sample should appear in DATA statements in lines 400–499. The output includes the value of the "pooled" estimate $\hat{\sigma}$, the computed value of t and the value of F to test $\sigma_1 = \sigma_2$. In the sample run the "ages" of 20 legislators have been compared with the "ages" of 15 legislators from the data of Exercise 5.1.3. The value of $F = 1.94$ and interpolating in Table B.V, we find $F(19,14)_{0.05} = 2.40$, indicating acceptance of $\sigma_1 = \sigma_2$ at the 10% level of significance. The t value of -0.142 and the P-VALUE of 0.888 suggest that

N2SAMP

```
10 REM THIS PROGRAM TESTS THE HYPOTHESIS: MU1 = MU2
20 REM AGAINST ONE OF THE ALTERNATIVES
30 REM 1) MU1 > MU2
40 REM 2) MU1 < MU2
50 REM 3) MU1 <> MU2
60 REM ASSUMING INDEPENDENT SAMPLES OF SIZE N1 AND N2
70 REM FROM TWO NORMAL POPULATIONS
80 REM THE SAMPLE SIZES ARE N1 AND N2
90 REM THE PROGRAM COMPUTES THE VALUE OF T ASSUMING EQUAL VARIANCES,
92 REM THE VALUE OF F TO TEST THE EQUALITY OF VARIANCES AND
95 REM OTHER STATISTICS OF INTEREST
100 PRINT "WHAT ARE THE VALUE OF N1 AND N2?"
110 INPUT N1,N2
120 PRINT "WHICH ALTERNATIVE, ";"1,2 OR 3?"
130 INPUT A
140 IF A = 1 THEN 190
150 IF A = 2 THEN 190
160 IF A = 3 THEN 190
170 PRINT "INCORRECT FORMAT ";"TRY AGAIN"
180 GO TO 110
190 FOR I = 1 TO N1
195 READ X
200 LET S = X + S
205 LET S1 = S1 +X†2
210 NEXT I
215 LET X1=S/N1
220 LET S1= SQR((S1-N1*X1†2)/(N1-1))
225 FOR I = 1 TO N2
230 READ X
235 LET T = T +X
240 LET T1 = T1 + X†2
245 NEXT I
250 LET X2=T/N2
255 LET S2=SQR((T1-N2*X2†2)/(N2-1))
260 LET S = SQR((((N1-1)*S1†2+(N2-1)*S2†2))/(N1+N2-2))
262 LET T = (X1-X2)/(S*SQR(1/N1+1/N2))
265 GOSUB 900
267 PRINT "ANALYSIS ASSUMING EQUAL VARIANCES"
270 PRINT "SAMPLE","MEAN","STD DEVIATION"
275 PRINT "  1   ",X1,S1
280 PRINT "  2   ",X2,S2
287 PRINT "STD DEV OF DIFF = ";S*SQR(1/N1+1/N2)
290 PRINT "THE CALCULATED VALUE OF T =";T;" ON ";N1+N2-2;" DOF"
295 PRINT "THE F VALUE = ";S1†2/S2†2;"ON";N1-1;" AND ";N2-1;" DOF"
300 IF A = 3 THEN 330
310 PRINT
320 PRINT "THE P-VALUE FOR ";"ALTERNATIVE ";A;" = ";P
325 GO TO 350
330 PRINT
340 PRINT "THE P-VALUE FOR ";"ALTERNATIVE ";A;" = ";P*2
350 PRINT
*
```

```
360 PRINT "ANALYSIS FOR UNEQUAL VARIANCES"
362 LET T =(X1-X2)/SQR(S1↑2/N1+S2↑2/N2)
364 LET D = ((S1↑2/N1)↑2/N1+(S2↑2/N2)↑2/N2)
366 LET D = (S1↑2/N1+S2↑2/N2)↑2/D
368 LET S =SQR(S1↑2/N1+S2↑2/N2)
370 PRINT "THE STD DEV OF DIFF = ";S
375 PRINT "THE CALUCULATED VALUE OF T = ";T
380 PRINT "WITH DOF = ";D
390 GO TO 1030
400 DATA 46, 44, 45, 40, 46, 52, 47, 48, 42, 44
410 DATA 57, 36, 53, 34, 63, 24, 62, 46, 61, 43
500 DATA 48, 50, 41, 41, 30, 51, 51, 43, 54, 55
510 DATA 41, 45, 48, 54, 54, 49, 46, 52, 56, 45
900 LET T1 = ABS(T)
905 LET C=1
915 LET D1=N1+N2-2
920 IF INT ((N1+N2)/2)=(N1+N2)/2 THEN 950
925 FOR I = 1 TO D1-2 STEP 2
930 LET C = C*(D1-I)/(D1-1-I)
935 NEXT I
940 LET C = C/(3.14159625*SQR(D1))
945 GO TO 970
950 FOR I = 1 TO D1-2 STEP 2
955 LET C =C*(D1-I)/(D1-1-I)
960 NEXT I
965 LET C = C/(2*SQR(D1))
970 DEF FNF(X)=C/(1+X↑2/D1)↑((D1+1)/2)
975 LET H = T1/100
980 FOR I = 1 TO 99
985 LET X =H*I
990 LET I1=I1+FNF(X)*H
995 NEXT I
1000 LET I1 = I1+(H/2)*(FNF(0)+FNF(T))
1010 LET P =0.5-I1
1020 LET P = 1E-4*(INT(1E4*P))
1025 RETURN
1030 END
*
 RUN
WHAT ARE THE VALUE OF N1 AND N2?
? 20, 15
WHICH ALTERNATIVE, 1, 2 OR 3?
? 3
ANALYSIS ASSUMING EQUAL VARIANCES
SAMPLE          MEAN              STD DEVIATION
   1            46.65             9.65333
   2            47.0667           6.9227
STD DEV OF DIFF = 2.93794
THE CALCULATED VALUE OF T =-0.141823  ON  33  DOF
THE F VALUE = 1.94448 ON 19  AND  14  DOF

THE P-VALUE FOR ALTERNATIVE  3  =  0.888

ANALYSIS FOR UNEQUAL VARIANCES
THE STD DEV OF DIFF = 2.80255
THE CALUCULATED VALUE OF T = -0.148674
WITH DOF = 34.9322

*
```

the hypothesis $\mu_1 = \mu_2$ should be retained. This is reassuring as the "samples" came from the same population. An analysis for the case $\sigma_1 \neq \sigma_2$ is also made.

A $(1 - \alpha)100\%$ confidence interval for $\mu_1 - \mu_2$ is found as in Sec. 8.2. Such a confidence interval has endpoints

$$\bar{X}_1 - \bar{X}_2 + \hat{\sigma}t_{\alpha/2,n_1+n_2-2}$$

In this case $\hat{\sigma} = 2.938$ and $\bar{X}_1 - \bar{X}_2 = -0.417$, so that a 95% confidence interval is given by $-0.417 \pm 2.938(1.991)$ or by the interval $[-6.267 , 5.433]$.

8.3.2   *The Case* $\sigma_1 \neq \sigma_2$

In the case that $\sigma_1^2 \neq \sigma_2^2$ with both variances unknown an approximate t test of H_0: $\mu_1 = \mu_2$ may be carried out using

$$t = \frac{\bar{X}_1 - \bar{X}_2}{\sqrt{S_1^2/n_1 + S_2^2/n_2}}$$

with approximate degrees of freedom

$$d = \frac{(S_1^2/n_1 + S_2^2/n_2)^2}{(S_1^2n_1)^2/n_1 + (S_2^2/n_2)^2/n_2}$$

In general d will not be an integer, but often the closest value of the degrees of freedom may be used. In some cases interpolation is required. In the sample run, assuming $\sigma_1 \neq \sigma_2$, we find $t = -0.149$ and $d = 34.9$. Using interpolation we find $t_{0.025,35} = 2.032$, so that again acceptance of $\mu_1 = \mu_2$ is clearly the appropriate decision.

Problems 8.3

1. (a) Show that $\hat{\sigma}^2 = ((n_1 - 1)S_1^2 + (n_2 - 1)S_2^2)/(n_1 + n_2 - 2)$.

    (b) Assuming $X_1 \sim N(\mu_1,\sigma^2)$ and $X_2 \sim N(\mu_2,\sigma^2)$, show that $E(\hat{\sigma}^2) = \sigma^2$. Is the normality necessary?

2. The table below gives data for 15 randomly selected boys and 10
    randomly selected girls of ages 9 through 10.

    Scores on a spelling test

i	Sample	n_1	$\bar{x}_i$	$\Sigma_j \, (x_{ij} - \bar{x}_i)^2$
1	Boys	15	81	1224
2	Girls	9	90	756

    Assume the observations come from normal distributions with
    common variance σ^2.

    (a) What is an unbiased estimate of σ^2?

    (b) Test the hypothesis that boys and girls score equally well
         against the alternative that girls have higher scores on
         the average. State clearly the rejection region for this
         test. Use $\alpha = 0.05$.

    (c) Find a 90% confidence interval for $\mu_2 - \mu_1$.

3. If we can assume $\sigma_1^2 = \sigma_2^2 = \sigma^2$, then a $(1 - \alpha)100\%$ confidence
    interval for μ_1 is given by $\bar{X}_1 \pm (\hat{\sigma}/\sqrt{n})t_{\alpha/2, n_1+n_2-2}$. Note the
    degrees of freedom. Find a 95% confidence interval for μ_1 and
    for μ_2 in Problem 8.3.2.

4. The starting lineups for the Iowa State versus Oklahoma State
    football game played at Stillwater, Oklahoma, on November 20,
    1976, were listed in the Des Moines Register of that date as
    follows:

Iowa State		*Oklahoma State*
Rogers, 216	LE	Blankenship 225
Petsch, 225	LT	Perrelli, 255
Greenwood, 240	LG	Ledford, 240
Boehm, 227	C	Gofourth, 250
Stoffel, 230	RG	Baker, 229
Cunningham, 228	RT	Hardaway, 329
Blue, 181	RE	Lisle, 188
Hardeman, 188	QB	Weatherbie, 184
Green, 171	LH	Miller, 189
Soloman, 179	RH	R. Taylor, 174
Cummings, 195	FB	S. Taylor, 193

(a) Test the hypothesis of no difference in average weight between the two lines against the alternative that the OSU line was heavier. Use $\alpha = 0.05$. Assume normality and that $\sigma_1 = \sigma_2$.

(b) Carry out the same comparison for the backfields.

5. In an article entitled "The Electrodermal Component of the Orienting Response in Blind and Deaf Individual," in the *British Journal of Psychology* (August, 1976), D. Carroll and P. G. Surtees report on comparisons between blind and sighted children. They report an average verbal IQ of 119.4 with a standard deviation of 8.13 for 11 blind children. The average IQ for 11 sighted children in a control group was 121.4 with a standard deviation of 7.4. Is there evidence of any significant differenc in IQ between these two groups? (Use $\alpha = 0.10$.)

Exercises 8.3

1. Suppose six plots of wheat are treated with fertilizer A and six plots with fertilizer B. The yields are recorded below:

A	26.3	28.6	25.4	29.2	27.6	25.6
B	28.5	30.0	28.8	25.3	28.4	26.5

(a) Use N2SAMP to test $\mu_1 = \mu_2$ versus $\mu_1 \neq \mu_2$ at $\alpha = 0.05$.
(b) Find a 95% confidence interval for $\mu_2 - \mu_1$.

2. Suppose 12 halfbacks are observed during spring training, six sampled from the freshmen and sophomores and six from the juniors and seniors. Each is given the opportunity of running four plays against the best defensive team. The resulting total yards for players is given below:

Freshmen and sophomores	10	9	5	11	16	9
Juniors and seniors	13	8	9	18	23	25

(a) Use N2SAMP to test the hypothesis that freshmen and sophomore backs are equally as good as junior and senior backs against the alternative that junior and senior backs are better. Use $\alpha = 0.05$.

(b)   Find a 95% confidence interval for $\mu_2 - \mu_1$.

3.   Use the data on the weights of men and women given in Exercise
     8.2.2 to answer the questions of that exercise using N2SAMP.

8.4   THE WILCOXON RANK SUM TEST

We again consider testing whether two random samples have come from
the same continuous population.   Here we assume that the X_1 population has a density function $f(x)$ while the X_2 population has a density
function $f(x - \Delta)$ for some real value Δ.   In Fig. 8.4.1 we display
two such density functions for which $\Delta > 0$.   We see that the effect
of $\Delta > 0$ is to "shift" the location of the second population to the
right of the first by an amount Δ.   Hence Δ is referred to as a shift
parameter.   We wish to test H_0: $\Delta = 0$ versus the natural alternatives
(1) $\Delta > 0$, (2) $\Delta < 0$, and (3) $\Delta \neq 0$.   The general idea of the test
is to obtain random samples of size n_1 and n_2 from the two populations.   We combine the observations in the two samples and rank the
$n_1 + n_2$ observations by magnitude from 1 to $n_1 + n_2$.   If the corresponding sum of ranks for the observations from sample 2 is either
"too large" or "too small," we tend to reject $\Delta = 0$ in favor of
$\Delta \neq 0$.

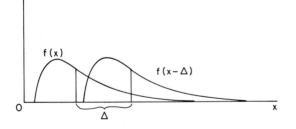

Fig. 8.4.1   Population 2 shifted to the right of 1

The actual statistic used in the small sample case is denoted as $U = \min(U_1, U_2)$, where

$U_1 \equiv \{$ the number of times observations from sample 1 exceed those from sample 2$\}$

$U_2 \equiv \{$ the number of times observations from sample 2 exceed those from sample 1$\}$

Under the hypothesis of continuous random variables X_1 and X_2 there can be no tied observations. Denoting $R_{X_{1i}}$ and $R_{X_{2j}}$, $i = 1, 2, \ldots,$ n_1 and $j = 1, 2, \ldots, n_2$ as the rank in the combined sample of the corresponding observation we see that

$$U_1 = \sum_{i=1}^{n_1} (R_{X_{1(i)}} - i) \quad \text{and} \quad U_2 = \sum_{j=1}^{n_2} (R_{X_{2(j)}} - j)$$

For example, for the ordered $X_{1(i)}$, the smallest observation in sample 1 exceeds $R_{X_{1(1)}} - 1$ observations in sample 2. Hence

$$U_1 = \sum_{i=1}^{n_1} R_{X_{1i}} - \frac{n_1(n_1 + 1)}{2} \quad \text{and} \quad U_2 = \sum_{j=1}^{n_2} R_{X_{2j}} - \frac{n_2(n_2 + 1)}{2}$$

It is actually necessary to compute only one of U_1 and U_2 as $U_1 + U_2 = n_1 n_2$. If we consider testing H_0: $\Delta = 0$ versus H_1: $\Delta \neq 0$, then the critical region will be of form $U \leq d$. Values of d such that $P(U \leq d) = \alpha''$ for α'' approximately equal to 0.10, 0.05, and 0.01 are given in Table B.VI.

For example suppose we have the following samples of heights for 12 boys and 10 girls (in inches):

Boys:	47.52	46.83	46.95	45.25	47.37
	(17)	(13)	(15)	(10)	(16)
	49.60	47.75	48.15	46.18	46.90
	(21)	(18)	(19)	(11)	(14)
	51.27	43.90			
	(22)	(4)			

Girls: 44.53   43.83   43.95   42.25   44.37
            (7)   (3)   (5)   (2)   (6)
            46.60   44.75   45.16   48.27   40.90
            (12)   (8)   (9)   (20)   (1)

The numbers in parentheses are the ranks of the corresponding values
in the combined sample. Tied observations are assigned ranks as in
the signed-rank test. Due to the format of the tables we should
always choose the smaller sample as "sample 1." We find $\sum_{i=1}^{10} R_{X_{1i}} =$
73; hence $U_1 = 73 - 55 = 18$ and $U_2 = 120 - 18 = 102$. The statistic
$U = 18$, and as $P(U \leq 22) = 0.009$ we certainly decide $\Delta \neq 0$.

As in the case of the Wilcoxon signed-rank test we can find a
large sample approximation to the distribution of U_1. As

$$U_1 = \sum_{i=1}^{n_1} R_{X_{1i}} - \frac{n_1(n_1 + 1)}{2} \quad \text{and} \quad E(R_{X_{1i}}) = \frac{(n_1 + n_2 + 1)}{2}$$

under H_0 for all i, we obtain $E(U_1) = n_1 n_2 / 2$. It can be further
shown that $Var(U_1) = n_1 n_2 (n_1 + n_2 + 1)/12$. For moderate and large
n_1 and n_2, the statistic

$$Z = \frac{U_1 - n_1 n_2 / 2}{\sqrt{n_1 n_2 (n_1 + n_2 + 1)/12}}$$

is approximately distributed as the standard normal variable under
H_0. Against the alternatives $\Delta > 0$, $\Delta < 0$, and $\Delta \neq 0$, the rejection
regions are $Z \leq -Z_\alpha$, $Z \geq Z_\alpha$, and $|Z| \geq Z_{\alpha/2}$, respectively.

Using the height data for the $n_1 = 12$ boys and the $n_2 = 10$
girls given in the example above, we find the computed value of
$Z = (102 - 60)/\sqrt{(10)(12)(23)/12} = 2.769$ and the corresponding P-VALUE
against a two-sided alternative of 0.0058. This calls for rejection
of $\Delta = 0$ in favor of $\Delta \neq 0$ at significance level $\alpha = 0.01$ (say). The
same data have been used with the program N2SAMP. The value of $t = $
3.376 with 20 degrees of freedom and a P-VALUE of 0.003 also strongly
suggests rejection of H_0: $\Delta = 0$. A 95% confidence interval for
$\mu_2 - \mu_1 = \Delta$ is given by $\bar{X}_2 - \bar{X}_1 \pm \hat{\sigma} t_{0.025,20} = -2.845 \pm 0.843(2.086)$
or by $[-4.604, -1.086]$. The true value of $\mu_1 = 48$ and $\mu_2 = 45$, so

that $\Delta = 3$ in this case. As the data actually do come from (simulated) normal populations the rank test is not as sensitive as the t test. As in the one sample case the Wilcoxon rank sum test is applicable to a much wider class of "shift" alternatives. Particularly if the underlying distributions have "long tails," the rank test is preferred to the parametric test.

Problems 8.4

1. Suppose that we denote population 1 by X and population 2 by Y. Assume $n_1 = 2$ and $n_2 = 3$. The null distribution of U_1 can be found by assigning equal probability to each of the $\binom{5}{2}$ permutations of 2 x's and 3 y's. For example, $U_1(xxyyy) = 0$, and hence $P(U_1 = 0) = 1/10$.

   (a) Find the null distribution of U_1.

   (b) Verify that $E(U_1) = n_1 n_2/2$ and $Var(U_1) = n_1 n_2(n_1 + n_2 + 1)/12$ in this case.

2. The $n_1 + n_2$ observations can be assumed to come from the same continuous population under H_0. The assignment of n_1 ranks to the X_1 observations can be considered to be a random selection of n_1 of the first $n_1 + n_2$ integers without replacement.

   (a) Using the results of Sec. 5.1 show that

   $$E\left(\sum_{i=1}^{n_1} R_{X_i} \right) = \frac{n_1(n_1 + n_2 + 1)}{2}$$

   and

   $$Var\left(\sum_{i=1}^{n_1} R_{X_i} \right) = \frac{n_1 n_2(n_1 + n_2 + 1)}{12}$$

   (b) Prove that $E(U_1) = n_1 n_2/2$ and $Var(U_1) = n_1 n_2(n_1 + n_2 + 1)/12$.

3. Use the yield data of Exercise 8.3.1 to test the hypothesis $\mu_1 = \mu_2$ versus $\mu_1 \neq \mu_2$ at $= 0.05$ using the Wilcoxon rank test based on the statistic U.

4. In the data below we give the batting averages of 15 right-hand-
    hitting left fielders and 10 left-hand-hitting left fielders.

R.H.	0.298	0.295	0.260	0.217	0.237
	0.224	0.294	0.220	0.283	0.290
	0.297	0.285	0.226	0.307	0.298
L.H.	0.224	0.259	0.312	0.255	0.261
	0.306	0.237	0.345	0.279	0.262

    (a) Use the large sample approximation to the statistic U_1 to
         test the hypothesis H_0: $\Delta = 0$ versus H_1: $\Delta \neq 0$ using
         $\alpha = 0.05$.
    (b) Use the program N2SAMP to test the same hypotheses as in
         (a). Compare the conclusions. Note: Ties to occur in
         practice. Assign the average of the ranks which would be
         assigned to the tied observations to each such observation.

5. Use the weights of the football players for Iowa State and
    Oklahoma State given in Problem 8.3.4 to test the hypothesis of
    no difference in average weight for the two teams. Use the rank
    test of this section with $\alpha = 0.10$.

6. Use the data of Exercise 8.3.2 on the yardage gained by freshmen
    and sophomore backs versus that gained by junior and senior backs
    to test the hypothesis stated in part (a) of that exercise. Use
    the Wilcoxon rank sum statistic U.

    Exercises 8.4

1. Use the program WIL2TEST with the age data of the example run of
    N2SAMP in Sec. 8.3. Compare your decision based on the rank sum
    statistic with that based on the parametric test.

2. Use the weight data of Exercise 8.2.2 to answer question 8.2.2(a).
    How does your decision compare with those of Exercises 8.2.2 and
    8.3.3?

WIL2TEST

```
10 REM THIS PROGRAM TESTS THE HYPOTHESIS THAT TWO POPULATIONS
20 REM HAVE THE SAME LOCATION PARAMETER VERSUS
30 REM 1) DELTA <0
40 REM 2) DELTA >0
50 REM 3) DELTA <> 0
60 REM ASSUMING INDEPENDENT SAMPLES OF SIZES N1 AND N2
70 REM FROM TWO CONTINUOUS POPULATIONS
80 REM THE PROGRAM COMPUTES THE VALUE OF THE WILCOXON RANK
85 REM SUM STATISTIC U1 AND THE CORRESPONDING Z AND P-VALUES
90 REM SAMPLE 1 AND SAMPLE 2 DATA APPEAR IN LINES 500-600 IN
95 REM IN THAT ORDER
100 PRINT "WHAT ARE THE VALUE OF N1 AND N2?"
110 INPUT N1,N2
120 PRINT "WHICH ALTERNATIVE,";"1, 2 OR 3?"
130 INPUT A
140 IF A = 1 THEN 190
150 IF A =2 THEN 190
160 IF A = 3 THEN 190
170 PRINT "INCORRECT FORMAT ";"TRY AGAIN"
180 GO TO 110
185 DIM X(200),T(200),Z(200)
190 FOR I = 1 TO N1+N2
200 READ X(I)
205 LET T(I)=I
210 NEXT I
215 FOR I = 1 TO N1
225 LET Z(I)=1
230 NEXT I
235 FOR J= 1 TO (N1+N2-1)
240 FOR I = 1 TO (N1+N2-J)
245 IF X(I)<=X(I+1) THEN 275
250 LET T=X(I+1)
255 LET T1=T(I+1)
260 LET X(I+1)=X(I)
262 LET T(I+1)=T(I)
265 LET X(I)=T
270 LET T(I)=T1
275 NEXT I
280 NEXT J
285 FOR I = 1 TO (N1+N2)
290 LET X(I)=I
295 NEXT I
300 FOR J = 1 TO (N1+N2-1)
305 FOR I = 1 TO (N1+N2-J)
310 IF T(I)<T(I+1) THEN 345
315 LET T =T(I+1)
320 LET T1=X(I+1)
325 LET T(I+1)=T(I)
330 LET X(I+1)=X(I)
335 LET T(I)=T
340 LET X(I)=T1
345 NEXT I
350 NEXT J
```
*

```
355 FOR I = 1 TO N1
360 LET T2=T2+X(I)*Z(I)
365 NEXT I
366 LET T2=T2-N1*(N1+1)/2
367 PRINT "THE VALUE OF U1 =";T2
370 LET Z =(T2-N1*N2/2)/SQR(N1*N2*(N1+N2+1)/12)
380 GOSUB 900
390 PRINT "THE CALCULATED VALUE OF Z = ";Z
400 IF A = 3 THEN 430
410 PRINT
420 PRINT "THE P-VALUE FOR ";"ALTERNATIVE ";A;" = ";P
425 GO TO 1030
430 PRINT
440 PRINT "THE P-VALUE FOR ";"ALTERNATIVE ";A;" = ";P*2
450 GO TO 1030
500 DATA 47.52, 46.83, 46.95, 45.25, 47.37, 49.6
510 DATA 47.75, 48.15, 46.18, 46.9, 51.27, 43.9
520 DATA 44.53, 43.83, 43.95, 42.25, 44.37
530 DATA 46.60, 44.75, 45.16, 48.27, 40.9
900 LET Z1= ABS(Z)
910 LET C=1/SQR(2)
920 LET C1= .14112821
930 LET C2= .08864027
940 LET C3=.02743349
950 LET C4= -.00039446
960 LET C5=.00328975
970 DEF FNZ(X)=1-1/(1+C1*X+C2*X↑2+C3*X↑3+C4*X↑4+C5*X↑5)↑8
980 LET P = .5+.5*FNZ(Z1*C)
990 LET P = 1E-4*(INT(1E4*P))
1010 LET P = 1-P
1020 RETURN
1030 END
*

 RUN
WHAT ARE THE VALUE OF N1 AND N2?
? 12,10
WHICH ALTERNATIVE, 1, 2 OR 3?
? 3
THE VALUE OF U1 = 102
THE CALCULATED VALUE OF Z =   2.7694

THE P-VALUE FOR ALTERNATIVE  3  =   5.79999 E-3

 *
```

```
    N2SAMP

EDIT LIST 400-510

400  DATA 47.52, 46.83, 46.95, 45.25, 47.37, 49.6
410  DATA 47.75, 48.15, 46.18, 46.9, 51.27, 43.9
500  DATA 44.53, 43.83, 43.95, 42.25, 44.37
510  DATA 46.6, 44.75, 45.16, 48.27, 40.9
*

   RUN
WHAT ARE THE VALUE OF N1 AND N2?
? 12,10
WHICH ALTERNATIVE, 1, 2 OR 3?
? 3
ANALYSIS ASSUMING EQUAL VARIANCES
SAMPLE          MEAN              STD DEVIATION
   1            47.3058            1.89613
   2            44.461             2.05202
STD DEV OF DIFF =  0.842565
THE CALCULATED VALUE OF T =  3.37639   ON  20   DOF
THE F VALUE =  0.853834 ON 11  AND  9   DOF

THE P-VALUE FOR ALTERNATIVE  3  =  0.003

ANALYSIS FOR UNEQUAL VARIANCES
THE STD DEV OF DIFF =  0.848933
THE CALUCLATED VALUE OF T =  3.35107
WITH DOF =  20.6016

   *
```

3. Use the grades for males and females given in Exercise 8.2.3
   with WIL2TEST to test the null hypothesis of no difference
   between the sexes with regard to statistics exam grades. Use
   $\alpha = 0.10$.

8.5   INFERENCE FOR MATCHED PAIRS

Consider a situation in which we want to determine whether there is
an effect on performance in an arithmetic exam due to birth order
of siblings. We consider it likely that the first born of siblings
will do better on an arithmetic exam given to each at age 6. We
collect data on the scores of five pairs of children on the test
(which has a maximum scores of 125).

Test score

Pair	Older	Younger
1	106	102
2	98	94
3	123	118
4	97	91
5	88	83

At first glance it may appear reasonable to use a two-sample
test. Using N2SAMP, we find t = 0.574 with 8 degrees of freedom
and a P-VALUE of 0.291 for testing H_0: $\mu_1 = \mu_2$ versus H_1: $\mu_1 > \mu_2$.
The F value of 0.983 indicates that the equal variance assumption
can be retained. The conclusion would be to accept $\mu_1 = \mu_2$. Alter-
natively, the Wilcoxon rank sum test statistic can be used. The
value of U = 9, which also leads to the conclusion $\Delta = \mu_2 - \mu_1 = 0$
at significance level 0.05.

We have however made an error in the statistical analysis of
these data. We have compared two groups of children, when actually
we want a test and confidence interval which is sensitive to the
differences of the scores within each of the five pairs. Using the
notation (X_{1i}, X_{2i}) for i = 1, 2, ..., n, we should consider a test
of whether the observed differences $D_i = X_{1i} - X_{2i}$ come from a

```
N2SAMP

EDIT LIST 400-500

400  DATA 106,98,123,97,88
500  DATA 102,94,118,91,83
*RUN
WHAT ARE THE VALUE OF N1 AND N2?
? 5,5
WHICH ALTERNATIVE, 1,2 OR 3?
? 1
ANALYSIS ASSUMING EQUAL VARIANCES
SAMPLE          MEAN           STD DEVIATION
   1            102.4             13.1643
   2             97.6             13.2778
STD DEV OF DIFF =  8.36182
THE CALCULATED VALUE OF T = 0.574038   ON  8   DOF
THE F VALUE =  0.982985 ON 4  AND  4   DOF

THE P-VALUE FOR ALTERNATIVE  1   =   0.2908

ANALYSIS FOR UNEQUAL VARIANCES
THE STD DEV OF DIFF =  8.36182
THE CALUCULATED VALUE OF T =  0.574038
WITH DOF =  9.99926

*OLD MTEST
*600 DATA 4,4,5,6,5
RUN
WHAT ARE THE VALUES OF N AND MU0?
? 5,0
WHICH ALTERNATIVE 1,2 OR 3?
? 1
WHAT IS THE VALUE OF ALPHA, 0.10,0.05,0.01?
? 0.10
THE CRITICAL REGION FOR A ONE=SIDED TEST
AT SIGNIFICANCE LEVEL =  0.05
IS T >=  2.1318

THE COMPUTED VALUE OF T = 12.8285

*
```

population D with $E(D) = \mu_1 - \mu_2 = 0$, versus the alternative that
$\mu_1 > \mu_2$. The appropriate test here is a one-sample test. Using the
program MTEST with the differences 4, 4, 5, 6, 5, we find t = 12.829
with 4 degrees of freedom. A test with significance level $\alpha = 0.05$
is given by the critical region $t \geq 2.132$. Hence we reject $\mu_1 = \mu_2$
in favor of $\mu_1 > \mu_2$. Calculations yield $\bar{d} = 4.8$ and $s_d^2 = 0.7$.
Hence we find a 95% confidence interval for $\mu_1 - \mu_2$ to be given by
$\bar{d} \pm s_d/\sqrt{5} t_{0.025,4} = 4.8 \pm (\sqrt{0.7}/\sqrt{5})2.776$ or by the interval
[3.761, 5.839]. The evidence for a better performance by the first
born is fairly strong in contrast to the decision based on the independent two-sample tests.

In the case in which there is a natural pairing of the elements
in the two groups and the question of interest is whether the members
of the pairs differ, we do not use an independent two-sample test.
The appropriate test statistic is based on

$$t = \frac{\bar{d}}{s_d/\sqrt{n}}$$

where the number of pairs is n. The appropriate degrees of freedom
for a confidence interval and for the test is n - 1 and not 2(n - 1)
as in the independent two-sample test. Many times we choose to
design a comparison so that the elements of each pair are as similar
as possible, except with respect to a difference we are attempting
to measure. The paired test is appropriate here. For example, we
may be interested in the effect of a drug. The drug is administered
at random to one of two members of each pair. We choose the members
in the pairs to be as similar as possible with respect to age, weight,
and possibly other variables. This is an elementary example of the
design of a statistical experiment. We collect the data in such a
way that the experiment provides the greatest possible information
about the question of interest. The field of experimental design
for complex statistical designs is an interesting and important area
of both mathematical and applied statistics.

Exercises 8.5

1. The following data[*] give the morning and evening heights of 10 individuals measured in centimeters. We wish to know if there is evidence for change of height of individuals during a day.

Subject	Morning	Evening
1	182.9	182.5
2	174.0	174.0
3	178.2	177.8
4	181.2	181.0
5	167.0	167.2
6	174.8	174.4
7	171.7	171.5
8	179.2	179.1
9	181.2	180.0
10	185.5	185.0

(a) Use the program N2SAMP to make a test using the independent two-sample test to decide between $\mu_1 = \mu_2$ and $\mu_1 \neq \mu_2$ using $\alpha = 0.05$.

(b) Use the program MTEST to make the same test.

(c) Using the appropriate design, find a confidence interval for $\mu_1 - \mu_2$.

2. A researcher is studying the effectiveness of two different methods of teaching spelling. He has a group of 20 children, who he matches as closely as possible in 10 pairs according to IQ, sex, and age. One of the members of each pair is assigned at random to method 1 and the other to method 2. The results of a test of the spelling of 16 words is reported below:

---

[*]Adapted from *Everyday Statistics*; G. W. Snedecor, 1950, W. C. Brown Company.

Pair	Method 1	Method 2
1	11	15
2	11	9
3	13	16
4	11	11
5	11	10
6	9	12
7	8	11
8	6	5
9	5	9
10	8	10

(a) Test the hypothesis H_0: $\mu_1 = \mu_2$ versus H_1: $\mu_1 \neq \mu_2$ using $\alpha = 0.05$.

(b) Find a 95% confidence interval for $\mu_1 - \mu_2$.

3. Nine pairs of universities were matches as closely as possible according to size, type of administrative control, and location. In each pair one institution had a faculty organized for collective bargaining, while the other was not so organized. The average difference in faculty compensation was $1600, with a standard deviation of the differences of $500. The average difference favored the organized institutions. Is the difference significant ($\alpha = 0.05$)? What is a 95% confidence interval for the difference?

4. Students in an elementary statistics course argued that their second hour exam was more difficult than their first. The following gives the scores of 15 students on the two exams, each of which had a maximum of 80.

Student	Hour exam 1	Hour exam 2
1	48	50
2	40	55
3	60	60
4	48	35
5	64	40
6	56	65
7	68	63
8	64	40
9	28	35
10	56	55
11	56	40
12	24	40
13	48	45
14	76	70
15	72	75

Use the matched-pairs test to test H_0: $\mu_1 = \mu_2$ versus H_1: $\mu_1 > \mu_2$. Is their evidence that exam scores on the second exam were lower? Does your answer to this question necessarily contradict the assertion made?

8.6 COMPARISON OF TWO POPULATION PROPORTIONS

The last two-sample situation which we consider is the comparison of two population proportions. We often have questions such as

1. Is the proportion of men and women in the adult population in favor of a certain governmental policy the same?
2. Is the proportion of defective items produced on two assembly lines the same?
3. Is the proportion of left-hand batters hitting at least 0.280 in the major leagues higher than that for right-hand batters?

We assume that from each of two populations we obtain random samples of size n_1 and n_2, respectively. Denote a proportion in population 1 by p_1 and in population 2 by p_2. Let X_1 and X_2 represent the numbers in the corresponding samples having a characteristic (e.g., favoring election of the president by popular vote rather than by the current electoral college method). We are interested in inference about p_1' and p_2 based on the estimators $\hat{p}_1 = X_1/n_1$ and $\hat{p}_2 = X_2/n_2$.

We first consider testing H_0: $p_1 = p_2$ in the moderate to large sample case. The natural alternatives are (1) $p_1 > p_2$, (2) $p_1 < p_2$, and (3) $p_1 \neq p_2$. If we consider the statistic $\hat{p}_1 - \hat{p}_2$, then $E(\hat{p}_1 - \hat{p}_2) = p_1 - p_2$ and $Var(\hat{p}_1 - \hat{p}_2) = Var(\hat{p}_1) + Var(\hat{p}_2) = p_1(1 - p_1)/n_1 + p_2(1 - p_2)/n_2$. Under the null hypothesis $p_1 = p_2 = p$ (say), we have $E(\hat{p}_1 - \hat{p}_2) = 0$ and $Var(\hat{p}_1 - \hat{p}_2) = p(1 - p)(1/n_1 + 1/n_2)$. The parameter p is unknown but can be estimated by a pooled estimate.

$$\hat{p} = \frac{X_1 + X_2}{n_1 + n_2}$$

Let

$$Z = \frac{\hat{p}_1 - \hat{p}_2}{\sqrt{\hat{p}(1 - \hat{p})(1/n_1 + 1/n_2)}}$$

It is clear that the appropriate rejection regions are $Z \geq Z_\alpha$, $Z \leq -Z_\alpha$, and $|Z| \geq Z_{\alpha/2}$ for the alternatives (1), (2), and (3).

To find a confidence interval for $p_1 - p_2$, we use $Var(\hat{p}_1 - \hat{p}_2) \doteq \hat{p}_1(1 - \hat{p}_1)/n_1 + \hat{p}_2(1 - \hat{p}_2/n_2 = s_{\hat{p}_1 - \hat{p}_2}^2$. Using Eq. (8.2.1) we find an approximate $(1 - \alpha)100\%$ confidence interval for $p_1 - p_2$.

$$\hat{p}_1 - \hat{p}_2 \pm s_{\hat{p}_1 - \hat{p}_2} Z_{\alpha/2} \qquad (8.6.1)$$

We cannot use the pooled estimate $\hat{p}$ here because the assumption that $p_1 = p_2$ is not being made here.

The program 2PTEST will test the hypothesis H_0: $p_1 = p_2$ versus one of the alternatives (1), (2), or (3). The program requests the values of n_1, n_2, x_1, x_2, and the appropriate alternative. The value of Z is computed and the corresponding P-VALUE is calculated. A 95% confidence interval for $p_1 - p_2$ is computed. In the sample

2PTEST

```
10 REM THIS PROGRAM TESTS THE HYPOTHESIS P1 = P2
20 REM AGAINST ONE OF THE ALTERNATIVES
30 REM 1) P1 > P2
40 REM 2) P1 < P2
50 REM 3) P1 <> P2
60 REM ASSUMING INDEPENDENT SAMPLES OF SIZE N1 AND N2
70 REM FROM TWO POPULATIONS WITH TRUE PROPORTIONS P1 AND P2
80 REM THE NUMBERS OF "SUCCESSES" IN THE TWO SAMPLES ARE X1 AND X2
90 REM THE PROGRAM COMPUTES Z, THE CORRESPONDING P- VALUE
95 REM AND A 95 PERCENT CONFIDENCE INTERVAL FOR P1 -P2
100 PRINT "WHAT ARE THE VALUE OF N1 AND N2?"
110 INPUT N1,N2
120 PRINT "WHICH ALTERNATIVE, ";"1,2 OR 3?"
130 INPUT A
140 IF A = 1 THEN 190
150 IF A = 2 THEN 190
160 IF A = 3 THEN 190
170 PRINT "INCORRECT FORMAT ";"TRY AGAIN"
180 GO TO 110
190 PRINT "WHAT ARE THE VALUES OF X1 AND X2?"
200 INPUT X1,X2
210 LET P1=X1/N1
220 LET P2 = X2/N2
230 LET P =(X1+X2)/(N1+N2)
235 LET P3=P
240 LET Z =(P1-P2)/SQR(P*(1-P)*(1/N1+1/N2))
250 LET S = SQR(P1*(1-P1)/N1+P2*(1-P2)/N2)
255 LET L = (P1-P2)-S*1.96
260 LET U = (P1-P2)+S*1.96
265 GOSUB 900
290 PRINT "THE CALCULATED VALUE OF Z = ";Z
350 PRINT "P1-HAT = ";P1,"P2-HAT = ";P2
370 PRINT "THE STD DEV OF DIFF = ";S
380 PRINT "THE POOLED ESTIMATE OF P = ";P3
390 PRINT "A 95 PERCENT CONF INT FOR P1- P2 IS GIVEN BY"
400 PRINT " L = ";L," U = ";U
405 IF A = 3 THEN 430
410 PRINT
420 PRINT "THE P-VALUE FOR ";"ALTERNATIVE ";A;" = ";P
425 GO TO 1030
430 PRINT
440 PRINT "THE P-VALUE FOR ";"ALTERNATIVE ";A;" = ";P*2
450 GO TO 1030
900 LET Z1= ABS(Z)
910 LET C=1/SQR(2)
920 LET C1= .14112821
930 LET C2= .08864027
940 LET C3=.02743349
950 LET C4= -.00039446
960 LET C5=.00328975
970 DEF FNZ(X)=1-1/(1+C1*X+C2*X↑2+C3*X↑3+C4*X↑4+C5*X↑5)↑8
980 LET P = .5+.5*FNZ(Z1*C)
990 LET P = 1E-4*(INT(1E4*P))
1010 LET P = 1-P
1020 RETURN
1030 END
*
```

```
 RUN
WHAT ARE THE VALUE OF N1 AND N2?
? 250, 450
WHICH ALTERNATIVE, 1, 2 OR 3?
? 3
WHAT ARE THE VALUES OF X1 AND X2?
? 120, 260
THE CALCULATED VALUE OF Z = -2.48528

THE P-VALUE FOR ALTERNATIVE  3  =  0.013
P1-HAT =  0.48                    P2-HAT =  0.577778
THE STD DEV OF DIFF =  3.92494 E-2
THE POOLED ESTIMATE OF P =  0.542857
A 95 PERCENT CONF INT FOR P1- P2 IS GIVEN BY
L = -0.174707   U = -0.020849

*
```

run 250 men and 450 women are asked whether they favor the retention
of the 55-mile/hr speed limit. The values $\hat{p}_1 = 0.480$, $\hat{p}_2 = 0.578$,
$Z = -2.488$, and P-VALUE of 0.013 suggest rejection of $p_1 \neq p_2$. A
95% confidence interval for $p_1 - p_2$ is given by $(-0.175, -0.021)$.
The evidence suggests that $p_1 < p_2$, i.e., the proportion of men in
favor of the 55-mile/hr limit is less than the proportion of women.

As a final example, consider a sample of the lifetime batting
averages of 140 major league hitters. This yielded 20 of 40 left-
hand batters with a lifetime average of at least 0.280, and 27 of 100
right-hand batters with lifetime averages of at least 0.280. Hence
$\hat{p}_1 = 0.5$, $\hat{p}_2 = 0.27$, $\hat{p} = 47/140 = 0.3347$, and $Z = 2.603$. As the
P-VALUE is less than 0.01, one would reject H_0: $p_1 = p_2$ in favor of
H_1: $p_1 > p_2$. In other words, left-hand batters seem to hit 0.280
or better with higher frequency than right-hand batters.

Exercises 8.6

1. Assume that of 200 sophomores and 250 seniors at a university,
    74 of the sophomores and 70 of the seniors voted for the same
    presidential candidate as their fathers in the last election.
    Assume that p_1 is the true proportion of sophomores voting for
    the same presidential candidate as their fathers and p_2 is the
    corresponding proportion for seniors.
    (a) Test $p_1 = p_2$ versus $p_1 > p_2$ at significance level 0.05.
    (b) Find a 95% confidence interval for $p_1 - p_2$.

2. Assume that during a given week that 190 of 220 arrivals of air-
    line A are "on time." An arrival is considered on time if a
    plane arrives within 10 min of its scheduled arrival time.
    During the same week 82 of 99 arrivals of airline B are on time.
    (a) Test the hypothesis that the proportions of on-time arrivals
         are the same for both airlines ($\alpha = 0.05$).
    (b) Find a 95% confidence interval for $p_A - p_B$.

3. In 1974, 2037 of 4832 chemistry majors applying to medical school
   were admitted. In the same year, 263 of 650 English majors applying to medical school were admitted. Is there statistical evidence
   that chemistry majors do better than English majors in applying
   to medical school?

ANALYSIS OF CLASSIFICATION DATA

9.1   INTRODUCTION

In many cases in which statistical inference is of importance we
measure the value of continuous random variables or random variables
which are assumed to be continuous.   For example, weights, lengths,
times, and distances are considered to be continuous random variables.
Test scores, ages, and words typed per minute, while actually measured
as discrete random variables, are treated as continuous random variables in questions of inference.   There are however many cases,
especially in the social sciences, in which observations can only be
classified into separate categories.   For example, a student is
graded as A, B, C, or failing.   A tire is classified as excellent,
acceptable with minor production flaws, or unacceptable.   We observe
the *number of observations* in the categories or classes in these
cases.   Data such as these are referred to as classification or
"count" data.   In this chapter we consider the elementary statistical
analysis of these type of data.

9.2   THE MULTINOMIAL DISTRIBUTION AND ITS RELATION TO THE CHI-SQUARE
DISTRIBUTION

In the case of the binomial distribution we consider n independent
trials of an "experiment" which are classified into one of two
classifications referred to as "success" or "failure."   With the

assumption of a constant probability p of success in each trial we
have obtained the distribution of the random variable $X \sim B(n;p)$
which gives the number of successes in the n trials. In the case of
the multinomial distribution we assume that at each trial an observation may be classified into one of $k \geq 2$ distinct classes. We
consider n independent observations with a constant probability p_i,
$i = 1, 2, \ldots, k$ of falling into class i (with $\Sigma\, p_i = 1$). Writing
$X_1, X_2, \ldots, X_k$ to denote the number of the observations in n trials
falling into classes $i = 1, 2, \ldots, k$, it is easy to see that $X_i \sim$
$B(n;p_i)$ for $i = 1, 2, \ldots, k$. However the joint probability of x_1
outcomes in class 1, x_2 outcomes in class 2, $\ldots$, x_{k-1} outcomes in
class k - 1 is given by the expression

$$P((x_1,x_2,\ldots,x_{k-1})) = \frac{n!}{x_1!x_2!\cdots x_k!}\, p_1^{x_1} p_2^{x_2} \cdots p_k^{x_k} \qquad (9.2.1)$$

The vector is of length k - 1, because the condition $\Sigma_{i=1}^{k} x_i = n$,
determines the value of x_k and the expression on the right can be
written with $n - \Sigma_{i=1}^{k-1} x_i$ replacing x_k, i.e., as a function of x_1,
$x_2, \ldots, x_{k-1}$ only. Hence X_k is not an additional random variable.
In the case k = 2, Eq. (9.2.1) reduces to

$$P(x_1) = \frac{n!}{x_1!(n - x_1)!}\, p_1^{x_1}(1 - p_1)^{n-x_1}$$

the familiar binomial expression.

As a simple example let us consider that in a game of matching
fair coins between David and Goliath, that each independently tosses
a penny n times. If both pennies are heads, David wins, and if both
pennies are tails, Goliath wins. If the coins differ, the outcome
is called a tie. Let X_1 represent the number of wins by David and
X_2 by Goliath in n trials. Then $p_1 = 1/4$, $p_2 = 1/4$, and $p_3 = 1/2$
and

$$P((x_1,x_2)) = \frac{n!}{x_1!x_2!x_3!}\, \left(\frac{1}{4}\right)^{x_1}\left(\frac{1}{4}\right)^{x_2}\left(\frac{1}{2}\right)^{x_3}$$

Assume $n = 10$, and we ask for the probability David and Goliath each win three times and the other four outcomes are ties. We find $P(3,3) = (10!/3!3!4!)(1/4)^3(1/4)^3(1/2)^4 = 525/8192 = 0.0641$. The probabilities of the other possible outcomes can be found by substituting all possible combinations of nonnegative integers satisfying $x_1 + x_2 + x_3 = 10$ into Eq. (9.2.1). We shall not pursue this further here, as our real interest is in inference concerning the values of the p_i jointly.

Suppose that we wish to test that the true values of the probabilities p_i are a certain fixed set of numbers. We write the null hypothesis as

$$H_0: \quad p_i = p_{i0} \quad i = i, 2, \ldots, k$$

where $\sum_{i=1}^{k} p_{i0} = 1$. The alternative to H_0 is generally stated just as the negation of H_0, namely

$$H_1: \quad p_i \neq p_{i0} \quad \text{for some } i$$

Under the hypothesis H_0, we have $E(X_i) = np_{i0}$ as the X_i are binomial variables. In order to test H_0 we observe the values of X_1, X_2, ..., X_k for a relatively large number of trials, n. The data can now be presented in a summary table as follows:

Class	1	2	· · ·	k	Total
Observed	x_1	x_2	· · ·	x_k	n
Expected	np_{10}	np_{20}	· · ·	np_{k0}	n

It is apparent that close agreement between the observed outcomes in each class with the corresponding expected values under H_0 suggests retention of H_0.

The statistic which is widely used in a test of the hypotheses H_0 and H_1 is

$$\chi^2 = \sum_{i=1}^{k} \frac{(x_i - np_{i0})^2}{np_{i0}} = \sum \frac{(\text{observed} - \text{expected})^2}{\text{expected}}$$

It can be shown that for multinomial data, if H_0 is true, the statistic χ^2 has asymptotically the chi-square distribution with k - 1 degrees of freedom. Rejection of H_0 occurs if the observed values are not close to the expected values, which results in a large value of the computed statistic χ^2. An approximate α level test is given by the critical region

$$\chi^2 > \chi^2_{\alpha,k-1}$$

as the value of χ^2 will exceed the $\chi^2_{\alpha,k-1}$ point, if H_0 is true, with probability approximately equal to α.

Let us consider the example of David and Goliath again. Suppose Goliath's coin is biased and falls heads with probability 2/3 and tails with probability 1/3, while David's coin is fair. Assuming independence of the two coins the true values of $p_1 = 1/3$, $p_2 = 1/6$, and $p_3 = 1/2$. Assuming fairness of the coins, we hypothesize

$$H_0: \quad p_1 = 1/4, \ p_2 = 1/4, \ p_3 = 1/2$$

Suppose that we observed the following results in 100 trials.

Outcome	2H	2T	1H, 1T	Total
Observed	38	13	49	100
Expected	25	25	50	100

We compute $\chi^2 = (38 - 25)^2/25 + (12 - 25)^2/25 + (49 - 50)^2/50 = 12.54$. As $\chi^2_{0.01,2} = 9.21$, we would reject H_0 at level of significance $\alpha = 0.01$. The hypothesized values appear to be incorrect, as in fact they are.

As another application of the large sample approximation of the chi-square distribution, let us consider testing $H_0: \ p = p_0$ versus $H_1: \ p \neq p_0$ in the binomial case in which $X \sim B(n;p)$ represents the number of successes. We observe

Outcome	Failures	Successes	Total
Observed	n - x	x	n
Expected	$n(1 - p_0)$	np_0	n

$$\chi^2 = \frac{(x - np_0)^2}{np_0} + \frac{(n - x - n(1 - p_0))^2}{n(1 - p_0)}$$

$$= \frac{(x - np_0)^2}{np_0(1 - p_0)}$$

For an α level test we reject H_0: $p = p_0$ if $\chi^2 > \chi^2_{\alpha,1}$.

In Chap. 7 we have seen that a test of

$$H_0: \quad p = p_0 \qquad \text{versus} \qquad H_1: \quad p \neq p_0$$

in the large sample case can be based on the statistic.

$$Z = \frac{x - np_0}{\sqrt{np_0(1 - p_0)}}$$

We reject H_0 if $|Z| \geq Z_{\alpha/2}$ for a test at significance level α. Note that the statistic $Z^2 = \chi^2$ as defined in the previous paragraph. Hence, under H_0

$$P(Z^2 \geq Z^2_{\alpha/2}) = \alpha \qquad \text{and} \qquad P(Z^2 \geq \chi^2_{\alpha,1}) \doteq \alpha$$

It should not be surprising then to find out that it can be proved that the square of a standard normal variable has the same distribution as a chi-square variable with one degree of freedom. Hence $Z^2_{\alpha/2} = \chi^2_{\alpha,1}$. For example, if $\alpha = 0.05$, $Z^2_{0.025} = (1.96)^2 = 3.841 = \chi^2_{0.05,1}$. The two tests will thus lead to identical decisions. For example, if we test that a coin is fair based on 400 tosses of which 220 are heads, we compute $Z = 2$ and $\chi^2 = 4$. At level $\alpha = 0.05$, we reject $p = 0.5$ as $Z \geq 1.96$ or $\chi^2 \geq 3.841$. It is sensible that the large sample tests of $p = p_0$ based on Z and χ^2 always yield the same decision.

As a final note we remark that the chi-square approximation is considered valid only if the expected value in each of the classes is not too small. A general rule is that the *expectation* should not be less than 1 in any class. Other statisticians use a minimum of 5, but we shall use the smaller bound in this chapter. If a class has expectation less than 1, it is combined with other classes until the expected value is at least 1.

Problems 9.2

1. Assume six fair dice are tossed. What is the probability of 2 twos, 2 fours, and 2 sixes?

2. A city has a population of registered voters, with 30% registered Republicans, 40% registered Democrats and 30% Independents. What is the probability of three Republicans and four Democrats in a sample of size 10, assuming independence of the selections.

3. Sketch a proof of Eq. (9.2.1).

4. Assume that (X_1, X_2) have the trinomial distribution with parameters p_1, p_2, p_3, and n.
   (a) What is the distribution of $X_1 + X_2$?
   (b) Use the fact that $\mathrm{Var}(X_1 + X_2) = \mathrm{Var}(X_1) + \mathrm{Var}(X_2) + 2\mathrm{Cov}(X_1, X_2)$ to find $\mathrm{Cov}(X_1, X_2)$.
   (c) Generalize the result in part (b) by finding $\mathrm{Cov}(X_i, X_j)$ in the general multinomial case.
   (d) Under the assumptions here, show that the conditional distribution of X_1, given $X_2 = j$, is $B(n - j;\ p_1/(p_1 + p_3))$.

5. Use the fact that the pdf of the standard normal variable is

$$f(z) = \frac{1}{\sqrt{2\pi}} e^{-z^2/2} \qquad -\infty < z < \infty$$

together with the result in Problem 8.4.6 to show that Z^2 has the chi-square distribution with 1 degree of freedom.

6. Use the table of the standard normal distribution to find

$$\chi^2_{\alpha,1} \qquad \text{for } \alpha = 0.1(0.1)(0.9)$$

7. Genetic theory predicts that 3/4 of hybrid crosses will display the dominant trait. Of 300 observations, 215 display the dominant trait. Test H_0: $p = 3/4$ versus H_1: $p \neq 3/4$ using the chi-square test. Use $\alpha = 0.05$.

8. A die is tossed 90 times with the following outcomes:

Number	1	2	3	4	5	6
Observed	20	17	13	14	16	10

    Use a chi-square test to decide whether to accept the assertion
    that the die is fair. (Let $\alpha = 0.05$.)

9. A faculty member who is grading "on a curve" is expected to
    distribute his grades as indicated below.

	A	*B*	*C*	*D*	*E*
Percent	10	20	30	20	10
Observations	19	28	41	8	2

    In a large class a faculty member gives the indicated grades.
    Do the grades actually given support the hypothesis that he is
    grading "on a curve"?

9.3   ANALYSIS OF CONTINGENCY TABLES

There are many cases in which observations are classified according
to two criteria. As an example, consider the question, "Are you in
favor of a voluntary armed service in the United States?" We may
be interested in whether the answer to this question is related to
party identification. Assume that we randomly sample 360 voters
and observe the following data:

Party identification/question	Yes	No	No opinion
Democrat	57	83	9
Republican	41	50	9
Independent	35	64	12

In this table each of the 360 observed voters have been classified
according to two criteria: opinion on the question and stated party
identification. Such a table is referred to as a *contingency table*.
The question which we would like to answer is whether the two criteria
of classification are independent or whether an association exists
between opinion and stated party preference.

We consider a general contingency table with r rows and c columns. The random variable X_{ij} represents the number of observations in a sample of n which fall into row class i and column class j. Such a table appears in Table 9.3.1. The notation $X_{i\cdot}$ is the sum of the number of observations in the ith row, and similarly, $X_{\cdot j}$ is the sum of the number of observations in the jth column. We see that

$$\sum_{i=1}^{r} X_{i\cdot} = \sum_{j=1}^{c} X_{\cdot j} = n$$

There are, in all, rc locations into which an observation may fall. For the table of the example we find $X_{1\cdot} = 149$, $X_{2\cdot} = 100$ and $X_{3\cdot} = 111$, that is, 149 Democrats, 100 Republicans, and 111 independents were sampled. Similarly, $X_{\cdot 1} = 133$, $X_{\cdot 2} = 197$, and $X_{\cdot 3} = 30$.

We further denote by p_{ij} the probability that a random observation from the population falls into the ijth location. By $p_{i\cdot}$ we shall mean the probability of falling into row class i and by $p_{\cdot j}$ the probability of falling into column class j. The null hypothesis of independence of the row and column classifications can be written as

$$H_0: \quad p_{ij} = p_{i\cdot} p_{\cdot j} \quad \quad i = 1, 2, \ldots, r; \, j = 1, 2, \ldots, c$$

Table 9.3.1   An r by c Contingency Table

	Column				
	X_{11}	X_{12}	$\cdots$	X_{1c}	$X_{1\cdot}$
	X_{21}	X_{22}	$\cdots$	X_{2c}	$X_{2\cdot}$
Row	$\cdots$	$\cdots$	$\cdots$	$\cdots$	$\cdots$
	X_{r1}	X_{r2}	$\cdots$	X_{rc}	$X_{r\cdot}$
	$X_{\cdot 1}$	$X_{\cdot 2}$	$\cdots$	$X_{\cdot c}$	n

The alternative, the negation of H_0, is $p_{ij} \neq p_{i\bullet}p_{\bullet j}$ for some pair (i,j). It is apparent that these probabilities are not known. However, the following estimates are appropriate:

$$\hat{p}_{i\bullet} = \frac{X_{i\bullet}}{n} \qquad i = 1, 2, 3, \ldots, r$$

$$\hat{p}_{\bullet j} = \frac{X_{\bullet j}}{n} \qquad j = 1, 2, \ldots, c$$

Under the null hypothesis, we would estimate p_{ij} by $\hat{p}_{ij} = \hat{p}_{i\bullet}\hat{p}_{\bullet j}$. With this estimate we can compute the estimated expected number of observations in location (i,j) under H_0 by

$$E_{ij} = n\hat{p}_{ij} = n\frac{X_{i\bullet}}{n}\frac{X_{\bullet j}}{n} = \frac{X_{i\bullet}X_{\bullet j}}{n}$$

The test will be based on the chi-square statistic

$$\chi^2 = \sum_{i=1}^{r} \sum_{j=1}^{c} \frac{(X_{ij} - E_{ij})^2}{E_{ij}} \qquad (9.3.1)$$

As in the previous section large values of χ^2 are inconsistent with H_0, and hence a critical region will be of form $\chi^2 \geq \chi^2_{\alpha,\nu}$, where ν is the appropriate degrees of freedom.

While it might seem that $rc - 1$ is the appropriate degrees of freedom from the discussion in the previous section, this is not so. The null hypothesis does not specify the hypothesized values of p_{ij} until the values of $\hat{p}_{i\bullet}$ and $\hat{p}_{\bullet j}$ are found. As

$$\sum_{i=1}^{r} \hat{p}_{i\bullet} = 1 \qquad \text{and} \qquad \sum_{j=1}^{c} \hat{p}_{\bullet j} = 1$$

there are $r - 1$ row probabilities estimated and $c - 1$ column probabilities. A theorem in mathematical statistics states that the statistic in Eq. (9.3.1) will have approximately a chi-square distribution for large n, where the degrees of freedom is given by

ν = number of locations - 1 - number of parameters estimated

Hence the value of ν for an r by c contingency table is
$rc - 1 - (r - 1) - (c - 1) = rc - r - c + 1 = (r - 1)(c - 1)$. Thus

for the example of this section the appropriate degrees of freedom
for the χ^2 statistic is $(3 - 1)(3 - 1) = 4$.

Using $E_{ij} = X_i \cdot Y_{\cdot j}/n$, we find the following table of expected
values:

Party identification/question	Yes	No	No opinion
Democrat	55.05	81.54	12.42
Republican	36.94	54.72	8.33
Independent	41.01	60.74	9.25

Notice that the numbers in any row or column sum to the corresponding
row or column sum in the original table. For example, for row 2 we
find $36.94 + 54.72 + 8.33 = 99.99$, which is 100 except for round-off
error. The computed value of χ^2 using Eq. (9.3.1) is found to be
3.814, and as $\chi^2_{0.05,4} = 9.49$, we would accept the null hypothesis.
The data do not support the hypothesis that opinion and party preference are associated. If we delete the information concerning independents and compare Democrats and Republicans, we find $\chi^2 = 1.204$.
The appropriate degrees of freedom is now 2. As $\chi^2_{0.05,2} = 5.99$,
again these data do not support the conclusion of association between
opinion and party preference.

As another example consider the following data on 200 countries,
which are classified according to the number of storks in the country
and the country's birthrate. The results are

	Countries with		
Birthrate	Few storks	Many storks	Totals
Low	42	10	52
High	130	18	148
	172	28	200

We compute the value of $E_{11} = 172(52)/200 = 44.72$, $E_{12} = 7.28$, $E_{21} = 127.28$, and $E_{22} = 20.72$. The computed chi-square is $\chi^2 = 1.597$,
and as $\chi^2_{0.05,1} = 3.841$. We conclude here that there is no statistical
evidence for the validity of the stork legend in these data.

CONTGY

```
10 REM THIS PROGRAM COMPUTES THE CHI-SQUARE STATISTIC AND THE
20 REM EXPECTED VALUES FOR AN R BY C CONTINGENCY TABLE.
25 REM THE P-VALUE FOR THE CALCULATED CHI-SQUARE IS ALSO COMPUTED.
30 REM DATA SHOULD BE PUT IN DATA STATEMENTS BEGINNING AT
40 REM LINE 400. THE DATA IS ENTERED BY ROWS.
50 REM THE COMPUTER WILL REQUEST THE VALUES OF R (ROWS)
60 REM AND C (COLUMNS).
100 DIM X(20,20),Y(20,20),R(20),C(20)
110 PRINT "WHAT ARE THE ";"VALUES OF R AND C?"
120 INPUT R,C
130 FOR I = 1 TO R
140 FOR J = 1 TO C
150 READ X(I,J)
155 LET R(I)=R(I)+X(I,J)
160 NEXT J
170 NEXT I
180 FOR J = 1 TO C
190 FOR I = 1 TO R
200 LET C(J)=X(I,J)+C(J)
210 NEXT I
220 NEXT J
230 FOR I = 1 TO R
240 LET N= N+R(I)
250 NEXT I
260 PRINT "ROW","COL","OBSERVED","EXPECTED"
270 PRINT
280 FOR I = 1 TO R
290 FOR J = 1 TO C
300 LET Y(I,J)=R(I)*C(J)/N
310 LET C2= C2 +(X(I,J)-Y(I,J))↑2/Y(I,J)
320 PRINT I,J,X(I,J),Y(I,J)
330 NEXT J
340 NEXT I
350 PRINT
360 PRINT "CHI-SQUARE = ";C2;"WITH ";(R-1)*(C-1);"DOF"
365 LET A = (R-1)*(C-1)
367 IF A = 1 THEN 660
370 GOSUB 500
375 PRINT "P-VALUE = ";I1
380 GO TO 800
400 DATA 57,83,9,41,50,9,35,64,12
500 IF INT (A/2)=A/2 THEN 545
505 LET T = INT((A-1)/2)
105LET C = 1
515 FOR I = 1 TO T
520 LET C = C*(A/2-I)
525 NEXT I
530 LET C = C*2↑(A/2)*SQR(3.14159263)
535 LET C = 1/C
540 GO TO 580
545 LET C = 1
550 LET T = INT(A/2-1)
555 FOR I = 1 TO T
560 LET C = C*(A/2-I)
565 NEXT I
570 LET C = C*2↑(A/2)
575 LET C=1/C
*
```

```
580 DEF FNF(X)=C*(X↑(A/2-1))*EXP(-X/2)
585 LET H = C2/100
590 LET N1 = 100
595 FOR I = 1 TO (N1-1)
600 LET X = H*I
605 LET I1 = I1+FNF(X)*H
610 NEXT I
615 LET I1 = I1 +(H/2)*(FNF(0)+FNF(C2))
620 IF ABS(I1-I2) < .001 THEN 645
625 LET I2 = I1
630 LET H = H/2
635 LET N1=2*N1
640 LET I1=0
642 GO TO 595
645 LET I1=1-I1
650 PRINT
655 RETURN
660 LET C1 = SQR(C2)
665 LET H = C1/100
670 DEF FNG(X)=SQR(2/3.14159263)*EXP(-X↑2/2)
675 LET N1 =100
680 FOR I = 1 TO (N1-1)
682 LET X = H*I
684 LET I1 = I1+FNG(X)*H
686 NEXT I
688 LET I1 = I1+(H/2)*(FNG(0)+FNG(C1))
690 IF ABS(I1-I2)<.001 THEN 702
692 LET I2=I1
694 LET H = H/2
696 LET N1 =2*N1
698 LET I1=0
700 GO TO 680
702 LET I1=1-I1
704 PRINT
706 GO TO 375
800 END
*
```

```
 RUN
WHAT ARE THE VALUES OF R AND C?
? 3,3
```

ROW	COL	OBSERVED	EXPECTED
1	1	57	55.0472
1	2	83	81.5361
1	3	9	12.4167
2	1	41	36.9444
2	2	50	54.7222
2	3	9	8.33333
3	1	35	41.0083
3	2	64	60.7417
3	3	12	9.25

```
CHI-SQUARE =   3.81441 WITH   4 DOF

P-VALUE =   0.431714

*RUN
WHAT ARE THE VALUES OF R AND C?
? 2,3
```

ROW	COL	OBSERVED	EXPECTED
1	1	57	58.6426
1	2	83	79.5863
1	3	9	10.7711
2	1	41	39.3574
2	2	50	53.4137
2	3	9	7.22892

```
CHI-SQUARE =   1.20428 WITH   2 DOF

P-VALUE =   0.547638

*400 DATA 42,10,130,18
RUN
WHAT ARE THE VALUES OF R AND C?
? 2,2
```

ROW	COL	OBSERVED	EXPECTED
1	1	42	44.72
1	2	10	7.28
2	1	130	127.28
2	2	18	20.72

```
CHI-SQUARE =   1.59689 WITH   1 DOF

P-VALUE =   0.206345

*
```

The program CONTGY has been written to calculate the value of chi-square for an r x c contingency table. The program requests the values of r and c. The observations should appear by rows in DATA statements in lines 400-499. The program computes the values of χ^2, E_{ij} for i = 1, 2, ..., r and j = 1, 2, ..., c, and also $P(\chi^2_{(r-1)(c-1)} \geq \chi^2)$, the P-VALUE for the contingency table. The program has been run for the examples of this section.

Problems 9.3

1. Prove the following for an r x c contingency table:

    (a) $\sum_{j=1}^{c} E_{ij} = X_{i\cdot}$ (b) $\sum_{i=1}^{r} E_{ij} = X_{\cdot j}$

2. Show that in a 2 x 2 contingency table

$$\chi^2 = \frac{n(X_{11}X_{22} - X_{12}X_{21})^2}{X_{1\cdot}X_{2\cdot}X_{\cdot 1}X_{\cdot 2}}$$

3. (a) Assume that r = c in a contingency table and that $X_{ii} = n_i$ for i = 1, 2, ..., r, while $X_{ij} = 0$ for i ≠ j. Defining $n = n_i$, show that $\chi^2 = (r - 1)n$.

    (b) The value (r - 1)n is the maximum value of χ^2 in a r x r contingency table with $\Sigma_{i=1}^{r}X_{i\cdot} = \Sigma_{\alpha=1}^{r}X_{\cdot j} = n$. Explain in words why this is reasonable.

4. (a) Prove that in a 2 x c contingency table for which the rows are proportional, that is, $X_{1j} = kX_{2j}$, that $\chi^2 = 0$. This is clearly the smallest value that χ^2 can assume.

    (b) Show that in a 2 x c contingency table.

$$\chi^2 = \sum_{\alpha=1}^{c} \frac{(X_{ij} - X_{\cdot j}\hat{p}_{1\cdot})^2}{X_{\cdot j}\hat{p}_{1\cdot}(1 - \hat{p}_{1\cdot})}$$

    (c) Explain why the χ^2 test provides a test of H_0: $p_{11} = p_{12} = p_{13} = \cdots = p_{1c}$ in this case, where p_{1j} can be interpreted as the probability of falling into the jth classification in sampling from population 1 and p_{2j} has the same interpretation for population 2.

Exercises 9.3

Use the program CONTGY as needed in the following:

1. In a statistics exam we classify the scores according to two
    criteria. The first is whether an individual's score is below
    the class median m or not. The second is whether a student is
    majoring in a physical science or mathematics or is a nonscience
    major. We observe the following:

	$< m$	$\geq m$
Science	31	38
Nonscience	37	30

    Test the hypothesis that there is no association between test
    performance and type of major. (Use $\alpha = 0.05$.)

2. A survey was conducted to evaluate the effectiveness of a new
    flu vaccine, administered in a small community. Some persons
    appeared only for one shot. The following data were observed:

	No vaccine	*One shot*	*Two shots*
Flu	48	17	16
No flu	270	105	550

    (a) Evaluate the effectiveness of the vaccine.
    (b) Combine the last two column classifications so that these
         classifications become "no vaccine" and "some vaccine."
         Do the data support your conclusion in part (a)?

3. A group of college students is asked to answer a question with
    possible responses 1, 2, 3, or 4. The respondents were also
    identified by sex, and the following bivariate frequency table
    was constructed from the data.

Responses	*1*	*2*	*3*	*4*	*Total*
Males	12	33	17	8	70
Females	42	87	4	7	140
Total	54	120	21	15	210

Test the null hypothesis that the sex of the respondent and the response are independent. Use $\alpha = 0.05$.

4. In the June 1976 issue of the *Journal of Advertising Research*, in an article by J. F. Dash, L. G. Schiffman, and Conrad Berenson entitled "Information Search and Store Choice," the following data concerning customers at a speciality store and a department store are reported.

Store Choice and Knowledge of Audio Equipment

| | Type of store | |
Level of product knowledge	Specialty	Department
Low	21%	63%
Medium	30%	27%
High	49%	10%
Number	263	156

Does the assertion that "customers shopping at the specialty store are more knowledgeable about audio equipment" appear to be justified? Use $\alpha = 0.05$.

5. Farley S. Brothwaithe reports on the relationship between teachers' class background and their career aspirations in the March–June 1976 issue of the *International Journal of Comparative Sociology* in an article entitled "Upward Mobility and Career (Value) Orientation." He reports the following data.

Career aspirations	*"Lower class background"*	*"Upper class background"*
Low	157	79
High	212	47

Is there an association between class background and career aspirations for this group of teachers? How would you describe the association? Use $\alpha = 0.01$.

9.4   GOODNESS OF FIT

An important question in statistics and in the scientific method
generally is whether data can be assumed to come from a particular
parametric family such as the binomial, Poisson, normal, or exponential families. As an example, consider the following data, the
number of calls received at an exchange in 100 1-min periods.

Number of calls (x)	0	1	2	3	4
Number of 1-min periods in which x calls arrive	40	30	18	10	2

Can we proceed under the assumption that the number of calls arriving is described by the Poisson distribution? A statistical answer
to this question is provided by the large sample distribution of
multinomial data introduced in Sec. 9.2. Assuming the Poisson model
does apply, the probability of x calls in a 1-min period is given
by $e^{-\mu}(\mu)^x/x! = p(x)$, where μ is the theoretical average number of
calls arriving in a 1-min period. If we estimate μ by $\hat{\mu}$, then the
expected number of 1-min periods in which x calls arrive is given
by $100\ p(x) = 100\ e^{-\mu}(\mu)^x/x!$. These expected numbers under the
Poisson assumption can be used with the observed numbers in a χ^2
test of

H_0:   the data are Poisson

against the alternative that they are not.

To carry out the test we would estimate μ by $\hat{\mu} = (40(0) +
30(1) + 18(2) + 10(3) + 2(4))/100 = 1.04$. The values of $p(x)$ may be
obtained by using the program POISSON. We find the following:

x	$p(x)$	$100\ p(x)$	*Observed*
0	0.353455	35.3455	40
1	0.367593	36.7593	30
2	0.191148	19.1148	18
3	0.066265	6.6265	10
4 or more	0.021540	2.1540	2

We compute the value of $\chi^2 = 3.649$. The number of classes here is
five and one parameter has been estimated. The correct degrees of
freedom would thus be $5 - 1 - 1 = 3$. Using $\chi^2_{0.05,3} = 7.81$, a decision in favor of H_0 would be made, i.e., the Poisson assumption
would be retained. This decision means that the data observed do
not strongly contradict the Poisson assumption. The test by no means
"proves" the data are Poisson, as it may be consistent with other
assumptions also. It does indicate that the Poisson assumption is
not strongly violated.

The same general method of testing goodness of fit may be used
to test whether observations come from a particular continuous distribution, such as the normal. If $(X_1, X_2, \ldots, X_n)$ represents a
random sample of n observations from $X \sim N(\mu, \sigma^2)$, then consider the
10 intervals $(Z_{(j-1)0.10}, Z_{j(0.10)})$ for $j = 1, 2, \ldots, 10$. Here
$Z_0 = -\infty$, $Z_{0.90} = -1.2817$, $\ldots$, $Z_{0.90} = 1.2817$, and $Z_{1.0} = \infty$. We see
that

$$P(Z_{(j-1)0.1} < \frac{X_i - \mu}{\sigma} \leq Z_{j(0.1)}) = 0.1$$

for each of these 10 intervals. Of course, in general μ and σ are
unknown but may be estimated by $\bar{x}$ and s, respectively. One can test
the hypothesis of normality by finding the number of the n transformed
observations $(x_i - \bar{x})/s$ falling into each of these 10 intervals. If
we denote this number by f_j, then the statistic

$$\chi^2 = \sum_{j=1}^{10} \frac{(f_j - n/10)^2}{n/10}$$

can be used to test the normality assumption. The appropriate degrees
of freedom is $10 - 1 - (2 \text{ parameters}) = 7$.

The program NORMFIT carries out the test described in the previous paragraph. The program requests the value of n as input.
Data should appear in lines 400-599. The program computes the values
of f_j and the chi-square statistic. In the first sample run, the
first 100 "ages" of legislators from the data of Exercise 5.1.3 has
been tested for normality. The value of $\chi^2 = 9.8$ on 7 degrees of

```
NORMFIT
10 REM THIS PROGRAM CARRIES OUT A GOODNESS OF FIT TEST
20 REM TO THE NORMAL DISTRIBUTION
30 REM SAMPLE DATA SHOULD APPEAR IN LINES 400-599
40 REM THE PROGRAM COMPUTES THE NUMBER OF STANDARDIZED OBSERVATIONS
50 REM IN EACH OF 10 EQUAL PROBABILITY INTERVALS UNDER THE
60 REM NORMALITY ASSUMPTION AND THE CORRESPONDING VALUE
70 REM OF CHI-SQUARE
100 PRINT "WHAT IS THE VALUE OF N?"
110 INPUT N
120 FOR I = 1 TO 9
130 READ Z(I)
140 NEXT I
145 DIM X(500)
150 FOR I = 1 TO N
160 READ X(I)
170 LET S = S +X(I)
180 LET S1 = S1+X(I)↑2
190 NEXT I
200 LET A = S/N
210 LET S = SQR((S1-N*A↑2)/(N-1))
220 FOR I = 1 TO N
230 FOR J = 1 TO 9
240 IF (X(I)-A)/S > Z(J) THEN 260
250 LET C(J)= C(J)+1
255 GO TO 290
260 IF J < 9 THEN 280
270 LET C(10)=C(10)+1
274 GO TO 290
280 NEXT J
290 NEXT I
300 PRINT "FREQ TABLE ";"FOR TRANSFORMED OBSERVATIONS"
310 PRINT
320 PRINT "CLASS","FREQUENCY","EXPECTED NO."
330 FOR I = 1 TO 10
340 PRINT I,C(I),N/10
350 LET C2 = C2+(C(I)-N/10)↑2/(N/10)
360 NEXT I
370 PRINT "CHI-SQUARE = ";C2;" ON 7 DOF"
380 DATA -1.2817,-0.8418,-0.5244,-0.2547,0,0.2547,0.5244,0.8418,1.2817
400 DATA 46, 44, 45, 40, 46, 52, 47, 48, 42, 44
410 DATA 57, 36, 53, 34, 63, 24, 62, 46, 61, 43
420 DATA 50, 48, 45, 62, 56, 46, 46, 40, 44, 52
430 DATA 43, 39, 49, 47, 42, 53, 53, 43, 41, 47
440 DATA 46, 34, 38, 44, 43, 48, 48, 45, 41, 61
450 DATA 42, 56, 40, 49, 52, 44, 49, 48, 46, 61
460 DATA 61, 39, 48, 54, 49, 46, 51, 65, 51, 42
470 DATA 47, 42, 56, 45, 42, 35, 48, 51, 58, 37
480 DATA 45, 55, 36, 49, 59, 51, 52, 43, 40, 43
490 DATA 43, 43, 46, 47, 48, 57, 50, 44, 54, 44
600 END
*
```

```
   RUN
WHAT IS THE VALUE OF N?
? 100
FREQ TABLE FOR TRANSFORMED OBSERVATIONS

CLASS              FREQUENCY          EXPECTED NO.
   1                  8                 10
   2                  8                 10
   3                 14                 10
   4                 12                 10
   5                 14                 10
   6                 13                 10
   7                  6                 10
   8                  7                 10
   9                  6                 10
  10                 12                 10
CHI-SQUARE =   9.8   ON 7 DOF

   *
```

```
400  DATA -.302375,-0.867488,1.96662,0.309971,-0.251166
405  DATA 0.716976,-0.991966,-0.451398,1.20585,0.6902
410  DATA 1.04625,-3.64143,-1.03501,-0.941516,0.36757
415  DATA -1.06065,0.127918,-0.45544,-15.7685,-0.853627
420  DATA -0.503163,4.44066,-2.03545,0.299753,-0.626281
425  DATA 1.82172,-2.73089,0.182603,-5.69252,5.24803
430  DATA -97.8301,-.0206872,-2.62331,-9.60167,2.19374
435  DATA 0.654189,-11.5311,2.59715,0.714333,-0.416665
440  DATA -0.859218,-1.25742,-0.101315,1.11337,-0.729777
445  DATA 0.772719,-3.1132,-12.036,0.0224493,-23.6451
   *
```

```
   RUN
WHAT IS THE VALUE OF N?
? 50
FREQ TABLE FOR TRANSFORMED OBSERVATIONS

CLASS              FREQUENCY          EXPECTED NO.
   1                  2                  5
   2                  1                  5
   3                  2                  5
   4                  1                  5
   5                  2                  5
   6                 24                  5
   7                 16                  5
   8                  2                  5
   9                  0                  5
  10                  0                  5
CHI-SQUARE =   120   ON 7 DOF

   *
```

freedom. The value of $\chi^2_{0.05,7} = 14.07$ and $\chi^2_{0.10,7} = 12.02$. Hence there is no statistical evidence to reject normality at levels of significance $\alpha = 0.05$ or $\alpha \overset{\text{o}}{=} 0.10$. The same discussion of the meaning of deciding for the Poisson distribution in the previous example applies to the decision to "accept" normality in this case.

In a second example, we consider 50 observations from the Cauchy distribution obtained by using the program CAUCHY. The program NORMFIT has been used with these data. The computed value of chi-square is found to be 120. As $\chi^2_{0.01,7} = 18.48$, the evidence is clear that the observations do not come from a normal population. This would suggest, for instance, that distribution-free methods of estimating parameters and testing hypotheses should be used with such data.

Exercises 9.4

Use the computer as required.

1. Use NORMFIT to test the hypothesis that the following data come from a normal distribution.

1.84	2.8	1.14	4.69	0.38	3.32	1.05	22.07	0.74	2.21
9.5	0.79	2.81	1.2	3.41	6.55	9.01	0.98	4.42	2.12
1.1	2.05	2.73	4.11	1.95	6.68	1.65	11.41	19.81	2.42
2.69	9.32	11.2	1.38	0.14	9.38	6.88	1.36	1.68	1.85
1.05	1.5	1.88	4.57	8.69	0.81	1.09	0.21	1.22	6.28

2. Consider the exponential distribution with pdf

$$f(x;\lambda) = \begin{cases} \dfrac{1}{\lambda}e^{-x/\lambda} & \text{for } x > 0 \\ \\ 0 & \text{elsewhere} \end{cases}$$

(a) Use the data of Exercise 9.4.1 to estimate λ unbiasedly by $\hat{\lambda}$.

(b) Solve the equation $y = 1 - e^{-x/\hat{\lambda}}$ for x given $y = 0.1(0.1)(0.9)$ to find the endpoints of 10 equal probability intervals for the data of Exercise 9.4.1, assuming the exponential family is appropriate.

(c) Carry out a chi-square goodness-of-fit test for the data
of Exercise 9.4.1 to test the hypothesis that the data
come from an exponential distribution. What is the appropriate number of degrees of freedom? (Use $\alpha = 0.05$.)

3. The following are "classical" data collected by Borkiewicz[*],
giving the number of deaths due to kicks from a horse in 200
Prussian army corps years.

Number of deaths	0	1	2	3	4
Number of corps years	109	65	22	3	1

Can the assumption be retained that these data are described by
the Poisson distribution?

4. Since 1895 the record for the mile run until the middle of 1975
has been reduced 23 times. The magnitudes of the reductions are
classified below. Time is measured in seconds.

Interval	Frequency
$0.0 < t \leq 0.5$	9
$0.5 < t \leq 1.0$	4
$1.0 < t \leq 1.5$	2
$1.5 < t \leq 2.0$	5
$2.0 < t \leq 2.5$	2
$2.5 < t \leq 3.0$	1

Test the hypothesis that the data are uniformly distributed on
$(0,3.0)$. (Use $\alpha = 0.05$.)

5. Extend the program ERRORS to calculate the chi-square test
statistic for the hypothesis that the observations are from a
uniform distribution on $(-0.5, 0.5)$. Carry out the test for
$N = 1000$ for several different runs using the RANDOMIZE command.

---

[*]M. Fisz, *Probability Theory and Mathematical Statistics*, Wiley,
New York, 1963, pp. 141-142.

SIMPLE LINEAR REGRESSION AND ASSOCIATION

10.1   INTRODUCTION (THE LEAST SQUARES LINE)

In almost every physical and social science it is of crucial importance to examine the relationship between variables. In many physical
sciences it has been possible to postulate a deterministic model to
describe the relationship between an independent variable and a dependent variable. For example, the distance traveled by a freely
falling body as a function of the independent variable time is given
by

$$s = f(t) = \frac{1}{2}gt^2$$

Here if s is measured in feet and t in seconds, then g is the accel-
eration constant equal to 32 ft/sec^2, approximately. This relationship is deterministic. For any value of t, there is a single value
of s given by this function. If t = 1, then s = f(1) = 16.   There
are many other situations in which a relationship exists between a
dependent variable and an independent variable, but the relationship
is not deterministic. As an example, we know that the weight of male
infants is related to the infants' age. Nevertheless, it is obviously
untrue that all 3-month-old male children have a given weight.   In
simple linear regression we wish to examine a relationship between
an independent real variable x and a dependent random variable Y.
Such a model is called *stochastic.*

In the simple linear regression model we assume that for every value of x, the expected value of Y can be written as a real-valued function of x. Further it is assumed that the particular function is linear in x. Hence

$$E(Y|x) = \alpha + \beta x$$

The line with unknown slope β and intercept α is called the *theoretical regression line*. In Fig. 10.1.1 we indicate that there is a population of possible values of Y for any value of x. For $x = x_1$, the expected value of Y is $\alpha + \beta x_1$ which in this case is less than the expected value of Y for $x = x_2$. In general, this relationship is indicated by the following:

$$E(Y_i) = \alpha + \beta x_i$$

For the example of infant male weights, the *expected* weight of x-month-old males is assumed to be a linear function of x.

One statistical problem is to obtain estimates of the unknown parameters α and β. In Fig. 10.1.2 we have plotted the weights in pounds of n = 5 male infants of different ages (measured in months) as given here:

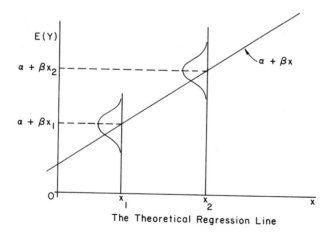

The Theoretical Regression Line

Fig. 10.1.1

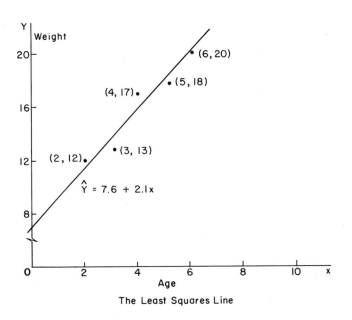

The Least Squares Line

Fig. 10.1.2

Infant(i)	Age(x_i)	Weight(y_i)
1	2	12
2	3	13
3	4	17
4	5	18
5	6	20

It is clear from Fig. 10.1.2 that there is no line through all five
points. We must choose a line which, according to some reasonable
criterion, fits all points well. The slope and intercept of this
line will be used to estimate α and β. A criterion which is often
used is to choose from all lines of form a + bx, the one for which
the quantity

$$\sum_{i=1}^{n} (y_i - (a + bx_i))^2 \qquad\qquad (10.1.1)$$

is least. The difference $y_i - (a + bx_i)$ is the vertical difference
between the observed value of Y at $x = x_i$ and the value of the arbitrary line $a + bx_i$ at $x = x_i$. There is a unique line called the
least squares line which minimizes expression (10.1.1). The intercept and slope of this line are given by

$$\hat{\alpha} = \bar{y} - \hat{\beta}\bar{x} \quad \text{and} \quad \hat{\beta} = \frac{\sum_{i=1}^n (x_i - \bar{x})(y_i - \bar{y})}{\sum_{i=1}^n (x_i - \bar{x})^2} \qquad (10.1.2)$$

We leave the proof of this to the problems. In the example here we
find $\bar{x} = 4$, $\bar{y} = 16$, $\sum_{i=1}^5 (x_i - \bar{x})(y_i - \bar{y}) = 21$, and $\sum_{i=1}^5 (x_i - \bar{x})^2 =$
10. Hence $\hat{\alpha} = 16 - 2.1(4) = 7.6$ and $\hat{\beta} = 21/10 = 2.1$. This line is
graphed in Fig. 10.1.2.

It is important to keep clearly in mind the fact that the line
in Fig. 10.1.1 is a theoretical line with intercept and slope satis-
fying $E(Y|x) = \alpha + \beta x$. The parameters α and β are unknown and after
estimation are still unknown, just as μ is unknown after estimation
by an observed value of $\bar{X}$. The least squares line in Fig. 10.1.2
is calculated from a single observation from each of the populations
of infant males of ages 2, 3, 4, 5, and 6 months, respectively. If
another such sample were taken, the observations $(2,y_1)$, $(3,y_2)$,
$(4,y_3)$, $(5,y_4)$, and $(6,y_5)$ would be different and the values of $\hat{\alpha}$
and $\hat{\beta}$ would be different. Hence the least squares line would also
be different. The estimates $\hat{\alpha}$ and $\hat{\beta}$ are random variables estimating
α and β, just as $\bar{X}$ is used to estimate μ. We turn to the statistical
properties of these estimators in the next section.

Problems 10.1

1. Suppose we observe the following pairs of observations (x_i, y_i),
   $i = 1, 2, \ldots, 5$: $(2,3)$, $(4,1)$, $(6,7)$, $(8,5)$, $(10,9)$. Find
   the equation of the least squares line.

2. Show that the least squares line defined by Eq. (10.1.2) passes
   through the point $(\bar{x}, \bar{y})$.

3. (a) Find the minimizing value of x for the quadratic function $px^2 + qx + r$ for $p > 0$.

   (b) Introduce the variable $a' = a + b\bar{x}$ in expression (10.1.1) and show that expression may be written as

$$\sum_{i=1}^{n} (y_i - a')^2 - 2b \sum_{i=1}^{n} (x_i - \bar{x})(y_i - \bar{y}) + b^2 \sum_{i=1}^{n} (x_i - \bar{x})^2$$

   (c) Show that the second two terms in part (b) are minimized by $\hat{\beta}$ as defined in Eq. (10.1.2), and this does not depend on the value of a'. Show that the first term is minimized by $a' = \bar{y}$, and hence $\hat{\alpha} = \bar{y} - \hat{\beta}\bar{x}$ and $\hat{\beta}$ minimize expression (10.1.1).

4. Assume that $E(Y|x) = \beta x$, i.e., the expected value of Y lies on a line through the origin. Find the equation of the least squares line through the origin based on a sample (x_i, y_i), $i = 1, 2, \ldots, n$.

5. In time series with an odd number of equally spaced time points, time is often considered an independent variable, and the values of a dependent random variable Y are measured at each time point. Assume time is coded as $-T$, $-(T - 1)$, $\ldots$, 0, $\ldots$, $T - 1$, T, and the observation at time t is denoted by y_t. Show that the intercept and slope of the least squares line are given by $\hat{\alpha} = \bar{y}$ and $\hat{\beta} = \sum_{t=-T}^{T} 3ty_t / T(T + 1)(2T + 1)$.

6. Use the following data on the civilian labor force in the United States, together with the previous problem, to find the least squares line coding 1969 as -2, etc.

Year	Civilian labor force(1000's)
1969	80,733
1970	82,715
1971	84,113
1972	86,542
1973	88,714

7. The following gives the scores of 10 persons on a verbal entrance
    examination and their grade point average (GPA) at the end of
    their freshman year. Use GPA as the dependent variable to find
    the least squares line given the following data:

    Verbal exam(x) 620 640 760 520 680 560 780 580 670 620
    GPA(Y)   3.28 3.16 3.40 2.88 3.64 3.24 3.80 2.80 3.54 2.96

10.2   INFERENCE IN THE LINEAR REGRESSION MODEL

10.2.1   *Statistical Properties of $\hat{\alpha}$ and $\hat{\beta}$*

The statistical properties of the estimators $\hat{\alpha}$ and $\hat{\beta}$ cannot be determined without further assumptions concerning the random variables
Y_i. The following assumptions, quite commonly made, permit inference concerning α and β.

1. $Y_i = \alpha + \beta x_i + e_i$ for real x_i, $i = 1, 2, \ldots, n$.
2. The $e_i \sim N(0, \sigma^2)$, $i = 1, 2, \ldots, n$.
3. The e_i and hence the Y_i are independent.

Assumptions 1 and 2 together imply that $E(Y_i) = \alpha + \beta x_i$ as before.
Notice that only e_i is random on the right-hand side of the equation
in 1. The same assumptions also imply that $Var(Y_i) = Var(e_i) = \sigma^2$
for all i, that is, the variance of the Y populations is the same
for all x. This is suggested by the symmetry of the density functions
in Fig. 10.1.1. The assumptions of normality of the e_i, and hence
of the Y_i, and the independence of the e_i, and hence of the Y_i, will
lead to confidence intervals and to tests of hypotheses for α and β.
The equation $Y_i = \alpha + \beta x_i + e_i$ means that the random variable e_i can
be thought of as a random error term to be added to the true expectation $\alpha + \beta x_i$ to yield values of Y_i. This is the reason for the
choice of notation e_i.

We will concentrate on the statistic $\hat{\beta}$, which estimates β. The
results for $\hat{\alpha}$ as an estimator of α are found analogously. The parameter β represents the theoretical average change in Y per unit increase in x. This clear from the fact that β is the slope of the

theoretical regression line (see Fig. 10.1.1). In the example β would represent the average change in weight per month for infant males. From Eq. (10.1.2) we have*

$$E(\hat{\beta}) = E\left[\frac{\Sigma (x_i - \bar{x})(Y_i - \bar{Y})}{\Sigma (x_i - \bar{x})^2}\right]$$

$$= \frac{\Sigma (x_i - \bar{x})(E(Y_i) - E(\bar{Y}))}{\Sigma (x_i - \bar{x})^2}$$

$$= \frac{\Sigma (x_i - \bar{x})(\alpha + \beta x_i - (\alpha + \beta\bar{x}))}{\Sigma (x_i - \bar{x})^2}$$

$$= \beta \left[\frac{\Sigma (x_i - \bar{x})^2}{\Sigma (x_i - \bar{x})^2}\right] = \beta$$

Thus $\hat{\beta}$ is an unbiased estimator of β. The variance of $\hat{\beta}$ may be found as follows:

$$Var(\hat{\beta}) = Var\left[\frac{\Sigma (x_i - \bar{x})Y_i}{\Sigma (x_i - \bar{x})^2}\right]$$

$$= \frac{\Sigma (x_i - \bar{x})^2 Var(Y_i)}{(\Sigma (x_i - \bar{x})^2)^2}$$

$$= \frac{\sigma^2}{\Sigma (x_i - \bar{x})^2} = \sigma_{\hat{\beta}}^2$$

The independence assumption has been used to write the variance of a sum as the sum of the variances. The fact that $\Sigma (x_i - \bar{x})(Y_i - \bar{Y}) = \Sigma (x_i - \bar{x})Y_i$ has also been used. Finally from Eq. (10.1.2) $\hat{\beta}$ is a sum of independent, normally distributed random variables. A theorem from mathematical statistics states that such a sum is again normally distributed. Hence

---

*All summations in the remainder of this chapter are for i = 1 to i = n, unless otherwise indicated.

$$Z = \frac{\hat{\beta} - \beta}{\sigma_{\hat{\beta}}} \qquad\qquad (10.2.1)$$

has the standard normal distribution.

Similar reasoning leads to the following results for α.

$$E(\hat{\alpha}) = \alpha \qquad Var(\hat{\alpha}) = \sigma^2 \left[\frac{1}{n} + \frac{\bar{x}^2}{\Sigma (x_i - \bar{x})^2} \right] = \sigma_{\hat{\alpha}}^2$$

and the random variable

$$Z = \frac{\hat{\alpha} - \alpha}{\sigma_{\hat{\alpha}}}$$

has the standard normal distribution also.

10.2.2 *Estimation of σ^2 and the ANOVA Table*

There remains one parameter of the model described in assumptions 1 through 3, namely σ^2, which has not been estimated. We define $\hat{Y}_i = \hat{\alpha} + \hat{\beta}x_i$, $i = 1, 2, \ldots, n$, which are referred to as the predicted values of the Y observations at $x = x_i$. It is left as a problem to show $\Sigma \hat{Y}_i/n = \bar{Y}$. We now introduce the concept of "partitioning the variability" of the Y observations. The following equality is valid and also left to the problems

$$\Sigma (Y_i - \bar{Y})^2 = \Sigma (\hat{Y}_i - \bar{Y})^2 + \Sigma (Y_i - \hat{Y}_i)^2 \qquad (10.2.2)$$

The left-hand side of this equation represents the "variation" of the Y observations about their average. The first term on the right-hand side represents the variation of the predicted Y observations about their average. This term is referred to as the *sum of squares due to regression* (SSR). The remaining term on the right in Eq. (10.2.2) is the sum of the squares of the (observed-predicted) Y values. The value $Y_i - \hat{Y}_i$ is referred to as the *ith residual*. The last term on the right is called the *sum of squares due to error* (SSE), or the sum of squares of the residuals. It is this sum of squares which is used to estimate σ^2 as we see in the following theorem.

THEOREM 10.2.1. Under the assumptions 1 through 3,

$$E\left[\frac{\Sigma (Y_i - \hat{Y}_i)^2}{n - 2}\right] = \sigma^2$$

Proof: We find the expectation of the first two sums of squares in in Eq. (10.2.2).

$$E(\Sigma (Y_i - \bar{Y})^2) = E(\Sigma (\alpha + \beta x_i + e_i - (\alpha + \beta\bar{x} + \bar{e}))^2)$$

$$= E(\beta^2 \Sigma(x_i - \bar{x})^2 + \Sigma (e_i - \bar{e})^2$$

$$+ 2\beta \Sigma (e_i - \bar{e})(x_i - \bar{x}))$$

$$= \beta^2 \Sigma (x_i - \bar{x})^2 + (n - 1)\sigma^2$$

using Sec. 5.3.1.

$$E(\Sigma (\hat{Y}_i - \bar{Y})^2) = E(\Sigma (\hat{\alpha} + \hat{\beta} x_i - \bar{Y})^2)$$

$$= E(\Sigma (\bar{Y} - \hat{\beta}\bar{x} + \hat{\beta} x_i - \bar{Y})^2) = E(\hat{\beta}^2 \Sigma (x_i - \bar{x})^2)$$

However as $\text{Var}(\hat{\beta}) = \sigma^2/\Sigma (x_i - \bar{x})^2 = E(\hat{\beta}^2) - \beta^2$, we find

$$E(\Sigma (\hat{Y}_i - \bar{Y})^2) = (\frac{\sigma^2}{\Sigma (x_i - \bar{x})^2} + \beta^2) \Sigma (x_i - \bar{x})^2$$

$$= \sigma^2 + \beta^2 \Sigma (x_i - \bar{x})^2$$

Taking expectations in Eq. (10.2.2) and canceling the common term, we have $(n - 1)\sigma^2 = \sigma^2 + E(\Sigma (Y_i - \hat{Y}_i)^2)$, from which the theorem follows directly. The statistic

$$\hat{\sigma}^2 = \frac{\Sigma (Y_i - \hat{Y}_i)^2}{n - 2}$$

is called the *mean square for error* (MSE) and is an unbiased estimate of σ^2.

Table 10.2.1   ANOVA Table

Source	Sum of squares	d.f.	MS	EMS
Regression	$\hat{\beta}^2 \Sigma (x_i - \bar{x})^2$	1	SSR	$\sigma^2 + \beta^2 \Sigma (x_i - \bar{x})^2$
Residual	$\Sigma (y_i - \hat{y}_i)^2$	$n - 2$	$\hat{\sigma}^2$	σ^2
Total	$\Sigma (y_i - \bar{y})^2$	$n - 1$		

[Incidentally the proof does not use the normality assumption, so that is only required that the e_i are independently distributed with $E(e_i) = 0$ and $Var(e_i) = \sigma^2$.]

The partitioning of the variation of the Y's and other related information is generally summarized in an analysis of variance (ANOVA) table. The format of the ANOVA table for the simple linear regression model is given in Table 10.2.1. The first column of the table gives a verbal description of the source of the variation of the corresponding sum of squares. The second column gives expressions for the sum of squares. It is left to the problems to show that the sum of squares for regression (SSR) can be written as $\hat{\beta}^2 \Sigma (x_i - \bar{x})^2$. The next column gives the degrees of freedom and the column labeled MS (mean square) is the quotient of the sum of squares divided by the appropriate degrees of freedom. The last column gives the expected values of the corresponding mean squares, which were derived in the proof of Theorem 10.2.1.

For the example of Sec. 10.1 we obtain the following table.

Source	SS	d.f.	MS	EMS
Regression	44.1	1	44.1	$\sigma^2 + 10\beta^2$
Residual	1.9	3	0.633	σ^2
Total	46	4		

The sum of squares for regression has been calculated as $\hat{\beta}^2 \Sigma (x_i - \bar{x})^2 = (2.1)^2(10) = 44.1$. The total sum of squares has been calculated as $\Sigma \, (y_i - \bar{y})^2 = 46$. The sum of squares of the residuals is generally computed as $\Sigma \, (y_i - \bar{y})^2 - \text{SSR} = 46 - 44.1 = 1.9$. The value of $\hat{\sigma}^2 = 1.9/3 = 0.633$ is the point estimate of σ^2.

10.2.3 Inference for α and β

As in Sec. 10.2.1 we shall concentrate on inference for β and indicate that analogous confidence intervals and tests can be found for the parameter α. We have seen in that section that the random variable still depends on the unknown parameter σ, as $\sigma_{\hat{\beta}} = \sigma / \sqrt{\Sigma \, (x_i - \bar{x})^2}$. It seems reasonable again to write $s_{\hat{\beta}} = \hat{\sigma} / \sqrt{\Sigma \, (x_i - \bar{x})^2}$ and consider the statistic

$$t = \frac{\hat{\beta} - \beta}{s_{\hat{\beta}}} \qquad (10.2.3)$$

The unknown parameter σ has been replaced by an estimate, the square root of the error mean square (MSE). It can be shown that under the assumptions 1 through 3 of this section that the random variable in Eq. (10.2.3) has Student's t distribution with $n - 2$ degrees of freedom. With this knowledge it is possible to construct confidence intervals and carry out tests for the unknown parameter β.

It is now straightforward to show that a $(1 - \alpha)100\%$ confidence interval for β has endpoints given by

$$\hat{\beta} \pm s_{\hat{\beta}} t_{\alpha/2,n-2}$$

In the example of this section we find a 95% confidence interval for the true average monthly weight gain to be given by $2.1 \pm \sqrt{0.633/10} \; t_{0.025,3}$ or by the interval $[1.299, 2.9012]$. Tests of the hypothesis H_0: $\beta = \beta_0$ versus the alternatives (1) $\beta > \beta_0$, (2) $\beta < \beta_0$, and (3) $\beta \neq \beta$ are based on the statistic $t = (\hat{\beta} - \beta_0)/s_{\hat{\beta}}$. The critical regions for α-level tests are found to be $t > t_{\alpha,n-2}$, $t < -t_{\alpha,n-2}$, and $|t| > t_{\alpha/2,n-2}$, respectively. Suppose we wish to test the hypotheses

$$H_0 \quad \beta \le 1.5 \qquad \text{versus} \qquad H_1 \colon \quad \beta > 1.5$$

for our example data. The critical region for a test with significance level 0.05 is given by t > $t_{0.05,3}$ = 2.353. As t =
(2.1 - 1.5)/0.2517 = 2.384, we would conclude that the true average
growth rate for male infants exceeds 1.5 lb/month. Similar confidence
intervals and tests for α (in the regression model) can be found by
replacing σ by $\hat{\sigma}$ in the expression for $\sigma_{\hat{\alpha}}$.

10.2.4   The Coefficient of Determination

A natural question to consider concerning the simple linear regression model is whether the model is actually satisfactory. We have
seen above that the relationship [see Eq. (10.2.3)]

$$\Sigma \ (y_i - \bar{y})^2 = SSR + SSE$$

is valid. Here SSR represents the sum of squares for regression and
SSE the sum of squares for error (or residual). The ratio r^2 =
SSR/$\Sigma \ (y_i - \bar{y})^2$ is called the *coefficient of determination*. As SSR
and SSE are nonnegative it is clear that $0 \leq r^2 \leq 1$. The statistic
r^2 can be interpreted as the "proportion of the variation in Y explained by regression." It is useful to consider the extreme cases
$r^2 = 1$ and $r^2 = 0$. If $r^2 = 1$, then SSE = $\Sigma \ (y_i - \hat{y}_i)^2 = 0$. Hence
$y_i = \hat{y}_i$ for i = 1, 2, ..., n, and since the $\hat{y}_i$ lie on a line, the
original observations do also.

   In the case that $r^2 = 0$, either $\hat{\beta} = 0$ or $\Sigma \ (x_i - \bar{x})^2 = 0$. The
later can occur only if we observe Y at one value of x. Clearly we
will not be able to make any statement about the change in Y per
unit change in x if we observe Y at one value of x. We exclude this
possibility. The situation in which $\hat{\beta} = 0$ implies that $\hat{y} = \hat{\alpha} = \bar{y}$.
This implies that knowledge of x does not assist in estimating E(Y).
For example, we would surely accept the hypothesis β = 0 as t = 0.
Thus we would retain the assumption E(Y|x) = α, i.e., the expected
value of Y does not depend *linearly* on x. This does not necessarily
imply that there is no relation between the observed values of Y and
x. Suppose we consider the (admittedly contrived) observations:
(-2,4), (-1,1), (0,0), (1,1), (2,4). We compute $\hat{\beta} = 0$ and $\hat{y} = 2$.

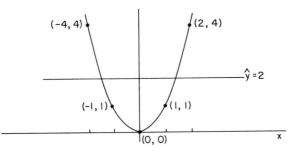

Fig. 10.2.1

As indicated in Fig. 10.2.1, the observations lie on the parabola $y = x^2$. However, no line fits the observed values well. Thus $r^2 = 0$ means that the data do not indicate a linear relationship of the form $E(Y|x) = \alpha + \beta x$, with $\beta \neq 0$. Hence the coefficient of determination is usually interpreted as a measure of the linearity of the relationship between Y and x.

10.2.5 A Program for the Linear Regression Model

The program LINREG computes the least squares line for n pointes (x_i, y_i) for $i = 1, 2, \ldots, n$. The data are placed in DATA statements in lines 500-599 in coordinate pairs. The program requests the number of points n. The program output includes the values of $\hat{\alpha}$, $\hat{\beta}$, $s_{\hat{\beta}}$, r^2, and $\hat{\sigma}$ and the t value for testing H_0: $\beta = 0$. In addition the analysis of variance table is printed out. The program has been run for the example of this section. As another example let us consider the data on the U. S. Gross National Product in Table 10.2.2. Considering time to be the independent variable coded as $1, 2, \ldots, 23$, where 1 represents 1950, we obtain the regression line $\hat{Y} = 309.86 + 19.487t$. The value of the coefficient of determination is $r^2 = 0.963$ indicating a substantial degree of linearity of real GNP with time. The annual growth of GNP in 1958 billions of dollars is estimated by $\hat{\beta} = 19.487$ with $s_{\hat{\beta}} = 0.837$. A 95% confidence interval for β, the "true" annual growth of GNP has endpoints $\hat{\beta} \pm s_{\hat{\beta}} t_{0.025, 21}$ and is given by the interval $[17.746, 21.228]$.

```
LINREG

5 DIM T(50)
10 REM THIS PROGRAM COMPUTES THE LEAST SQUARES LINE,
20 REM THE ANOVA TABLE, AND RELATED STATISTICS FOR
30 REM THE SIMPLE LINEAR REGRESSION MODEL
40 REM DATA PO INTS SHOULD APPEAR IN LINES 500-599 IN PAIRS
50 REM I.E. X(1),Y(1),X(2),Y(2),...,X(N),Y(N)
60 REM THE PROGRAM WILL REQUEST THE VALUE OF N.
70 PRINT "WHAT IS THE VALUE OF N?"
80 INPUT N
81 DATA 12.7 062,4.3027,3.1824,2.7764,2.5706
82 DATA 2.4469,2.3646,2.3060,2.2622,2.2281
83 DATA 2.2010,2.1788,2.1604,2.1448,2.1315
84 DATA 2.1199,2.1098,2.1009,2.0930,2.0860
85 DATA 2.0796,2.0739,2.0687,2.0639,2.0595
86 DATA 2.0555,2.0518,2.0484,2.0452,2.0423
87 DATA 2.0395,2.0369,2.0345,2.0322,2.0301
88 DATA 2.0281,2.0262,2.0244,2.0227,2.0211
89 DATA 2.0195,2.0181,2.0167,2.0154,2.0141
90 DATA 2.0129,2.0117,2.0106,2.0096,2.0086
91 FOR I = 1 TO 50
92 READ T(I)
93 NEXT I
100 FOR I = 1 TO N
110 READ X,Y
120 LET X1=X1+X              sum of x's
130 LET X2= X2+X↑2           sumsqp
140 LET Y1 = Y1 +Y           sum of y's
150 LET Y2 = Y2 +Y↑2         sumsqy
160 LET Z1= Z1+X*Y
170 NEXT I
180 LET M1=X1/N              mean of x
190 LET M2=Y1/N              mean of y
200 LET S1= X2-N*M1↑2
210 LET S2= Y2-N*M2↑2
220 LET C = Z1-N*M1*M2
230 LET B = C/S1
240 LET A = M2-B*M1
250 PRINT "VARIABLE","MEAN","STD DEV"
260 PRINT "X",M1,SQR(S1/(N-1))
270 PRINT "Y",M2,SQR(S2/(N-1))
280 PRINT
290 PRINT "FOR THE LEAST ";"SQUARES LINE"
300 PRINT "INTERCEPT","SLOPE"
310 PRINT A,B
320 PRINT
330 PRINT "ANOVA TABLE"
340 PRINT "SOURCE","S.S.","DOF","M.S."
350 LET S3= S2-(B↑2)*S1      Residual
360 PRINT "REGRESSION",(B↑2)*S1,N-(N-1),(B↑2)*S1
370 PRINT "RESIDUAL",S3,N-2,S3/(N-2)
380 PRINT "TOTAL",S2,N-1
390 PRINT
400 LET F = (B↑2)*S1*(N-2)/S3
410 PRINT "STD DEV OF B","T-VALUE","F VALUE"
420 LET T = SQR(F)
430 PRINT SQR(S3/(S1*(N-2))),T,F
440 PRINT "SIGMA-HAT","COEFFICIENT OF DET"
```

```
450 PRINT SQR(S3/(N-2)),(B↑2)*S1/S2
460 PRINT "SAMPLE CORRELATION COEFFICIENT R = ";B*SQR(S1/S2)
464 PRINT "DO YOU WISH AN ESTIMATE OF THE AVERAGE VALUE OF Y?"
468 PRINT "ANSWER 1 IF YES AND 0 IF NO"
471 INPUT Q
474 IF Q = 0 GO TO 700
477 PRINT "AT WHAT VALUE OF X?"
481 INPUT X1
484 PRINT "THE ESTIMATE IS ";A+B*X1
487 PRINT "A 95 PERCENT CONFIDENCE INTERVAL FOR THE AVERAGE Y IS"
491 LET V =SQR(S3/(N-2)*(1/N+(X1-M1)↑2/S1))
493 PRINT "L = ";A+B*X1-V*T(N-2);" U= ";A+B*X1+V*T(N-2)
497 GO TO 464
500 DATA 2,12,3,13,4,17,5,18,6,20
700 END
*
 RUN
WHAT IS THE VALUE OF N?
 5
 VARIABLE        MEAN              STD DEV
X               4                 1.58114
Y               16                3.39116

FOR THE LEAST SQUARES LINE
INTERCEPT       SLOPE
 7.6            2.1

ANOVA TABLE
SOURCE          S.S.              DOF          M.S.
REGRESSION      44.1              1            44.1
RESIDUAL        1.9               3            0.633333
TOTAL           46.0              4

STD DEV OF B    T-VALUE           F VALUE
 0.251661       8.34455           69.6316
SIGMA-HAT       COEFFICIENT OF DET
 0.795822       0.958696
SAMPLE CORRELATION COEFFICIENT R =   0.97913
DO YOU WISH AN ESTIMATE OF THE AVERAGE VALUE OF Y?
ANSWER 1 IF YES AND 0 IF NO
 0

*
```

```
500 DATA 1,355.3,2,383.4,3,395.1,4,412.8,5,407.0
510 DATA 6,438.0,7,446.1,8,452.5,9,447.3,10,475.9
520 DATA 11,487.7,12,497.2,13,529.8,14,551.0,15,581.1
530 DATA 16,617.8,17,658.1,18,675.2,19,706.6,20,725.6
540 DATA 21,725.5,22,745.4,23,790.7
*

 RUN
WHAT IS THE VALUE OF N?
? 23
VARIABLE          MEAN              STD DEV
X                 12                6.78233
Y                 543.7             134.702

FOR THE LEAST SQUARES LINE
INTERCEPT         SLOPE
 309.858          19.4869


ANOVA TABLE
SOURCE            S.S.              DOF            M.S.
REGRESSION        384295            1              384295.
RESIDUAL          14886.5           21             708.881
TOTAL             399181.           22


STD DEV OF B      T-VALUE           F VALUE
 0.836944          23.2833           542.114
SIGMA-HAT         COEFFICIENT OF DET
 26.6248           0.962707
SAMPLE CORRELATION COEFFICIENT R =   0.981177
DO YOU WISH AN ESTIMATE OF THE AVERAGE VALUE OF Y?
ANSWER 1 IF YES AND 0 IF NO
? 1
AT WHAT VALUE OF X?
? 24
THE ESTIMATE IS   777.542
A 95 PERCENT CONFIDENCE INTERVAL FOR THE AVERAGE Y IS
L =   753.678    U = 801.407
DO YOU WISH AN ESTIMATE OF THE AVERAGE VALUE OF Y?
ANSWER 1 IF YES AND 0 IF NO
? 0


*
```

Table 10.2.2   U. S. Gross National Product (1958 billions of dollars)[a]

Year	GNP	Year	GNP
1950	355.3	1961	497.2
1951	383.4	1962	529.8
1952	395.1	1963	551.0
1953	412.8	1964	581.1
1954	407.0	1965	617.8
1955	438.0	1966	658.1
1956	446.1	1967	675.2
1957	452.5	1968	706.6
1958	447.3	1969	725.6
1959	475.9	1970	725.5
1960	487.7	1971	745.4
		1972	790.7

[a]Economic Report of the President, 1974.

Problems 10.2

1. (a) For the data given in Problem 10.1.1, find the analysis of variance table and the values of r^2, $\hat{\sigma}^2$, and $s_{\hat{\beta}}$.

   (b) Test the hypothesis H_0: $\beta = 0$ under the assumptions 1 through 3.

2. Suppose that we have data on U. S. personal disposable income (x) and personal consumption (Y) over a period of 10 years. The least squares line is found to be

   $$\hat{Y} = 7.0 + 0.9x \quad \text{and} \quad s_{\hat{\beta}} = 0.1$$

   Both x and Y are measured in terms of constant 1954 dollars (billions).

   (a) Find a 95% confidence interval for β. Give a verbal description of the meaning of β.

   (b) If the average personal consumption was $295 (in constant 1954 billions), what was the average personal income (in constant 1954 billions)?

3. In a simple linear regression x represents the number of depend-
    ents declared by an individual on his income tax return and Y
    represents the income tax due (in thousands). A sample of 52
    such returns yields

$$Y = 2.5 - 0.4x \qquad \sum_{i=1}^{52} (x_i - \bar{x})^2 = 1600 \qquad \sum_{i=1}^{52} (y_i - \bar{y})^2 = 1506$$

(a) Construct the ANOVA table for these data.
(b) Find the value of the coefficient of determination r^2.
(c) Find a 95% confidence interval for β.
(d) Test the hypothesis: H_0: $\beta \geq -0.2$ versus H_1: $\beta < -0.2$
     at level of significance $\alpha = 0.05$.

4. The heights (in inches) of different tomato plants (Y) was
    observed after varying periods of time (x) after being trans-
    planted to the garden (3-9 weeks). The following summary data
    are available: $n = 10$, $\bar{x} = 6$, $\bar{y} = 12$,

$$\sum (x_i - \bar{x})^2 = 32 \qquad \sum (x_i - \bar{x})(y_i - \bar{y}) = 160$$
$$\sum (y_i - \bar{y})^2 = 1000$$

(a) Find the equation of the least squares line.
(b) What are the units β?
(c) Construct the ANOVA table and find r^2.

5. Replace $\hat{\alpha}$ by $\bar{Y} - \hat{\beta}\bar{x}$ in the expression for $\hat{Y}_i$ to prove that
    $\sum \hat{Y}_i/n = \bar{Y}$.

6. Show, using Problem 10.2.5, that $\sum (Y_i - \hat{Y}_i) = 0$. (The sum of
    the observed residuals is 0.)

7. (a) Writing $\sum (Y_i - \bar{Y})^2 = \sum (Y_i - \hat{Y}_i + \hat{Y}_i - \bar{Y})^2$ show that the
         cross product term $2 \sum (Y_i - \hat{Y}_i)(\hat{Y}_i - \bar{Y}) = 0$ by replacing
         $\hat{Y}_i$ by $\bar{Y} - \hat{\beta}(x_i - \bar{x})$.
    (b) Prove that the basic equality of the ANOVA table in Eq.
         (10.2.2) is correct.

8. (a) Using the fact that $(\bar{x}, \bar{y})$ lies on the least squares line,
         show that $\hat{\beta} = (\hat{y}_i - \bar{y})/(x_i - \bar{x})$ for $i = 1, 2, \ldots, n$.

(b) Prove that the sum of squares for regression (SSR) may be written as $\hat{\beta}^2 \Sigma (x_i - \bar{x})^2$.

9. (a) Using the ANOVA table (Table 10.2.1) show that the SSR and $\hat{\sigma}^2$ are both unbiased estimates of σ^2 if $\beta = 0$.

(b) What is the relationship between E(SSR) and σ^2 if $\beta \neq 0$ and the x_i are not all equal?

10. (a) If $\beta = 0$, the statistic $MSR/\hat{\sigma}^2 = F$ has the F distribution with 1 and n - 2 degrees of freedom under the assumptions of this section. A test at level α of H_0: $\beta = 0$ versus H_1: $\beta \neq 0$ can be made using the critical region F > $F > F(1, n - 2)_\alpha$.

(b) Show that $t^2_{0.025,3} = F(1,3)_{0.05}$.

(c) Show that the value of t for testing H_0: $\beta = 0$ versus H_1: $\beta \neq 0$ satisfies $t^2 = F$ as defined in part (a).

(d) Demonstrate numerically that the tests of H_0: $\beta = 0$ versus H_1: $\beta \neq 0$ are identical for the infant weight example of this section (for $\alpha = 0.05$). In fact, the square of a t variable with ν degrees of freedom (t_ν^2) has the same distribution as an $F(1,\nu)$ random variable, so these two tests are equivalent for any given value of α.

11. Show that $E(\hat{\alpha}) = \alpha$, using the fact that $E(\hat{\beta}) = \beta$.

12. In the Quarterly Review of Economics and Business (Spring 1976), Paul C. Nystrom and Clarke C. Johnson present interesting data in an article entitled "Labor's Share: New Evidence on an Old Controversy," The variable "labor's share" was the dependent variable and the time measured in years was the independent variable. The model used was $Y_t = \alpha + \beta t + e_t$. The following regression lines were found in the indicated industries.

Industry	Regression line	r^2
Food	$\hat{Y}_t = 0.7777 - 0.00246t$	0.31
Petroleum	$\hat{Y}_t = 0.5598 - 0.00815t$	0.49
Machinery	$\hat{Y}_t = 0.7138 + 0.00349t$	0.08

(a) In this case, t = 1, 2, 3, ..., 18. Estimate labor's
     share in each of the industries at time t = 20.

(b) Using the fact that the t statistic may be written as
     $t = (r/\sqrt{1 - r^2})\sqrt{n - 2}$ with r having the sign of $\hat{\beta}$, test
     the hypothesis H_0: $\beta = 0$ versus H_1: $\beta \neq 0$ for each
     industry. ($\alpha = 0.05$.)

(c) In which of these three industries is there statistically
     significant evidence that labor's share is changing?

Exercises 10.2

1. The following table gives the total U. S. farm employment for
    1950-1972 in thousands*.

Year	Farm employment	Year	Farm employment
1950	9926	1962	6700
1951	9546	1963	6518
1952	9149	1964	6110
1953	8864	1965	5610
1954	8651	1966	5214
1955	8381	1967	4903
1956	7852	1968	4749
1957	7600	1969	4749
1958	7503	1970	4596
1959	7342	1971	4523
1960	7057	1972	4373
1961	6919		

(a) Find the equation of the least squares line using the pro-
     gram LINREG (code 1950 as 0, etc.).

(b) What is the value of the coefficient of determination.

(c) Find a 95% confidence interval for β, the average change
     per year in total U. S. farm employment over this period.

---

*Source: Economic Report of the President, 1974.

2. The following table gives the per capita disposable income in the United States from 1950-1972 in constant dollars*.

Year	Per capita disposable income	Year	Per capita disposable income
1950	1646	1962	1969
1951	1657	1963	2015
1952	1678	1964	2126
1953	1726	1965	2239
1954	1714	1966	2335
1955	1795	1967	2403
1956	1839	1968	2486
1957	1844	1969	2534
1958	1831	1970	2610
1959	1881	1971	2680
1960	1883	1972	2767
1961	1909		

(a) Find the equation of the least squares line coding time as in Exercise 10.2.1.

(b) What is r^2?

(c) Find a 95% confidence interval for β, the average change per year in per capita income in the United States over this period.

10.3   PREDICTION USING THE REGRESSION MODEL

We consider here the prediction of the average value of Y at a given level of x. The natural predictor of the expectation $\mu_{Y|x}$ is given by

$$\hat{\mu}_{Y|x} = \hat{\alpha} + \hat{\beta}x \qquad (10.3.1)$$

Under the model assumptions made in the previous section we see that $E(\hat{\mu}_{Y|x}) = E(\hat{\alpha} + \hat{\beta}x) = \alpha + \beta x$, so that the estimator is unbiased. The

---

*Source: Economic Report of the President, 1974.

value of x at which the prediction is made may be a value at which
Y has not been observed. The statistic given by Eq. (10.3.1) is,
of course, a random variable. Under the assumptions of the previous
section we can find the variance of this estimator as follows:

$$\text{Var}(\hat{\alpha} + \hat{\beta}x) = \text{Var}(\bar{Y} + \hat{\beta}(x - \bar{x}))$$
$$= \text{Var}(\bar{Y}) + \text{Var}(\hat{\beta})(x - \bar{x})^2$$
$$= \frac{\sigma^2}{n} + \frac{\sigma^2(x - \bar{x})^2}{\Sigma \ (x_i - \bar{x})^2}$$

The independence of $\bar{Y}$ and $\hat{\beta}$ has been used here but will not be proved.
Again $\mu_{Y|x}$ is a normally distributed variable, so that

$$\frac{\hat{\mu}_{Y|x} - (\alpha + \beta x)}{\sigma \sqrt{1/n + (x - \bar{x})^2 / \Sigma \ (x_i - \bar{x})^2}}$$

has the standard normal distribution. It is also true that if σ is
replaced by $\hat{\sigma}$, then

$$t = \frac{\hat{\mu}_{Y|x} - \mu_{Y|x}}{\hat{\sigma} \sqrt{1/n + (x - \bar{x})^2 / \Sigma \ (x_i - \bar{x})^2}}$$

has Student's t distribution with n - 2 degrees of freedom.

As usual a confidence interval for $\mu_{Y|x}$ is more informative than
a point estimate. The usual $(1 - \alpha)100\%$ confidence interval for
$\mu_{Y|x}$ is given by

$$\hat{\alpha} + \hat{\beta}x \pm \hat{\sigma} \sqrt{\frac{1}{n} + \frac{(x - \bar{x})^2}{\Sigma \ (x_i - \bar{x})^2}} \ t_{\alpha/2, n-2} \qquad (10.3.2)$$

It is important to note that for a fixed value of the confidence
coefficient, $(1 - \alpha)100\%$, and observations (x_i, y_i), i = 1, 2, ..., n,
the length of these intervals varies. It is least for x = $\bar{x}$ and
increases as x gets further away from $\bar{x}$. In Fig. 10.3.1 we consider
again the data for infant weights. The endpoints of 95% confidence
intervals are graphed for various values of x. It is clearly more
risky to estimate $\mu_{Y|x}$ at values of x far away from $\bar{x}$ than close to

```
LINREG

RUN
WHAT IS THE VALUE OF N?
? 5
VARIABLE          MEAN            STD DEV
X                 4               1.58114
Y                 16              3.39116

FOR THE LEAST SQUARES LINE
INTERCEPT         SLOPE
 7.6              2.1


ANOVA TABLE
SOURCE            S.S.            DOF              M.S.
REGRESSION        44.1            1                44.1
RESIDUAL          1.9             3                0.633333
TOTAL             46              4


STD DEV OF B     T-VALUE        F VALUE
 0.251661         8.34455         69.6316
SIGMA-HAT         COEFFICIENT OF DET
 0.795822         0.958696
SAMPLE CORRELATION COEFFICIENT R =   0.97913
DO YOU WISH AN ESTIMATE OF THE AVERAGE VALUE OF Y?
ANSWER 1 IF YES AND 0 IF NO
? 1
AT WHAT VALUE OF X?
? 7
THE ESTIMATE IS  22.3
A 95 PERCENT CONFIDENCE INTERVAL FOR THE AVERAGE Y IS
L =   19.6438   U=   24.9562
DO YOU WISH AN ESTIMATE OF THE AVERAGE VALUE OF Y?
ANSWER 1 IF YES AND 0 IF NO
? 0


*
```

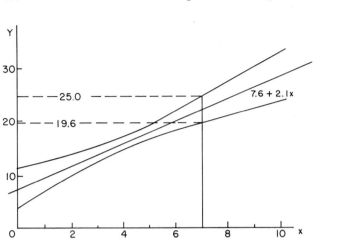

95 Percent Confidence Intervals for Average Weight

Fig. 10.3.1

$\bar{x}$. For x = 7, we find $\hat{\mu}_{Y|7}$ = 7.6 + 2.1(7) = 22.3. The endpoints of a 95% confidence interval for the average weight of 7-month-old male infants are given by expression (10.3.2) as 22.3 $\pm$ $\sqrt{0.6333(1/5 + 9/10)}$ $t_{0.025,3}$. The interval is [19.64, 24.96]. The program LINREG has been written to estimate the expectation of Y for a given x and to provide a 95% confidence interval for the estimate.

We indicate here, by example, two causes of erroneous predictions using the linear regression model. Let us again consider the example of infant weights. The regression line was found to be Y = 7.6 + 2.1x. If we estimate E(Y|x = 100), we find $\hat{\mu}_{Y|100}$ = 217.6 lb for an 8 1/3-year-old child. The data which produced this prediction were taken over the period of the first 6 months after birth. While the linear model may be appropriate over a short period of time (say up to 10 months), growth is not linear over a long period of time. The model is hence deficient for long-term predictions. Consider the example of relating GNP to time. The least squares equation found in the previous section, $\hat{Y}$ = 309.86 + 19.487t, would predict GNP for 1973 (t = 24) to be $777.55B. The actual GNP for 1973 was $839.2B (Source: *Survey of Current Business*, February 1975).

This estimate is again unsatisfactory. With economic time series, predictions based on long historical data are often not as accurate as those which weight more heavily the most recent past. Even with these methods, unforeseen economic events can (and do) lead to erroneous forecasts. One should generally limit the use of the simple linear regression model for forecasting to cross-sectional data (i.e., not a time series) and for values of x not too far from $\bar{x}$. Elaborations of the simple linear regression model and other methods are more appropriate for time series data.

Problems 10.3

1. Using the information given in Problem 10.2.3 with the additional information that $\bar{x}$ = 2.3, find the predicted average income tax due for an individual with two dependents. Find a 95% confidence interval for this prediction.

2. Using the information in Problem 10.2.4, find the predicted average height of tomato plants 8 weeks after being transplanted and also a 95% confidence interval for this prediction.

3. Use LINREG with the data on infant weights of Sec. 10.1 to pre-dict the weight of 6-month-old infants.

10.4 ASSOCIATION AND CORRELATION

In the previous sections of this chapter we have considered that the expected value of a random variable Y was a function of a real variable x, that is, $E(Y|x) = f(x)$. It is quite often the case that we observe the values of a bivariate random vector (X,Y) at n points, (x_i, y_i) i = 1, 2, ..., n. However it is not possible to state which random variable is dependent and which is independent. For example, we may observe

a. The height and weight of n individuals.
b. The scores on a verbal test and a mathematics test for n persons.
c. The inventory turnover (cost of goods sold/value of inventory) and the earnings as a percent of sales of n companies.

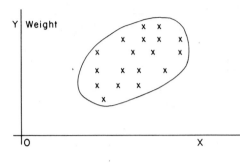

Fig. 10.4.1

We are aware that the random variables X and Y are often "correlated" or "associated." In example a. we would assume that higher weights and heights would occur together and lesser weights and heights would occur together. In Fig. 10.4.1 we illustrate this graphically. We would like to have a statistical measure of the intensity of the association between two random variables.

10.4.1   The Correlation Coefficient

We shall use the following notation in this section: $E(X) = \mu_x$, $E(Y) = \mu_y$, $Var(X) = \sigma_x^2$, and $Var(Y) = \sigma_y^2$. We now consider measures of the population association between X and Y. We have previously (in Problem 3.2.14) introduced the covariance of X and Y which is defined by

$$Cov(X,Y) = E((X - \mu_x)(Y - \mu_y)) \qquad (10.4.1)$$

If large values of X and Y occur together and small values of X and Y occur together the product $(X - \mu_x)(X - \mu_y)$ will tend to be positive and so will its expectation. We denote $Cov(X,Y)$ by σ_{xy} and present several properties of the covariance in the following theorem. The proof is left to the problems.

THEOREM 10.4.1. If σ_{xy} is defined, then

a. $\sigma_{xy} = E(XY) - \mu_x\mu_y$.

b. $\sigma_{xy} = \sigma_{yx}$.

c. $\text{Cov}(aX,bY) = ab\sigma_{xy}$ for any constants a and b.

d. $\sigma_{xx} = \sigma_x^2$.

e. $\text{Cov}(X, Y + Z) = \text{Cov}(X,Y) + \text{Cov}(X,Z)$.

It is clear from part c of this theorem that the covariance itself cannot be used as a measure of association for if one of the random variables, say X is multiplied by a constant, then σ_{xy} is also multiplied by the same constant. We would like to have a measure of association which is free of the units of measurement.

The population parameter which is often used as a measure of the association between X and Y is the theoretical correlation coefficient between X and Y defined by

$$\rho = \frac{\sigma_{xy}}{\sigma_x \sigma_y}$$

It is easy to see that ρ is free of the units of measurement of X and Y. Another important property of ρ is given in the following theorem.

THEOREM 10.4.2. If ρ is defined for two random variables X and Y, then

$$-1 \le \rho \le 1$$

Proof: Consider the standardized random variables $X' = (X - \mu_x)/\sigma_x$ and $Y' = (Y - \mu_y)/\sigma_y$. Both X' and Y' have expectation 0 and variance 1. Clearly $\rho = E(X'Y')$. Consider $E(X' + Y')^2$. We have

$$E(X' + tY')^2 = E(X')^2 + 2E(X'Y')t + t^2(E(Y'))^2$$
$$= 1 + 2\rho t + t^2$$

for any real t. However this quadratic in t is nonnegative for all t as it is the expectation of a nonnegative random variable. Hence the discriminant of this quadratic cannot be positive and $4\rho^2 - 4 \leq 0$ or $\rho^2 \leq 1$, which is equivalent to $-1 \leq \rho \leq 1$.

It is natural to ask for a statistic, which can be calculated from a random sample of n independent observations (x_i, y_i), i = 1, 2, ..., n, from the bivariate population (X,Y) to *estimate* ρ. If we estimate the covariance σ_{xy} by

$$\hat{\sigma}_{xy} = \frac{\Sigma (x_i - \bar{x})(y_i - \bar{y})}{n - 1}$$

then a natural estimator of ρ is

$$r = \frac{\hat{\sigma}_{xy}}{s_x s_y} = \frac{\Sigma (x_i - \bar{x})(y_i - \bar{y})}{\sqrt{\Sigma (x_i - \bar{x})^2 \Sigma (y_i - \bar{y})^2}} \tag{10.4.2}$$

The statistic r is called the *sample correlation coefficient*. The value of r is in fact numerically equal to the square root of the coefficient of determination, with the sign of $\hat{\beta}$ prefixed, calculated in the regression of Y on x. We can see this by writing

$$r = \frac{\Sigma (x_i - \bar{x})(y_i - \bar{y})}{\Sigma (x_i - \bar{x})^2} \frac{s_x}{s_y} = \frac{\hat{\beta} s_x}{s_y} \tag{10.4.3}$$

Hence

$$r^2 = \frac{\hat{\beta}^2 s_x^2}{s_y^2} = \frac{\hat{\beta}^2 \Sigma (x_i - \bar{x})^2}{\Sigma (y_i - \bar{y})^2} = \frac{SSR}{\Sigma (y_i - \bar{y})^2}$$

The last expression is the coefficient of determination. Hence r, the sample correlation coefficient is the square root of the coefficient of determination, and by Eq. (10.4.3), r has the sign of $\hat{\beta}$.

Inference for ρ may only be based on the knowledge of the distribution of the bivariate random variable (X,Y). If (X,Y) has a bivariate normal distribution*, it can be shown that

$$E(Y|x) = \mu_y + \rho \frac{\sigma_y}{\sigma_x}(x - \mu_x)$$

The expectation of Y given x is then linear in x with slope $\rho = \beta(\sigma_y/\sigma_x)$. It is clear that $\rho = 0$ if and only if $\beta = 0$. Hence it should not be too surprising that, in the bivariate normal case, a test of $\rho = 0$ is based on the same (numerical) t statistic used to test $\beta = 0$. It can be shown that in the bivariate normal case

$$t = \frac{r}{\sqrt{1 - r^2}} \sqrt{n - 2} \tag{10.4.4}$$

has Student's t distribution with $n - 2$ degrees of freedom if $\rho = 0$. Hence against alternatives (1) $\rho < 0$, (2) $\rho > 0$, and (3) $\rho \neq 0$, we have the critical regions $t < -t_{\alpha,n-2}$, $t > t_{\alpha,n-2}$, and $|t| > t_{\alpha/2,n-2}$, where the statistic t is defined in Eq. (10.4.4).

We consider a small example to illustrate the computation of r and a test of $\rho = 0$. Assume that we obtain the following data from a sample of five firms.

Company	Cost of goods sold/inventory	Earnings as a percent of sales
1	3	10
2	4	8
3	5	12
4	6	14
5	7	16

We find $\bar{x} = 5$, $\bar{y} = 12$, $\Sigma (x_i - \bar{x})(y_i - \bar{y}) = 18$, $\Sigma (x_i - \bar{x})^2 = 10$, and $\Sigma (y_i - \bar{y})^2 = 40$. Hence $r = 18/\sqrt{400} = 0.9$. We consider a test of H_0: $\rho = 0$ versus H_1: $\rho > 0$ assuming bivariate normality. We

---

*The definition requires the multivariate calculus, see Draper and Smith (1966).

calculate $t = (0.9/\sqrt{1 - 0.81})\sqrt{3} = 3.576$. As $t_{0.05,3} = 2.132$, we
would decide that $\rho > 0$ at the 5% level of significance. There is
an apparent positive correlation between capital utilization and
profits.

A confidence interval for ρ can be found by using a transformation of r, namely $z = T(r) = 0.5 \log((1 + r)/(1 - r))$, called *Fisher's
z* after Sir Ronald Fisher, the famous British statistician. This
statistic has approximately a normal distribution with $E(z) =$
$0.5 \log((1 + \rho)/(1 - \rho))$ and $Var(z) = 1/(n - 3)$. Hence, approximately, $P(z - 1.96/\sqrt{n - 3} \le 0.5 \log((1 + \rho)/(1 - \rho)) \le z +$
$1.96/\sqrt{n - 3}) = 0.95$. The transformation $T(r)$ is monotonic so that
$P(T^{-1}(z - 1.96/\sqrt{n - 3}) \le \rho \le T^{-1}(z + 1.96/\sqrt{n - 3})) = 0.95$, where
T^{-1} is the inverse transformation of T. The program COR has been
written to compute the correlation coefficient r and to calculate
an approximate 95% confidence interval for ρ.

10.4.2 A Distribution-free Measure of Association

In this section we shall again assume that we are sampling from a
continuous bivariate distribution (X,Y). Assume that (X_1,Y_1) and
(X_2,Y_2) represent two independent observations from this distribution. We define a parameter which reflects association between X
and Y as follows. First let us define

$$\pi_c = P(X_1 > X_2 \text{ and } Y_1 > Y_2) + P(X_1 < X_2 \text{ and } Y_1 < Y_2)$$

This parameter represents the probability that two independent observations from (X,Y) are "concordant." Using the continuity assumption, we define $\pi_d = 1 - \pi_c$ as the probability that two independent
observations from (X,Y) are "discordant." As a measure of association we define $\tau = \pi_c - \pi_d = 2\pi_c - 1$. Note that if X and Y are
independent, then $\pi_c = 1/2$ (why?) and $\tau = 0$. Clearly as $0 \le \pi_c \le 1$,
$-1 \le \tau \le 1$.

As an estimator of τ based on a sample of n independent observations (X_i,Y_i), i = 1, 2, ..., n, we use

$$t = \sum_{i=1}^{n-1} \sum_{j=i+1}^{n} \frac{\text{sgn}(X_i - X_j)\,\text{sgn}(Y_i - Y_j)}{\binom{n}{2}}$$

Assuming that X and Y are continuous variables, the probability $X_i = X_j$ for any pair (i,j) is 0 and similarly for Y_i and Y_j. Hence $\text{sgn}(X_i - X_j)\,\text{sgn}(Y_i - Y_j) = \pm 1$. The value is +1 if the ith and jth pairs are concordant and -1 otherwise. Hence each of the $\binom{n}{2}$ terms in the sum has expectation $\pi_c - \pi_d$ and $E(t) = \tau$. If X and Y are independent, $E(t) = 0$. Similarly, under the independence assumption, we can find $\text{Var}(t) = (4/9)(n + 5/2)/n(n - 1)$. A test of $H_0: \tau = 0$ can be made for small n using the tables available in the *Handbook of Tables for Probability and Statistics* (1966). These tables give the percentile points for the statistic t under the hypothesis of independence for n = 4 to 10. For larger values of n, one can use

$$Z = \frac{t}{\sqrt{\text{Var}(t|H_0)}}$$

For the alternatives $\tau > 0$, $\tau < 0$, and $\tau \neq 0$, the usual critical regions are appropriate.

More than a word of caution is required about the interpretation of a "statistically significant" correlation between two observed random variables. Such a correlation need not imply a causal relationship. For example, the length of feet of children is correlated with their ability to do arithmetic problems, but this is due to the fact that with increasing age, both the length of feet and arithmetical ability increase. Often common trends in time of two time-series data yield a "spurious" correlation. The value of alcoholic consumption and the value of razor blades sold may appear correlated due to a common inflationary trend. On the other hand, in many cases, the evidence of statistical correlation may, in fact, imply a true association between two random variables.

Problems 10.4

1. Prove Theorem 10.4.1 using the definition

$$Cov(X,Y) = E((X - \mu_x)(Y - \mu_y))$$

2. Prove that if two random variables are linearly related, that is,
    $Y = aX + b$, then $\rho = 1$ if $a > 0$ and $\rho = -1$ if $a < 0$.

3. In a correlation analysis we find $\Sigma (x_i - \bar{x})^2 = 100$, $\Sigma (y_i - \bar{y})^2 = 64$, and $\Sigma (x_i - \bar{x})(y_i - \bar{y}) = 48$ with $n = 18$.
    (a) Find the value of r.
    (b) Test the hypothesis H_0: $\rho = 0$ versus H_1: $\rho \neq 0$ assuming
         bivariate normality. (Use $\alpha = 0.05$.)

4. Suppose in a sample of 29 adult males we measure the height X
    in inches and weight Y in pounds of each individual. We find
    $\Sigma (x_i - \bar{x})^2 = 100$, $\Sigma (y_i - \bar{y})^2 = 400$, and $\Sigma (x_i - \bar{x})(y_i - \bar{y}) = 100$.
    (a) Find the value of r.
    (b) Carry out the test of Problem 10.4.3 part (b).

5. Suppose that two judges rank seven contestants in a diving
    competition. Test the hypotheses

    H_0: the judges rankings are independent.
    H_1: positive association exists between the rankings.

Contestant	A	B	C	D	E	F	G
First judge	2	1	4	5	3	7	6
Second judge	3	4	2	5	1	6	7

    (Use the statistic t and $\alpha = 0.07$.) Note $P(t > 0.524) = 0.07$
    if $n = 7$.

6. In the absence of ties the statistic t is based on the ranks of
    the X observations and Y observations. It is clear from the
    definition of t that if the ranks of the X observations and Y
    observations replace the observations themselves, the value of
    t is not altered.

(a) Use the 24 possible permutations of the Y ranks together
with the X ranks 1, 2, 3, 4 to find the null distribution
of t for n = 4.

(b) Verify that E(t) = 0 and Var(t) = $(4/9)(n + 5/2)/n(n - 1)$,
in this case.

7. Lawrence G. Hrebiniak has studied the relationship between
hospital size (measured by number of beds) and degree of professionalism (measured by man-hours of work done by certain
professional categories). His work appears in *The Academy of
Management Journal*, December 1976, in an article entitled "Size
and Staff Professionalism." He reports the following correlations between size and staff professionalism.

Staff	*Public*	*Private (profit)*	*Private (nonprofit)*
Psychiatrists	0.790	0.421	0.395
Nurses (RN's)	0.762	0.781	0.888
Administrative staff	0.340	0.165	0.390

The number of public hospitals was 47, private (profit) hospitals
was 76, and private (nonprofit) hospitals was 172.

(a) Test whether each of the calculated correlations implies
that H_0: $\rho = 0$ should be rejected in favor of H_1: $\rho > 0$
at significance level $\alpha = 0.05$.

(b) Find 95% confidence intervals for the theoretical correlation coefficients.

8. In an article entitled "Higher Education Tax Allowances: An
Analysis," by Larry L. Leslie in the *Journal of Higher Education*
(September/October 1976), the following correlations are reported.

Simple Correlation Coefficients between Appropriations and
Sociometric Variables

	Appropriations to all public institutions per capita	
	1960	*1969*
Industrialization	−0.57	−0.28
Affluence	0.51	0.48
Median school years	0.50	0.48
College educated percent	0.28	0.40
Personal income	0.03	0.17
Corporate income	0.02	0.26
College age population	0.09	0.27

The study involved observations from the 50 states. Which correlations appear significant at $\alpha = 0.05$ level? Were there significant correlations for any sociometric variables in 1960 which were not significant in 1969?

Exercises 10.4

1. For 17 Miss America winners from 1959–1975 we give their weight and height.

Year	Height	Weight	Year	Height	Weight
1959	65	114	1968	69	135
1960	67	120	1969	67	125
1961	67	116	1970	65.5	110
1962	65.5	118	1971	68	121
1963	65	115	1972	67	118
1964	66.5	124	1973	68	120
1965	66	124	1974	69	125
1966	67	115	1975	68	119
1967	66	116			

(a) Use COR to find the value of r.

(b) Use the confidence interval to test $\rho = 0$ versus $\rho = 0$ at level $\alpha = 0.05$.

```
COR

10 REM THIS PROGRAM COMPUTES THE PEARSON CORRELATION COEFFICIENT
20 REM BASED ON A SAMPLE OF N POINTS FROM A BIVARIATE POPULATION.
30 REM A 95 PERCENT CONFIDENCE INTERVAL FOR RHO IS FOUND
40 REM USING FISHER'S APPROXIMATION.
45 REM THE DATA SHOULD APPEAR IN LINES 500-699 IN PAIRS,
50 REM I.E. X(1),Y(1),X(2),Y(2),...,X(N),Y(N)
60 REM THE PROGRAM WILL REQUEST THE VALUE OF N.
70 PRINT "WHAT IS THE VALUE OF N?"
80 INPUT N
100 FOR I = 1 TO N
110 READ X,Y
120 LET X1=X1+X
130 LET X2= X2+X↑2
140 LET Y1 = Y1 +Y
150 LET Y2 = Y2 +Y↑2
160 LET Z1= Z1+X*Y
170 NEXT I
180 LET M1=X1/N
190 LET M2=Y1/N
200 LET S1= X2-N*M1↑2
200 LET S2= Y2-N*M2↑2
220 LET C = Z1-N*M1*M2
240 LET R = C/SQR(S1*S2)
400 LET U = 0.5*LOG((1+R)/(1-R)) + 1.96*SQR(1/(N-3))
410 LET L = 0.5*LOG((1+R)/(1-R)) -1.96*SQR(1/(N-3))
420 LET U = (EXP(2*U)-1)/((1+EXP(2*U)))
430 LET L = (EXP(2*L)-1)/((1+EXP(2*L)))
440 PRINT " THE VALUE OF R="; R
445 PRINT
450 PRINT "A 95 PERCENT CONFIDENCE INTERVAL FOR RHO IS GIVEN BY:"
460 PRINT "L = ";L,"U = ";U
500 DATA 2,5,3,5,6,5,4,3,5,4,3,1,1,2,4,4,5,5,6,6
700 END
*RUN
WHAT IS THE VALUE OF N?
? 5
 THE VALUE OF R= 0.8

A 95 PERCENT CONFIDENCE INTERVAL FOR RHO IS GIVEN BY:
L = -0.279664  U =  0.986197

*
```

2. (a) The data below* give the 1970 rank of the states by popula-
         tion and by land area. Find the value of r and test H_0:
         $\rho = 0$ versus H_1: $\rho \neq 0$ using the Student t statistic.
         (Use $\alpha = 0.05$.)

State	Rank by population	Rank by area	State	Rank by population	Rank by area
Alabama	21	29	Montana	43	4
Alaska	50	1	Nebraska	35	15
Arizona	33	6	Nevada	47	7
Arkansas	32	27	New Hampshire	41	44
California	1	3	New Jersey	8	46
Colorado	30	8	New Mexico	37	5
Connecticut	24	48	New York	2	30
Delaware	46	49	North Carolina	12	28
Florida	9	22	North Dakota	45	17
Georgia	15	21	Ohio	6	35
Hawaii	40	47	Oklahoma	27	18
Idaho	42	13	Oregon	31	10
Illinois	5	24	Pennsylvania	3	33
Indiana	11	38	Rhode Island	39	50
Iowa	25	25	South Carolina	26	40
Kansas	28	14	South Dakota	44	16
Kentucky	23	37	Tennessee	17	34
Louisiana	20	31	Texas	4	2
Maine	38	39	Utah	36	11
Maryland	18	42	Vermont	48	43
Massachusetts	10	45	Virginia	14	36
Michigan	7	23	Washington	22	20
Minnesota	19	12	West Virginia	34	41
Mississippi	29	32	Wisconsin	16	26
Missouri	13	19	Wyoming	49	9

---

*Source: *The World Almanac*, 1975.

TAU

```
10 REM THIS PROGRAM COMPUTES THE TAU STATISTIC
20 REM BASED ON A SAMPLE OF SIZE N FROM A CONTINUOUS
30 REM BIVARIATE DISTRIBUTION
40 REM DATA SHOULD APPEAR IN LINES 500-699 IN PAIRS,
50 REM I.E. X(1),Y(1),X(2),Y(2),....,X(N),Y(N).
60 REM THE PROGRAM COMPUTES KENDALL'S TAU, THE ASSOCIATED
70 REM P-VALUE,AND THE NUMBER OF PAIRS CONTRIBUTING 0
75 REM TO THE CALCULATION OF (CONCORDANT-DISCORDANT).
80 PRINT "WHAT IS THE VALUE OF N?"
90 INPUT N
100 DIM X(100),Y(100)
110 FOR I = 1 TO N
120 READ X(I),Y(I)
130 NEXT I
140 FOR I = 1 TO N-1
150 FOR J = I+1 TO N
160 IF (X(I)-X(J))*(Y(I)-Y(J))< 0 THEN 210
170 IF (X(I)-X(J))*(Y(I)-Y(J))= 0 THEN 200
180 LET C=C+1
190 GO TO 210
200 LET T1=T1+1
210 NEXT J
220 NEXT I
230 LET T =(2*C-N*(N-1)/2)/(N*(N-1)/2)
240 PRINT "TAU = ",T
250 PRINT "NUMBER OF TIES = ",T1
260 LET Z = T/SQR((4/9)*(N+5/2)/(N*(N-1)))
270 GOSUB 900
280 PRINT "IS THE ALTERNATIVE ONE-SIDED? 1 IF YES 0 IF NO"
290 INPUT A
300 IF A= 1 THEN 320
310 PRINT "P-VALUE FOR TWO-SIDED ALTERNATIVE = ";2*P
315 GO TO 1030
320 PRINT "P-VALUE FOR ONE-SIDED ALTERNATIVE = ";P
330 GO TO 1030
500 DATA 2,3,1,4,4,2,5,5,3,1,7,6,6,7
900 LET Z1= ABS(Z)
910 LET C=1/SQR(2)
920 LET C1= .14112821
930 LET C2= .08864027
940 LET C3=.02743349
950 LET C4= -.00039446
960 LET C5=.00328975
970 DEF FNZ(X)=1-1/(1+C1*X+C2*X↑2+C3*X↑3+C4*X↑4+C5*X↑5)↑8
980 LET P = .5+.5*FNZ(Z1*C)
990 LET P = 1E-4*(INT(1E4*P))
1010 LET P = 1-P
1020 RETURN
1030 END
*
```

```
 RUN
WHAT IS THE VALUE OF N?
?  7
TAU =              0.428571
NUMBER OF TIES =                0
IS THE ALTERNATIVE ONE-SIDED 1 IF YES 0 IF NO
? 1
P-VALUE FOR ONE-SIDED ALTERNATIVE =    0.0883

*
```

(b) The program TAU computes t and the corresponding Z value for a large sample test of H_0: $\tau = 0$. Use TAU to test H_0: $\tau = 0$ for the data given in part (a). (Use $\alpha = 0.05$.)

3. The following gives the final batting averages and winning percentages for the National League teams in 1974. Use the program TAU to find the value of t and to test H_0: $\tau = 0$ versus H_1: $\tau > 0$ ($\alpha = 0.025$).

	Team batting average	*Team winning percentage*
Pittsburg	0.274	0.543
St. Louis	0.265	0.534
Philadelphia	0.261	0.494
Montreal	0.254	0.491
New York	0.235	0.438
Chicago	0.251	0.407
Los Angeles	0.272	0.630
Cincinnati	0.260	0.605
Atlanta	0.248	0.543
Houston	0.263	0.500
San Francisco	0.252	0.444
San Diego	0.229	0.370

THE ANALYSIS OF VARIANCE

11.1   INTRODUCTION

In Chap. 8 we considered methods for testing whether there existed a
a difference between the expectations of two populations. In statistical language we say that we are comparing the effects of two
treatments. One compares the test performance of students taught
by two teaching methods, the weight gain of animals on two diets,
the life expectancy after two medical treatments, and so forth.
It is quite natural to think of the number of different treatments
being greater than two. Again we wish to compare the effects of the
treatments and to make statistical statements concerning differences
among such treatments. This topic is a part of the body of statistical knowledge known as the analysis of variance and is considered
in Sec. 11.2.

In Chap. 10, in simple linear regression, we considered a means
of ascertaining the influence of the real variable x upon the average
value of a random variable Y. There are also the situations in which
the level of a categorical variable or variables influences the value
of a random variable Y. For example, we may consider that political
attitude, measured by degree of conservatism on a 0 to 100 scale,
may be affected by both the categorical variables of region of residence and income class. The regions could be east, south, midwest,
and west, while the income classes could be low, medium and high.

We wish to analyze the effect of these two factors, region of residence and income class, upon political attitudes. In Sec. 11.3 we consider the randomized block design in this regard. These two areas of the analysis of variance present only a brief introduction to the vast and challenging statistical field of the design and analysis of experiments.

11.2   SINGLE-FACTOR ANALYSIS OF VARIANCE

11.2.1   The One-Way Classification Model

In Chap. 8 we considered two-sample situations. In Sec. 8.3 we specifically considered testing hypotheses concerning the difference of the true means of two normal populations and finding a confidence interval for this difference. A natural extension is to compare the true means of $t \geq 2$ normal populations. Thus we assume that X_i
$X_i \sim N(\mu_i, \sigma^2)$, $i = 1, 2, \ldots, t$. One hypothesis of interest is

$$H_0: \quad \mu_1 = \mu_2 = \cdots = \mu_t \qquad (11.2.1)$$

Note that, as all theoretical variances are equal, this is equivalent to stating that each of the t populations have the same distribution. We might also be interested in testing a hypothesis of form

$$H_0: \quad \frac{\mu_1 + \mu_2}{2} = \mu_3 \qquad \text{versus} \qquad H_1 : H_0 \text{ is false}$$

This null hypothesis states that the average "effect" of populations 1 and 2 is the same as the "effect" of population 3. A confidence interval for $(\mu_1 + \mu_2)/2 - \mu_3$ might also be required. In this section we consider a method of answering such questions which is referred to as the one-way classification model in the analysis of variance.

The populations defined in the preceding paragraph are referred to as treatments. Initially, let us consider observing n observations from each of the t treatments, nt observations in all. X_{ij} will refer to the jth observation on the ith treatment. Our model then is

$$X_{ij} = \mu_i + e_{ij} \quad i = 1, 2, \ldots, t; \; j = 1, 2, \ldots, n$$

with $e_{ij} \sim NID(0, \sigma^2)$. The abbreviation NID means *normally and independently distributed*. Hence each observation is independent of others in the same treatment and of all observations from the other treatments. By defining $\mu = \Sigma_{i=1}^{t} \mu_i / t$ and $\mu_i = \mu + \alpha_i$, we write the basic model as

$$X_{ij} = \mu + \alpha_i + e_{ij} \quad e_{ij} \sim NID(0, \sigma^2) \tag{11.2.2}$$

$i = 1, 2, \ldots, t$ and $j = 1, 2, \ldots, n$. It may be easily checked that $\Sigma_{i=1}^{t} \alpha_i = 0$. Thus μ can be thought of as an "overall effect" common to all the populations and $\alpha_i = \mu_i - \mu$ as an "effect" due to the ith population. The hypothesis (11.2.1) is then restated as $H_0: \alpha_i = 0$ for all i.

The observed values of X_{ij} are usually recorded in a table such as Table 11.2.1. Clearly tests about the theoretical means will be based on the sample means $\bar{x}_i$, so that these are calculated. Let $\bar{x}$ be the overall mean and consider the identity

$$x_{ij} = \bar{x} + (\bar{x}_i - \bar{x}) + (x_{ij} - \bar{x}_i)$$

or

$$x_{ij} - \bar{x} = (\bar{x}_i - \bar{x}) + (x_{ij} - \bar{x}_i) \tag{11.2.3}$$

which is obtained by mimicking the model equation (11.2.2). Squaring both sides of Eq. (11.2.3), summing first on j and then on i, yields

$$\sum_{i=1}^{t} \sum_{j=1}^{n} (x_{ij} - \bar{x})^2 = n \sum_{i=1}^{t} (\bar{x}_i - \bar{x})^2 + \sum_{i=1}^{t} \sum_{j=1}^{n} (x_{ij} - \bar{x}_i)^2 \tag{11.2.4}$$

The identity (11.2.4) is the basic identity in the one-way classification. It is symbolically written as

$$SST = SSA + SSE \tag{11.2.4'}$$

and read "the sum of squares for total equals the sum of squares among treatments plus the sum of squares for error."

Table 11.2.1   Observed Values of X_{ij}

Treatment	Observations	Mean
1	x_{11} x_{12} $\cdots$ x_{1n}	$\bar{x}_1$
2	x_{21} x_{22} $\cdots$ x_{2n}	$\bar{x}_2$
3	$\cdots\cdots\cdots\cdots\cdots$	
t	x_{t1} x_{t2} $\cdots$ x_{tn}	$\bar{x}_t$
		$\bar{x} = \Sigma_{i=1}^{t}\bar{x}_i/t$

Table 11.2.2   Analysis of Variance Table for the One-Way Classification

Source of variation	SS	d.f.	MS	EMS
Treatments	SSA	$t - 1$	$SSA/(t - 1)$	$\sigma^2 + n\,\Sigma_{i=1}^{t}\alpha_i^2/(t - 1)$
Error	SSE	$t(n - 1)$	$SSE/t(n - 1)$	σ^2
Total	SST	$nt - 1$		

The information in Eq. (11.2.4) is summarized in an analysis of variance table such as that given in Table 11.2.2. This analysis of variance table is similar to the analysis of variance table of Sec. 10.2 in the regression case. Here the total sum of squares is divided into two parts, that due to variation among the treatments and that due to error. The degrees of freedom are integers chosen so that the respective mean squares are unbiased estimates of σ^2, under the hypothesis H_0: $\alpha_i = 0$ for all i. This is clear from the expressions given for the expected mean squares, whose derivations are left to the problems. Note that the mean square for error (MSE = $\hat{\sigma}^2$) is an unbiased estimate of σ^2 whether or not H_0 is true.

Now we consider a test of

$$H_0: \quad \alpha_i = 0 \quad i = 1, 2, \ldots, t$$

versus   (11.2.5)

$$H_1: \quad \alpha_i \neq 0 \quad \text{for at least one i}$$

The statistic which is used in this case is MSA/MSW. It is clear from the EMS column of Table 11.2.2 that under H_0 both numerator and denominator estimate σ^2 unbiasedly, but under H_1 the numerator will tend to be larger. Hence we should reject H_0 if MSA/MSW is large. It can be shown that MSA and MSW are independent χ^2 random variables divided by their degrees of freedom under H_0 (although a proof is beyond the scope of this text). Hence MSA/MSW = F has an F distribution with $t - 1$ and $t(n - 1)$ degrees of freedom. Thus an α level test of statement (11.2.5) is given by the critical region $F > F(t - 1, t(n - 1))_\alpha$.

Let us consider an example of the one-way classification analysis. Assume that a football coach is interested in whether the statement "college grades increase with class" (i.e., freshman, sophomore, junior, senior) is true for his football players. He randomly selects five players from each class and obtains their current GPA, which is measured on a 0.0 to 5.0 scale. The observations are recorded in Table 11.2.3. The program ONEWAY has been written to perform the analysis of variance computations for the one-way classification. Data are entered according to the instructions in lines 100 through 230. Here we see that the analysis of variance table given in Table 11.2.4 is appropriate.

Table 11.2.3   Observed GPA

Class	Observations					$\bar{x}_i$
Freshman	3.64	2.79	3.21	2.98	3.48	3.22
Sophomore	4.23	3.53	3.62	4.98	3.50	3.97
Junior	4.10	3.87	3.87	4.29	4.04	4.03
Senior	4.50	4.24	4.36	3.77	4.24	4.22

Table 11.2.4   ANOVA Table

Source	SS	d.f.	MS	EMS
Classes	2.917	3	0.972	$\sigma^2 + 5 \Sigma_{i=1}^{4} \alpha_i^2 / 3$
Error	2.536	16	0.1585	σ^2
Total	5.453	19		

The calculated value of the F = MSA/MSW = 0.972/0.1585 = 6.13. Hence if we wish to test the null hypothesis of no difference in grade averages among the classes (that is, H_0: $\alpha_i = 0$, i = 1, 2, 3, 4) versus the alternative that such differences exist, we reject H_0 if F > $F(3,16)_\alpha$. If α = 0.05, the critical value is 3.24, and if α = 0.01, the critical value is 5.29. As the calculated F value exceeds both of these, there is clear statistical evidence for the view that grade averages differ with class (i.e., freshman, sophomore, junior, senior). Of course we are more interested in how these grade averages differ. We shall take up this question later in the section when we discuss contrasts.

It is often the case that the number of observations on a treatment are not the same. For example in an educational experiment to compare the effectiveness of three teaching methods, we may observe the scores on a standardized exam for the students in each of the classes. It often will be the case that the number of students in the classes will differ. (This may happen even though we plan to have the same number in each class for example, if some students leave the course.) The appropriate model in such cases is given by

$$X_{ij} = \mu_i + e_{ij} \qquad i = 1, 2, \ldots, t; \ j = 1, 2, \ldots, n_i$$

with $e_{ij} \sim NID(0,\sigma^2)$. Using the following,

$$\bar{x}_i = \sum_{j=1}^{n_i} \frac{x_{ij}}{n_i} \qquad \bar{x} = \sum_{i=1}^{t} \sum_{j=1}^{n_i} \frac{x_{ij}}{N} \qquad N = \sum_{i=1}^{t} n_i$$

we obtain

$$\sum_{i=1}^{t} \sum_{j=1}^{n_i} (x_{ij} - \bar{x})^2 = \sum_{i=1}^{t} n_i(\bar{x}_i - \bar{x})^2 + \sum_{i=1}^{t} \sum_{j=1}^{n_i} (x_{ij} - \bar{x}_i)^2$$

or SST = SSA + SSE. This corresponds to Eq. (11.2.4) and reduces to that expression if n_i = n for i = 1, 2, ..., t.

The appropriate analysis of variance table in the case of unequal sample sizes is given in Table 11.2.5. In this case the test of H_0: α_i = 0, i = 1, 2, ..., t is given by the rejection region F > F(t - 1, N - t)$_\alpha$, where the calculated statistic F = MSA/EMS, as before. Again the mean square for error is an unbiased estimate of σ^2 whether H_0 is true or not.

As an example, assume the following are final exam scores of students from three different colleges in a statistics course.

College	Scores
Agriculture	112, 76, 124, 64, 132, 112, 88, 100, 80
Liberal Arts	84, 100, 56, 108, 96, 64, 92, 124, 136, 132
Home Economics	80, 96, 96, 92, 116

Here we see that n_1 = 9, n_2 = 10, and n_3 = 5. Using the program ONEWAY, the analysis of variance table is constructed in Table 11.2.6. The calculated F = 0.0327 and F(2,21)$_{0.05}$ = 3.49. Hence there is no statistical indication of difference in performance due to college of registration.

Table 11.2.5   ANOVA Table One-Way Classification (Unequal n_i)

Source	SS	d.f.	MS	EMS
Treatments	SSA	t - 1	SSA/(t - 1)	$\sigma^2 + \sum_{i=1}^{t} n_i \alpha_i^2/(t - 1)$
Error	SSE	N - t	SSE/(N - t)	σ^2
Total	SST	N - 1		

Table 11.2.6 ANOVA Table

Source	SS	d.f.	MS	EMS
Colleges	35.733	2	17.8667	$\sigma^2 + \Sigma_{i=1}^{3} n_i \alpha_i^2/2$
Error	11481.6	21	546.7428	σ^2
Total	11517.333	23		

11.2.2 Comparisons or Contrasts

It is often the case, as we have noted above, that in a statistical analysis using the one-way classification we wish to find out more information about treatment means than just deciding to reject or accept the hypothesis in Eq. (11.2.1). We often desire to compare two particular treatment means or one treatment mean with all the others. Such comparisons can be expressed as statistical hypotheses of the form

$$H_0: \quad \sum_{i=1}^{t} \lambda_i \mu_i = 0 \quad \text{where} \quad \sum_{i=1}^{t} \lambda_i = 0 \quad (11.2.6)$$

Such a linear combination of the μ_i is called a *comparison* or *contrast*. Consider the first example of Sec. 11.2.1. Suppose we wish to compare the grade averages of juniors and seniors. We test $H_0: \quad \mu_3 - \mu_4 = 0$ versus $H_1: \quad \mu_3 - \mu_4 \neq 0$. Evidence for H_1 suggests the GPAs of juniors and seniors differ. The contrast $\mu_1 - (\mu_2 + \mu_3 + \mu_4)/3$ can be used to compare freshman with upper classmen.

Let us consider the general case of testing the null hypothesis that a contrast is zero, i.e., Eq. (11.2.6). The same linear combination of the observed means $\bar{x}_i$, $L = \Sigma_{i=1}^{t} \lambda_i \bar{x}_i$ is an unbiased estimate of the contrast. Assuming an equal number of observations for each treatment, we see that $\text{Var}(L) = (\sigma^2/n) \Sigma_{i=1}^{t} \lambda_i^2$. It is natural to consider the standardized statistic $(L - E(L))/\sqrt{\text{Var}(L)}$. Under Eq. (11.2.6), $E(L) = 0$ and $Z = L/\sqrt{\text{Var}(L)}$ would have the standard normal distribution. However, σ^2 is not known and is replaced by

ONEWAY

```
00100 REM DESCRIPTION: COMPUTES THE ANALYSIS OF VARIANCE TABLE
00110 REMFOR A ONE-WAY COMPLETELY RANDOMIZED DESIGN.
00120 REM INSTRUCTION: ENTER DATA IN LINE 900 AND FOLLOWING.
00130 REM ENTER DATA IN THE FOLLOWING ORDER:
00140 REM 1) THE TOTAL NUMBER OF OBSERVATIONS
00150 REM 2) M, THE NUMBER OF DIFFERENT TREATMENTS
00160 REM 3) N(1),...,N(M), WHERE N(J) IF THE NUMBER OF
00170 REM OBSERVATIONS ON TREATMENT J.
00180 REM 4) AND THEN, THE OBSERVATIONS THEMSELVES, FIRST FOR
00190  REM TREATMENT 1) THEN TREATMENT 2, ETC.
00200 REM IF ANY N(J)> 20, CHANGE THE DIMS IN LINE 240
00210 REM IF M > 10,CHANGE THE DIMS IN LINE 240.
00220 REM SAMPLE DATA IS IN LINES
00230  REM      900 THROUGH 903
00240 DIM X(20,10),N(10),T(10),S(10)
00250 READ A,M
00260  MAT S = ZER
00270  LET U = 0
00272  LET R = 0
00274  LET V = 0
00280  MAT T = ZER
00290  FOR I = 1 TO M
00300  READ N(I)
00310  NEXT I
00320  FOR J = 1 TO M
00330  FOR I = 1 TO N(J)
00340  READ X(I,J)
00350  NEXT I
00360  NEXT J
00370  FOR J = 1 TO M
00380  FOR I = 1 TO N(J)
00390  LET T(J)=T(J)+X(I,J)
00410   LET S(J)=S(J)+X(I,J)*X(I,J)
00420  NEXT I
00430  LET U = U + T(J)
00440  LET R = R+ S(J)
00450  LET V = V+T(J)*T(J)/N(J)
00460  NEXT J
00470  LET C = U*U/A
00480  LET W = V-C
00490  LET E = R- V
00500 PRINT "ANOVA TABLE"
00510 PRINT
00520 PRINT "    ITEM    "
00530 PRINT "GRAND TOTAL","SS=",R,"DF=",A
00540  PRINT "GRAND MEAN","SS=",C,"DF=","1"
00550 PRINT "TREATMENTS","SS=",W,"DF=",M-1,"MEAN SQUARE=",W/(M-1)
00560  PRINT " ERROR    ","SS=",E,"DF=",A-M,"MEAN SQUARE=",E/(A-M)
00570  LET F = W/(M-1)/(E/(A-M))
00580  PRINT
00590 PRINT "F=",F,"ON",M-1,"AND",A-M,"DEGREES","OF FREEDOM"
00900  DATA 20,4,5,5,5,5
00901  DATA 3.64,2.79,3.21,2.98,3.48
00902  DATA 4.23,3.53,3.62,4.98,3.50
00903  DATA 4.1,3.87,3.87,4.29,4.04,4.5,4.24,4.36,3.77,4.24
01000  END
*
```

```
   RUN
ANOVA TABLE

   ITEM
GRAND TOTAL     SS=              303.754      DF=              20
GRAND MEAN      SS=              298.301      DF=        1
TREATMENTS      SS=              2.91724      DF=              3
MEAN SQUARE=    0.97241
   ERROR        SS=              2.53628      DF=              16
MEAN SQUARE=    0.15852

F=              6.13443     ON                3           AND
   16          DEGREES     OF FREEDOM
```

*

the MSE an unbiased estimate. It is true that the resulting statistic has Student's t distribution with t(n - 1) degrees of freedom.
Hence

$$t = \frac{L}{\sqrt{(MSE/n) \ \Sigma_{i=1}^{t} \lambda_i^2}}$$

is computed. The rejection regions are the obvious ones with
$|t| > t_{\alpha/2, t(n-1)}$ the appropriate region for a two-sided test with
significance level α.

Let us compare the grade averages of juniors and seniors using
the data given in Sec. 11.2.1. The GPA for five juniors was 4.03
and for five seniors was 4.22. From Table 11.2.4 we find the mean
square for error to be 0.1585. Hence

$$t = \frac{4.03 - 4.22}{\sqrt{(0.1585/5)(0^2 + 0^2 + 1^2 + (-1)^2)}} = \frac{-0.19}{0.2518} = -0.75$$

As $t_{0.025,16} = 2.12$ there is no significant difference at the 5%
level of significance. Suppose we wish to compare freshmen with all
other classes. We test H_0: $\mu_1 - (\mu_2 + \mu_3 + \mu_4)/3 = 0$ versus
H_1: $\mu_1 - (\mu_2 + \mu_3 + \mu_4)/3 < 0$. Here we obtain

$$t = \frac{3.22 - (3.97 + 4.03 + 4.22)/3}{\sqrt{(0.1585/5)(4/3)}} = -4.13$$

Against H_1 we reject H_0 if $t < -t_{0.05,16} = -1.746$. The statistical
evidence suggests that upperclassmen have higher GPAs on the average.

To obtain a confidence interval for a contrast we proceed as
previously and find

$$\sum_{i=1}^{t} \lambda_i \bar{x}_i \pm \sqrt{\frac{MSE}{n} \ \sum_{i=1}^{t} \lambda_i^2} \ t_{\alpha/2, t(n-1)}$$

to be the endpoints for a $(1 - \alpha)100\%$ confidence interval for the
corresponding contrast. The confidence interval for $(\mu_2 + \mu_3 + \mu_4)/3$
$- \mu_1$ in the example would be

$$0.85 \pm \sqrt{\frac{0.1585}{5} \frac{4}{3}} \, 2.12 \quad \text{or} \quad 0.85 \pm 0.436$$

with confidence coefficient 0.95.

Finally if we have an unequal number of observations on a treatment, the estimate of a contrast $\Sigma_{i=1}^{t} \lambda_i \mu_i$ is still $L = \Sigma_{i=1}^{t} \lambda_i \bar{x}_i$. However as $\text{Var}(L) = \sigma^2 \Sigma_{i=1}^{t} \lambda_i^2 / n_i$ the appropriate t statistic to test Eq. (11.2.1) is

$$t = \frac{L}{\sqrt{\text{MSE} \, \Sigma_{i=1}^{t} \lambda_i^2 / n_i}}$$

The correct degrees of freedom for the test is $N - t$. Similarly a $(1 - \alpha)100\%$ confidence interval for the contrast becomes

$$L \pm \sqrt{\text{MSE} \, \Sigma_{i=1}^{t} \frac{\lambda_i^2}{n_i}} \, t_{\alpha/2, N-t}$$

Problems 11.2

1. Consider the one-way classification model for which $X_{ij} = \mu_i + e_{ij}$, $i = 1, 2, \ldots, t$; $j = 1, 2, \ldots, n$, with $e_{ij} \sim \text{NID}(0, \sigma^2)$.
   (a) Verify that by defining $\mu = \Sigma_{i=1}^{t} \mu_i / t$ and $\alpha_i = \mu - \mu_i$ that $\Sigma_{i=1}^{t} \alpha_i = 0$.
   (b) Prove that the identity (11.2.4) is correct.

2. In the model of Problem 11.2.1 prove
   (a) $E(\text{SSE}) = \sigma^2 (n - 1)t$.
   (b) $E(\text{SSA}) = \sigma^2 (t - 1) + n \Sigma_{i=1}^{t} \alpha_i^2$.
   Note in part (b) that $\bar{x}_{i.} - \alpha_i - \bar{x}$ are $\text{NID}(0, \sigma^2/n)$.

3. Assume $t = 2$ in the model defined in Problem 11.2.1. Show the following:
   (a) $\text{SSA} = n(\bar{x}_1 - \bar{x}_2)^2 / 2$.
   (b) $\text{MSE} = \hat{\sigma}^2$, the pooled estimate of the common variance defined in Sec. 8.3.

(c) MSA/MSE = $F = t^2$, where t is the test statistic for the
     two-sample test defined in Sec. 8.3 with $n_1 = n_2 = n$.

(d) Using the result stated in Problem 10.2.10 concerning the
     relationship between the t and F distributions, show that
     the t test for H_0: $\mu_1 = \mu_2$ versus H_1: $\mu_1 \neq \mu_2$ gives the
     same result as the F test in the one-way classification
     for t = 2 treatments and $n_1 = n_2 = n$.

Exercises 11.2

Use the program ONEWAY as needed.

1. Current expenditures per pupil in six eastern, southern, and
    midwestern states are given below. Is there statistical evidence
    of differential expenditure in these regions (α = 0.05)?

East	Expenditure($)	South	Expenditure($)
Conneticut	1241	Alabama	590
Maine	840	Florida	885
Massachusetts	1090	Georgia	782
New Jersey	1294	Mississippi	689
New York	1584	North Carolina	802
Rhode Island	1113	South Carolina	751

Midwest	Expenditure($)
Illinois	1234
Indiana	855
Iowa	1055
Kansas	969
Missouri	861
Nebraska	953

Source: The World Almanac 1975, data for 1972-1973.

2. In the previous question state, using the appropriate parameters,
    the following hypotheses:

(a)   There is no difference in expenditure between eastern and midwestern states.

(b)   There is no difference between southern and midwestern states.

(c)   There is no difference between southern states and the other areas.

Test each of these hypothesis at level of significance $\alpha = 0.05$.

3.   It is argued by a large department at a university that its class sizes in its senior undergraduate courses is increasing over time. The following are the class sizes of five randomly selected senior classes in each of four years.

Year	Class sizes
1974	38, 19, 24, 52, 23
1975	30, 36, 15, 54, 24
1976	41, 21, 12, 63, 29
1977	31, 35, 20, 58, 28

(a)   Use the analysis of variance technique to test the null hypothesis of equal average class size in these 4 years ($\alpha = 0.05$).

(b)   Use the appropriate contrast to test the hypothesis that class sizes in 1974 and 1977 are the same on the average versus the alternative that they have increased.

4.   A study is made to compare the costs of elementary calculus texts published by companies A, B, and C. Company A publishes three such texts, Company B publishes four, and Company C publishes five. The following are the retail prices.

Company	Prices
A	10.95, 12.95, 9.95
B	12.95, 14.95, 16.00, 8.95
C	10.95, 13.95, 15.95, 16.95, 18.95

(a)   Test the null hypothesis of equal average price ($\alpha = 0.05$).

(b)   Test H_0: $(\mu_A + \mu_B)/2 = \mu_C$ and find a 95% confidence interval for $\mu_C - (\mu_A + \mu_B)/2$.

5. It is argued by some that the Central Division of the National
    Football Conference is a low-scoring one. Data are given below
    for the total points scored by each team in the NFC in 1973.

Eastern Division	Points	Western Division	Points
Dallas	382	Los Angeles	388
Washington	325	Atlanta	318
Philadelphia	310	San Francisco	262
St. Louis	286	New Orleans	163
New York	226		

Central Division	Points
Minnesota	296
Detroit	271
Green Bay	202
Chicago	195

(a) Test the hypothesis of equal scoring for each of the three
     divisions ($\alpha = 0.05$).

(b) Test the hypothesis that teams in the Central Division score
     as well as teams in the other divisions against the alterna-
     tive of lower scoring ($\alpha = 0.05$).

11.3   THE RANDOMIZED BLOCK DESIGN

Often in statistics it is of interest to compare the effect of treatments, where we assume that we make observations on all the treatments
within homogeneous groups called *blocks*. For example, we may wish
to compare five institutions of higher education with regard to the
compensation of their faculty. We might then observe the following
table giving the average compensation at the five institutions in
thousands of dollars. Clearly it would be inappropriate to compare
the level of compensation of instructors at one institution with the
level of compensation of professors at another. In this case the
blocks are the instructors, assistant professors, associate professors,
and professors.

	Institution				
	1	2	3	4	5
Instructors	11.8	11.8	12.4	19.1	11.6
Assistant professors	16.0	15.9	15.1	23.4	15.3
Associate professors	19.2	19.0	20.9	30.0	17.8
Professors	27.1	24.8	29.1	37.4	22.5

The observations in the general randomized block design in which we have t reatements and b blocks can be modeled as follows:

$$X_{ij} = \mu + \alpha_i + \beta_j + e_{ij} \qquad i = 1, 2, \ldots, t; \ j = 1, 2, \ldots, b$$

$$\sum_{i=1}^{t} \alpha_i = 0 \qquad \sum_{j=1}^{b} \beta_j = 0 \qquad e_{ij} \sim NID(0, \sigma^2)$$

Here X_{ij} represents the observation of the ith treatment in the jth block. The expectation $E(X_{ij}) = \mu + \alpha_i + \beta_j$. Thus the expectation is assumed to be the sum of an overall effect, an effect due to the treatment i and an effect due to the block j. We denote the actual observations as x_{ij} and make the following definitions:

$$\bar{x} = \sum_{i=1}^{t} \sum_{j=1}^{b} \frac{x_{ij}}{tb} \qquad \bar{x}_{i\bullet} = \sum_{j=1}^{b} \frac{x_{ij}}{b} \qquad \bar{x}_{\bullet j} = \sum_{i=1}^{t} \frac{x_{ij}}{t}$$

for the overall mean, the ith treatment mean, and the jth block mean. In general we are interested in testing hypotheses about the treatment effects. Thus, as in the one-way classification, we wish to test H_0: $\alpha_i = 0$, $i = 1, 2, \ldots, t$ and to make inference about contrasts of the form $\sum_{i=1}^{t} \lambda_i \alpha_i$.

The method of proceeding is analogous to the one-way classification. We consider the identity

$$x_{ij} - \bar{x} = (\bar{x}_{i\bullet} - \bar{x}) + (\bar{x}_{\bullet j} - \bar{x}) + (x_{ij} - \bar{x}_{i\bullet} - \bar{x}_{\bullet j} + \bar{x})$$

which again results from the mimicking the model equation. Squaring and summing first on j and then on i yields

$$\sum_{i=1}^{t} \sum_{j=1}^{b} (x_{ij} - \bar{x})^2 = b \sum_{i=1}^{t} (\bar{x}_{i\bullet} - \bar{x})^2 + t \sum_{j=1}^{b} (\bar{x}_{\bullet j} - \bar{x})^2$$

$$+ \sum_{i=1}^{t} \sum_{j=1}^{b} (x_{ij} - \bar{x}_{i\bullet} - \bar{x}_{\bullet j} + \bar{x})^2$$

or

SST = SSA + SSB + SSE

Again this is read "the total sum of squares equals the sum of squares for treatments plus the sum of squares for blocks plus the sum of squares for error."

This information is summarized in an analysis of variance table given in Table 11.3.1. The appropriate expected mean squares have been found. Note that $E(MSA) = \sigma^2 + b \sum_{i=1}^{t} \alpha_i^2/(t-1)$ involves σ^2 and the treatment effects *only* and not the block effects. This makes possible again a test of H_0: $\alpha_i = 0$, $i = 1, 2, \ldots, t$ using $F = MSA/MSE$ as the appropriate statistic. The rejection region for an α level test is of the form $F > F(t-1, (t-1)(b-1))_\alpha$.

Let us consider the compensation data presented above for the faculties at five universities. The program RBLOCK performs the required analysis for the randomized block design. In Table 11.3.2 we present the analysis of variance table for the example data. The F test for the hypothesis of equal institutional compensation is

Table 11.3.1   ANOVA Table for the Randomized Block Design

Source of Variation	SS	d.f.	MS	EMS
Treatments	SSA	t − 1	SSA/(t−1)	$\sigma^2 + b \sum_{i=1}^{t}\alpha_i^2/(t-1)$
Blocks	SSB	b − 1	SSB/(b−1)	$\sigma^2 + t \sum_{j=1}^{b}\beta_j^2/(b-1)$
Error	SSE	(t−1)(b−1)	SSE/(t−1)(b−1)	σ^2
Total	SST	bt − 1		

RBLOCK

```
00010   REM DESCRIPTION: COMPUTES THE ANALYSIS OF VARIANCE TABLE
00020   REM FOR A RANDOMIZED BLOCK DESIGN.
00030   REM INSTRUCTIONS: ENTER DATA IN LINE 900 AND FOLLOWING
00040   REM ENTER DATA IN THE FOLLOWING ORDER:
00050   REM 1) B, THE NUMBER OF BLOCKS
00060   REM 2) T, THE NUMBER OF TREATMENTS
00070   REM 3) THE DATA THEMSELVES BY TREATMENTS, I.E. FIRST OBS ON
00080   REM TREATMENT 1, BLOCKS 1,2,...,B, THEN TREATMENT 2,
00090   REM BLOCKS 1,2,...,B,   ETC.
00095   REM IF B > 10 OR T >10 CHANGE DIMS IN LINE 100
00098   REM SAMPLE DATA IS IN LINES 900-902
00100   DIM X(10,10),V(10),W(10),S(10)
00110   READ B,T
00120   MAT X = ZER
00130   MAT V = ZER
00140   MAT S = ZER
00150   MAT W = ZER
00160   FOR J= 1 TO T
00170   FOR I = 1 TO B
00180   READ X(I,J)
00190   NEXT I
00200   NEXT J
00210   FOR J = 1 TO T
00220   FOR I = 1 TO B
00230   LET V(J) = V(J)+X(I,J)
00240   LET S(J) = S(J)+X(I,J)*X(I,J)
00250   NEXT I
00260   LET A = A +V(J)
00270   LET E = E+S(J)
00275   LET C = C+V(J)*V(J)/B
00280   NEXT J
00290   FOR I = 1 TO B
00300   FOR J = 1 TO T
00310   LET W(I)=W(I)+X(I,J)
00320   NEXT J
00330   LET D = D +W(I)*W(I)/T
00340   NEXT I
00350   LET C1 = A*A/(B*T)
00360   LET C = C - C1
00370   LET D = D - C1
00380   LET T1 = E - C1
00390   LET E1 = T1 - C - D
00400   PRINT "   ANOVA TABLE"
00410   PRINT "     SOURCE "
00420   PRINT "TREATMENTS","SS=",C,"DF=",T-1,"MS=",C/(T-1)
00430   PRINT "BLOCKS   ","SS=",D,"DF=",B-1,"MS=",D/(B-1)
00440   PRINT " ERROR    ","SS=",E1,"DF=",(T-1)*(B-1),"MS=",E1/((T-1)*(B-1))
00450   PRINT "TOTAL(CORR)","SS=",T1,"DF=",B*T-1
00460   PRINT "F FOR TREATMTS=",C/(T-1)/(E1/((T-1)*(B-1)))
00470   PRINT "DOF ARE", T-1, " AND",(T-1)*(B-1)
00900   DATA 4,5
00901   DATA 11.8,16.0,19.2,27.1,11.8,15.9,19.0,24.8,12.4,15.1,20.9,29.1
00902   DATA 19.1,23.4,30.0,37.4,11.6,15.3,17.8,22.5
01000   END
*
```

```
RUN
  ANOVA TABLE
     SOURCE
TREATMENTS        SS=              292.788          DF=              4
MS=              73.197
BLOCKS            SS=              606.758          DF=              3
MS=              202.253
  ERROR           SS=              27.0919          DF=             12
MS=              2.25766
TOTAL(CORR)       SS=              926.638          DF=             19
F FOR TREATMTS= 32.4216
DOF ARE           4                AND               12


*
```

Table 11.3.2   ANOVA Table

Source	SS	d.f.	MS	EMS
Institutions	292.788	4	73.197	$\sigma^2 + 4 \Sigma \alpha_i^2$
Ranks	606.758	3	202.253	$\sigma^2 + 5 \Sigma \beta_j^2/3$
Error	27.092	12		σ^2
Total	926.638	19		

based on $F = MSA/MSE = 73.197/2.258 = 32.42$. As the critical value $F(4,12)_{0.05} = 3.26$, the evidence for differing levels of compensation is overwhelming. The F test for blocks would clearly also be significant at the 5% level. This test is not important however because we expect that there will be different levels of compensation for the professorial ranks. The fact that the model permits a test of treatment differences in the face of assumed block differences makes it a useful design in the analysis of variance.

If we consider a test of contrast in the randomized block design of form

$$H_0 : \quad \sum_{i=1}^{t} \lambda_i \alpha_i = 0 \qquad \sum_{i=1}^{t} \lambda_i = 0 \qquad\qquad (11.3.1)$$

we proceed generally as before. The statistic $L = \Sigma_{i=1}^{t} \lambda_i \bar{x}_i$ has expectation $\Sigma_{i=1}^{t} \lambda_i \alpha_i$ and $Var(L) = (\sigma^2/b)\Sigma_{i=1}^{t} \lambda_i^2$. Thus

$$t = \frac{L}{\sqrt{(MSE/b)(\Sigma_{i=1}^{t} \lambda_i^2)}}$$

has Student's distribution with $(t - 1)(b - 1)$ degrees of freedom under the hypothesis in Eq. (11.3.1). Thus if we wish to test the hypothesis that the contrast $\alpha_3 - \alpha_2 = 0$, we obtain

$$t = \frac{19.375 - 17.875}{\sqrt{2.258(2/4)}} = 1.411$$

As $t_{0.025,12} = 2.179$ institutions 2 and 3 cannot be said to differ
with regard to faculty compensation at the 5% level.

Exercises 11.3

Use the program RBLOCK as required.

1. Three publishers are to be compared with regard to cost of math-
    ematics and statistics texts. The costs of books in different
    fields for each publisher is given below.

Text	Publisher A	B	C
Calculus	12.95	19.25	15.95
Elementary statistics	7.95	10.95	10.95
Time series	12.95	20.50	12.95
Statistical methods	16.95	19.00	15.00
Group theory	9.95	12.00	11.00

   (a) Do the data indicate a true difference in the cost of books
        produced by the three publishers? Carry out a test at level
        $\alpha = 0.05$.

   (b) The first publisher is not located in the United States and
        the other two are. Test the null hypothesis that the cost
        of books published by the foreign publisher does not differ
        from the U. S. publishers ($\alpha = 0.05$).

2. The average monthly payment to various categories of individuals
    for assistance is given below for five states for December 1973.

Group	Alabama	Georgia	State Massachusetts	Utah	Wisconsin
Old age	72.92	58.56	121.21	46.73	89.49
Blind	102.54	75.36	156.53	87.19	92.63
Disabled	78.88	69.25	153.61	82.45	103.08
AFDC	21.72	32.22	95.93	61.86	95.42
General assistance	12.50	28.40	106.49	73.01	56.66

Source: *The World Almanac and Book of Facts*, 1975.

(a) Do the data suggest a difference in the level of assistance
     payments among these states ($\alpha = 0.05$).
(b) Compare the two southern states with the others ($\alpha = 0.05$).
     Compare Massachusetts with all others ($\alpha = 0.05$).

A.1   AN INTRODUCTION TO BASIC

BASIC (beginner's all-purpose symbolic instruction code) is a computer programming language developed initially at Dartmouth College in 1963 and 1964 by Professors John G. Kemeny and Thomas E. Kurtz (1972). It was designed to be used in the time-sharing mode, i.e., by users located at teletype terminals remote from the central computer. Such a system permits the apparent simultaneous communication by many users with the computer. The language was developed in order to permit easy access to the power of the digital computer by persons without advanced training in computer technology. Hence the grammar, format, and statements in the language are quite simple and easy to learn. The language has proved to be quite popular for these reasons and is now widely used.

It will be assumed that the reader has some familiarity with a programming language. However, we review here certain elementary definitions:

DEFINITION A.1.1. A computer *program* is a list of instructions, which when executed causes the computer to perform a sequence of operations in a given order.

DEFINITION A.1.2. A *statement* is one line in a program, requiring
the computer to take a specific action.

DEFINITION A.1.3. A *variable*, written in BASIC as a single letter
(e.g., T) or a single letter followed by a digit (e.g., T2), can
take on various values. A variable is thus similar to a variable in
algebra.

DEFINITION A.1.4. A *constant* is a fixed numerical value.

DEFINITION A.1.5. A symbolic arithmetic *operation*, for example, *
causes two variables, two constants, or a variable and a constant to
be combined arithmetically. For example, A*B means to multiply A
by B.

To illustrate these definitions and some elementary BASIC grammar, consider the following simple program in BASIC.

```
10 PRINT "N","A"
20 LET N=1
30 LET A=(1+1/N)↑N
40 PRINT N,A
50 LET N=N+1
60 IF N>10 THEN 80
70 GO TO 30
80 END
```

There are several properties of this simple program which are common
to all BASIC programs. First, all statements in the program are
numbered. They will be executed by the computer according to increasing statement numbers. Secondly, the last statement is END,
which is true in all BASIC programs. We see that the variables N
and A have given values in the program. The constant 1 has been

used in statements 20, 30, and 50. Finally the algebraic operators
+, /, and ↑ standing for addition, division, and exponentiation,
respectively, have been employed.

Before making a systematic study of the types of statements
used in BASIC, let us consider what the preceding program achieves.
When the program is given the command RUN, the following output
results

N	A
1	2
2	2.25
3	2.370370
4	2.441406
5	2.48832
6	2.521626
7	2.546500
8	2.565785
9	2.581174
10	2.593742

The program has computed $(1 + 1/N)^N$ for N = 1, 2, ..., 10. This
sequence converges to e = 2.718282 as N → ∞. (Note that the conver-
gence is slow, so that this is not an efficient method of approximating e.)

Observing the statements in the program in order, we see that
the PRINT statement using quotation marks causes the printing of the
letters N and A. The LET statments in 20 and 30 assign numerical
values to the variables N and A. The PRINT statement in 40 causes
the *current values* of N and A to be printed. The statement LET N =
N + 1 shows that the equal sign should be interpreted as a replacement. The variable N is replaced by the current value of N plus 1.
The IF ... THEN statement in 60 causes the statement 80 END to be
executed if N exceeds 10. If N is less or equal to 10, control passes
to statement 70. This GO TO statement transfers control to statement 30. Note that if statement 60 were omitted, then a never-ending
loop would be created.

A.2　　ELEMENTARY STATEMENTS AND OPERATIONS IN BASIC

The elementary statements in BASIC can be classified into three types:
(1) those causing the computer to receive input or to produce output,
(2) those requiring the computer to perform a calculation, and (3)
those affecting the sequence in which the computer executes statements in a program.

A.2.1　　*Input and Output Statements*

1.　The READ...DATA Statements:　The READ statement causes a variable
or variables to be assigned a numerical value or values given in a
DATA statement.　For example,

　　10 READ X, Y, Z
　　100 DATA 5, 9, 17

assigns the variable X the value 5, Y the value 9, and Z the value
17.　The READ statement causes the values in DATA statements which
have not been read previously to be assigned to variables.　Thus if
the value 5 in statement 100 had already been used as data and the
statement

　　110 DATA 4, 2, 1

followed statement 100, then X would be given the value 9, Y the
value 17, and Z the value 4.　The READ statement must always be used
with a DATA statement.　If a numerical value is not available to be
assigned to a variable, the program will terminate with the statement OUT OF DATA.

2.　The INPUT Statement:　The INPUT statement causes the computer
to stop executing the program and request the value or values of
variables to be typed in from the terminal.　For example, the statements

　　10 PRINT "WHAT IS YOUR AGE AND WEIGHT?"
　　20 INPUT A, W

when executed result in the following:

WHAT IS YOUR AGE AND WEIGHT?
?

The program will not continue until the programmer respons with two
numerical values, e.g., ? <u>24,160</u>. (The underlined values indicate
that these are provided at the terminal.) Following transmission
of these values the variable A is assigned the value 24 and the
variable W is assigned the value 160.

3. The LET Statement: The LET statement directly assigns a value
to a variable. For example, LET T2 = 3.14159 assigns T2 the indicated
value.

4. The PRINT Statement: The PRINT statement causes either the
printing of the value of a variable, a constant, or some text.
Suppose the program in Sec. A.2.1.2 is completed with

        10 PRINT "WHAT IS YOUR AGE AND WEIGHT?"
        20 INPUT A, W
        30 PRINT A, W
        40 END

We have seen that line 10 causes the words in quotation marks to be
printed out. The PRINT statement in line 30 would cause the values
24 and 160 to be printed. The PRINT statement alone causes a line
to be skipped in the output.

A.2.2   Statements Causing a Calculation to be Performed

The symbols in BASIC which are used to perform the elementary arithmetic operations are indicated below:

Symbol	Expression	Meaning
+	X + Y	Add X to Y
−	X − Y	Subtract Y from X
*	X * Y	Multiply X by Y
/	X / Y	Divide X by Y
↑	X ↑ Y	Raise X to power Y

These symbols may be used to combine variables and constants. The most common use is in a LET statement in which a variable is given a value in terms of variables currently assigned values. The arithmetic operations are allowed only on the right-hand side of the equal sign. For example, consider

10 LET Z = (X↑2 + Y↑2)/2

The variable Z is assigned one-half of the sum of the squares of the values currently assigned to X and Y.

The hierarchy of the operations is exponentiation, multiplication and division, and addition and subtraction. This means that exponentiation will be carried out first. Multiplication and division follow and will be performed in the order in which the symbols appear. Addition and subtraction are performed last. As examples, consider

Expression	*Value*
3*2↑3+1	25
(3*2)↑3+1	217
2↑4/4	4
2/4↑2	0.125
2*4↑2	32
(2*3)↑2-4*2	28
2↑(-2)	0.25
3/4*8	6
3*4-2	10
3*(4-2)	6

The reader should check to see that these expressions are correctly evaluated. If doubt exists as to how an expression is written, then sufficient parentheses should be used to ensure that the correct expression is being computed. Furthermore, it is not allowable to use two symbolic operators adjacent to one another in an expression. Hence in writing 3 to the power -4 in BASIC, 3↑-4 is not permissible. The correct expression is 3↑(-4).

A.2.3   *Statements Controlling the Sequence in Which the Computer Executes Program Statements*

As remarked above, all statements in a BASIC program are numbered. The statements will be executed by the program in the order of the statement numbers. There need not be a statement for each number and, in fact, this is a bad idea, as program corrections will often involve the insertion of additional statements. It is of fundamental importance to computing that a program can contain statements which allow the determination of the order in which statements within the program are executed. This permits the repetition of steps in a program by the computer without a repetition of the statements in the program.

1. The GO TO Statement: The GO TO statement requires the computer to break its normal sequence of execution and to execute the statement indicated. In the example program of Sec. A.1, the statement 70 GO TO 30 requires statement 30 to be executed following this instruction. Thereafter the computer would execute statements in numerical order following 30.

2. THE IF...THEN... Statement: The IF...THEN... statement can be thought of as a conditional GO TO statement. If the terms of the conditional statement are met, then the control passes to the indicated statement. If the terms are not met, the next statement in numerical order is executed. In line 60 of the program in Sec. A.1 we have

        60 IF N > 10 THEN 80

The conditional statement is $N > 10$. If this condition is met by the current value of N, then the next statement executed is 80 END. Otherwise, statement 70 is executed next.

The relational operators which may appear in the conditional portion of the IF...THEN... statement are as follows:

Symbol	Meaning
=	Equal
<	Less than
>	Greater than
<=	Less or equal
>=	Greater or equal
<>	Unequal

3. The FOR..., NEXT... Statements: The ability of the digital computer to perform repetitive calculations very swiftly and accurately is one of the major factors in its widespread use. The statements which permit the repeated execution of a certain sequence of statements are the FOR and NEXT statements. We illustrate these in the following program.

```
10 PRINT "APPROXIMATION,"NEXT TERM"
20 READ N
30 LET T = 1
40 LET S = 0
50 FOR I = 1 TO N
60 LET S = S+T
70 LET T = T/I
80 PRINT S,T
90 NEXT I
100 DATA 10
110 END
```

Statements 50 through 90 constitute the FOR...,NEXT... "loop" in this program. The variable I is set equal to 1 the first time that statement 50 is executed. Statements 60 to 80 are then executed. At statement 90 if I is less than N, control is returned to statement 50, I is increased by 1, and the loop continued until I = N at statement 90. At this point control passes to the next statement in numerical order following the NEXT statement.

Let us look at the output from this program.

Approximation	Next term
1	1
2	0.5
2.5	0.166667
2.66667	0.0416667
2.70833	0.00833333
2.71667	0.00138889
2.71806	0.000198413
2.71825	0.0000248016
2.71828	0.00000275573
2.71828	0.000000275573

It is easy to check that this program computes the Maclaurin expansion approximations to e, that is,

$$\sum_{k=0}^{n-1} \frac{1}{k!} \quad \text{for } n = 1, 2, \ldots, 10$$

It is known that the error is less than 3/n! for such approximation. Thus we see that for n = 10, the error is less than 3(0.000000275573) = 0.000000826719. This program does provide an efficient method of approximating the constant e, in contrast with the program of Sec. A.1.

The FOR..., NEXT... statements provide greater flexibility in looping than has been indicated. The step size for the "stepping variable" I was 1 in the example program. However, this need not be the case. The statements

```
10 FOR X = 0.0 TO 1.0 STEP 0.05
   .
   .
   .
40 NEXT X
```

cause the loop to be executed 21 times with X taking the values 0.0, 0.05, 0.10, ..., 1.00, successively. This is particularly useful when it is desired to evaluate and print out the value of an expression depending upon X over a range of values of X. If the STEP portion of a FOR command is omitted, the step size is assumed to be 1.

A.3   ADDITIONAL INFORMATION ABOUT BASIC

A.3.1   The REM Statement

In order to make a BASIC program more readable, a remark in English
text may be included at any point in the program. For example, the
statement

　　5 REM COMPUTES AN APPROXIMATION OF e

might be the first statement of the program in Sec. A.2.3.3. The
computer ignores remark statements in the execution of a program.

A.3.2   Functions Available in BASIC

There are some functions such as the elementary trigonometric functions (the sine, cosine, and tangent) which are used so frequently
that BASIC provides the means of calculating these functions in a
single statement. For example the program

```
10 LET X = 1
20 LET Y = EXP(X)
30 PRINT X,Y
40 END
```

produces the output

　　1　　2.7182818

The function e^X has been evaluated for X = 1.

　　The following functions are available in BASIC:

Function	Mathematical meaning
ABS(X)	$\lvert x \rvert$
ATN(X)	Arctan x
COS(X)	Cosine x
COT(X)	Cotangent x
EXP(X)	e^X
INT(X)	$[x]$
LOG(X)	ln x

SGN(X)	−1 if x < 0, 1 if x > 0, and 0 if x = 0
SIN(X)	Sine x
SQR(X)	$\sqrt{x}$
TAN(X)	Tangent x

Such a function has an argument in parentheses which must be either a constant or a variable whose value is defined. The arguments of the trigonometric functions are assumed to be in radians.

A.3.3   Subscripted Variables

When it is improtant to have a vector of numbers available for computation the BASIC system allows the definition of a subscripted variable. Consider the following example:

```
10 READ N
20 FOR I = 1 TO N
30 READ X(I)
40 NEXT I
50 DATA 8
60 DATA 4, 8, 7, 16, 2, 1, 2, 5
```

This portion of a program will cause the subscripted variable X to take on the values $X(1) = 4$, $X(2) = 8$, $X(3) = 7$, $X(4) = 16$, $X(5) = 2$, $X(6) = 1$, $X(7) = 2$, and $X(8) = 5$. These values can then be used in subsequent calculations, for example, for the mean, variance, and standard deviation of the N numbers.

A subscripted variable may also have two subscripts. For example, $T(I,J)$ is a variable which can store an array or table of values. The first subscript represents the row in the table and the second subscript represents the column. The loop causing the reading of a table with R rows and C columns would be

```
10 FOR I = 1 TO R
20 FOR J = 1 TO C
30 READ T(I,J)
40 NEXT J
50 NEXT I
```

Note that the table is read by rows. If the index for a variable
with a single subscript takes on values exceeding 10, or if either
of the indexes exceed 10 for a double subscripted variable, it is
necessary to include a statement giving the maximum index or indexes.
The statement is of form

    10 DIM X(100), T(20,20)

The dimension statement DIM may appear at any point in the program.

A.3.4 Definition of Functions

In some programs a certain function, not among those available in
BASIC, is used so frequently that it becomes efficient to *define* a
new function in the program. This is accomplished by the DEF command.
For example, suppose the calculation of the function $f(x) = x + \sin x$
$+ \sqrt{x^2 + 1}$ is required repeatedly in a program. We use the following
statement

    10 DEF FNF(X) = X + SIN(X) + SQR(X↑2+1)

to define a new function FNF. The statement does not require the
calculation of any quantity, and the argument X is a dummy argument.
If the statement

    20 Y = FNF(0)

appeared later in the program, then Y would be given the value of
$f(0) = 1$. The functions which can be defined in a program are of
form FN_, where the last character is a letter, e.g., FNA, FNT, FNW.

A.3.5 The GOSUB and RETURN Statements

We illustrate the use of these statements in the following example:

    10 DEF FNF(X) = X↑2
    20 READ A,B
    30 GOSUB 200
    40 PRINT "LOWER ENDPOINT","UPPER ENDPOINT","INTEGRAL APPROX"

```
 50 PRINT A,B,S
 60 GO TO 20
190 REM SUBROUTINE FOR TRAPEZOIDAL RULE
200 LET H = (B-A)/100
210 LET S = 0
220 FOR I = 1 TO 99
230 LET S = S + FNF(A+H*I)
240 NEXT I
250 LET S = S + 0.5 * (FNF(A) + FNF(B))
260 LET S = H*S
270 RETURN
280 DATA 0,1
290 DATA 1,2
300 DATA 2,3
310 END
```

The effect of the GOSUB statement is to transfer control to the
statement indicated in the GOSUB command. In this case the statements in lines 200 through 260 carry out an approximate integration
of the function FNF(X) on the interval with endpoints given by A
and B, using the trapezoidal rule (with 100 subintervals). When the
statement 270 RETURN is encountered, control is returned to the
statement immediately following the GOSUB statement. In the example
this would be the PRINT statement in line 40.

This program computes approximations to the integral of x^2 on
the intervals [0, 1], [1, 2], and [2, 3]. The values obtained from
a RUN of the program were 0.33335, 2.33335, and 6.33335. These are
approximations of 1/3, 7/3, and 19/3, the exact values. It is clear
that a simple change of line 10 would allow approximate integration
of other functions. The GOSUB... RETURN statements are useful when
a particular algorithm is to be used frequently in the program. The
values of any variables in the subroutine portion of the program
must either have been assigned in the main program itself or be
assigned in the subroutine.

APPENDIX B: TABLES

Table B.I   Cumulative Normal Distribution[a]

$$F_Z(z) = \int_{-\infty}^{z} \frac{1}{\sqrt{2\pi}} e^{-z^2/2} dz$$

Example:

$F_Z(z)$

z	.00	.01	.02	.03	.04	.05	.06	.07	.08	.09
.0	.5000	.5040	.5080	.5120	.5160	.5199	.5239	.5279	.5319	.5359
.1	.5398	.5438	.5478	.5517	.5557	.5596	.5636	.5675	.5714	.5753
.2	.5793	.5832	.5871	.5910	.5948	.5987	.6026	.6064	.6103	.6141
.3	.6179	.6217	.6255	.6293	.6331	.6368	.6406	.6443	.6480	.6517
.4	.6554	.6591	.6628	.6664	.6700	.6736	.6772	.6808	.6844	.6879
.5	.6915	.6950	.6985	.7019	.7054	.7088	.7123	.7157	.7190	.7224
.6	.7257	.7291	.7324	.7357	.7389	.7422	.7454	.7486	.7517	.7549
.7	.7580	.7611	.7642	.7673	.7704	.7734	.7764	.7794	.7823	.7852
.8	.7881	.7910	.7939	.7967	.7995	.8023	.8051	.8078	.8106	.8133
.9	.8159	.8186	.8212	.8238	.8264	.8289	.8315	.8340	.8365	.8389
1.0	.8413	.8438	.8461	.8485	.8508	.8531	.8554	.8577	.8599	.8621
1.1	.8643	.8665	.8686	.8708	.8729	.8749	.8770	.8790	.8810	.8830
1.2	.8849	.8869	.8888	.8907	.8925	.8944	.8962	.8980	.8997	.9015
1.3	.9032	.9049	.9066	.9082	.9099	.9115	.9131	.9147	.9162	.9177
1.4	.9192	.9207	.9222	.9236	.9251	.9265	.9279	.9292	.9306	.9319
1.5	.9332	.9345	.9357	.9370	.9382	.9394	.9406	.9418	.9429	.9441
1.6	.9452	.9463	.9474	.9484	.9495	.9505	.9515	.9525	.9535	.9545
1.7	.9554	.9564	.9573	.9582	.9591	.9599	.9608	.9616	.9625	.9633
1.8	.9641	.9649	.9656	.9664	.9671	.9678	.9686	.9693	.9699	.9706
1.9	.9713	.9719	.9726	.9732	.9738	.9744	.9750	.9756	.9761	.9767
2.0	.9772	.9778	.9783	.9788	.9793	.9798	.9803	.9808	.9812	.9817
2.1	.9821	.9826	.9830	.9834	.9838	.9842	.9846	.9850	.9854	.9857
2.2	.9861	.9864	.9868	.9871	.9875	.9878	.9881	.9884	.9887	.9890
2.3	.9893	.9896	.9898	.9901	.9904	.9906	.9909	.9911	.9913	.9916
2.4	.9918	.9920	.9922	.9925	.9927	.9929	.9931	.9932	.9934	.9936
2.5	.9938	.9940	.9941	.9943	.9945	.9946	.9948	.9949	.9951	.9952
2.6	.9953	.9955	.9956	.9957	.9959	.9960	.9961	.9962	.9963	.9964
2.7	.9965	.9966	.9967	.9968	.9969	.9970	.9971	.9972	.9973	.9974
2.8	.9974	.9975	.9976	.9977	.9977	.9978	.9979	.9979	.9980	.9981
2.9	.9981	.9982	.9982	.9983	.9984	.9984	.9985	.9985	.9986	.9986
3.0	.9987	.9987	.9987	.9988	.9988	.9989	.9989	.9989	.9990	.9990
3.1	.9990	.9991	.9991	.9991	.9992	.9992	.9992	.9992	.9993	.9993
3.2	.9993	.9993	.9994	.9994	.9994	.9994	.9994	.9995	.9995	.9995
3.3	.9995	.9995	.9995	.9996	.9996	.9996	.9996	.9996	.9996	.9997
3.4	.9997	.9997	.9997	.9997	.9997	.9997	.9997	.9997	.9997	.9998

[a]Adapted from Table 1 of E. S. Pearson and H. O. Hartley (eds.)
(1970), *Biometrika Tables for Statisticians*, Vol. 1, 3rd Ed.,
Cambridge University Press, Cambridge, England, with permission of
the Biometrika Trustees.

Table B.II   Student's t Distribution[a]

An entry in the table gives the area or probability $P(t > t_{\alpha,\nu}) = \alpha$, where t has a Student's t distribution with ν = degrees of freedom.

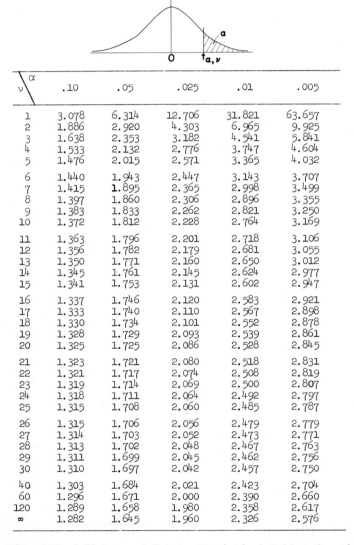

ν \ α	.10	.05	.025	.01	.005
1	3.078	6.314	12.706	31.821	63.657
2	1.886	2.920	4.303	6.965	9.925
3	1.638	2.353	3.182	4.541	5.841
4	1.533	2.132	2.776	3.747	4.604
5	1.476	2.015	2.571	3.365	4.032
6	1.440	1.943	2.447	3.143	3.707
7	1.415	1.895	2.365	2.998	3.499
8	1.397	1.860	2.306	2.896	3.355
9	1.383	1.833	2.262	2.821	3.250
10	1.372	1.812	2.228	2.764	3.169
11	1.363	1.796	2.201	2.718	3.106
12	1.356	1.782	2.179	2.681	3.055
13	1.350	1.771	2.160	2.650	3.012
14	1.345	1.761	2.145	2.624	2.977
15	1.341	1.753	2.131	2.602	2.947
16	1.337	1.746	2.120	2.583	2.921
17	1.333	1.740	2.110	2.567	2.898
18	1.330	1.734	2.101	2.552	2.878
19	1.328	1.729	2.093	2.539	2.861
20	1.325	1.725	2.086	2.528	2.845
21	1.323	1.721	2.080	2.518	2.831
22	1.321	1.717	2.074	2.508	2.819
23	1.319	1.714	2.069	2.500	2.807
24	1.318	1.711	2.064	2.492	2.797
25	1.315	1.708	2.060	2.485	2.787
26	1.315	1.706	2.056	2.479	2.779
27	1.314	1.703	2.052	2.473	2.771
28	1.313	1.702	2.048	2.467	2.763
29	1.311	1.699	2.045	2.462	2.756
30	1.310	1.697	2.042	2.457	2.750
40	1.303	1.684	2.021	2.423	2.704
60	1.296	1.671	2.000	2.390	2.660
120	1.289	1.658	1.980	2.358	2.617
∞	1.282	1.645	1.960	2.326	2.576

[a]From Table III of Fisher and Yates, *Statistical Tables for Biological Agricultural and Medical Research*, published by Longman Group, Ltd., London, 1974. (Previously published by Oliver and Bond, Edinburgh) with permission of the authors and publishers.

Table B.III   The χ^2 Distribution[a]

An entry in the table gives the area or probability $P(\chi^2 > \chi^2_{\alpha,\nu}) = \alpha$, where χ^2 has a chi-square distribution with ν = degrees of freedom.

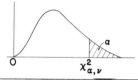

P ν	0.995	0.990	0.975	0.950	0.050	0.025	0.010	0.005
1	0.0000393	0.0001571	0.0009821	0.0039321	3.84146	5.02389	6.63490	7.87944
2	0.0100251	0.0201007	0.0506356	0.102587	5.99147	7.37776	9.21034	10.5966
3	0.0717212	0.114832	0.215795	0.351846	7.81473	9.34840	11.3449	12.8381
4	0.206990	0.297110	0.484419	0.710721	9.48773	11.1433	13.2767	14.8602
5	0.411740	0.554300	0.831211	1.145476	11.0705	12.8325	15.0863	16.7496
6	0.675727	0.872085	1.237347	1.63539	12.5916	14.4494	16.8119	18.5476
7	0.989265	1.239043	1.68987	2.16735	14.0671	16.0128	18.4753	20.2777
8	1.344419	1.646482	2.17973	2.73264	15.5073	17.5346	20.0902	21.9550
9	1.734926	2.087912	2.70039	3.32511	16.9190	19.0228	21.6660	23.5893
10	2.15585	2.55821	3.24697	3.94030	18.3070	20.4831	23.2093	25.1882
11	2.60321	3.05347	3.81575	4.57481	19.6751	21.9200	24.7250	26.7569
12	3.07382	3.57056	4.40379	5.22603	21.0261	23.3367	26.2170	28.2995
13	3.56503	4.10691	5.00874	5.89186	22.3621	24.7356	27.6883	29.8194
14	4.07468	4.66043	5.62872	6.57063	23.6848	26.1190	29.1413	31.3193
15	4.60094	5.22935	6.26214	7.26094	24.9958	27.4884	30.5779	32.8013
16	5.14224	5.81221	6.90766	7.96164	26.2962	28.8454	31.9999	34.2672
17	5.69724	6.40776	7.56418	8.67176	27.5871	30.1910	33.4087	35.7185
18	6.26481	7.01491	8.23075	9.39046	28.8693	31.5264	34.8053	37.1564
19	6.84398	7.63273	8.90655	10.1170	30.1435	32.8523	36.1908	38.5822
20	7.43386	8.26040	9.59083	10.8508	31.4104	34.1696	37.5662	39.9968
21	8.03366	8.89720	10.28293	11.5913	32.6705	35.4789	38.9321	41.4010
22	8.64272	9.54249	10.9823	12.3380	33.9244	36.7807	40.2894	42.7956
23	9.26042	10.19567	11.6885	13.0905	35.1725	38.0757	41.6384	44.1813
24	9.88623	10.8564	12.4011	13.8484	36.4154	39.3641	42.9798	45.5585
25	10.5197	11.5240	13.1197	14.6114	37.6525	40.6465	44.3141	46.9278
26	11.1603	12.1981	13.8439	15.3791	38.8852	41.9232	45.6417	48.2899
27	11.8076	12.8786	14.5733	16.1513	40.1133	43.1944	46.9630	49.6449
28	12.4613	13.5648	15.3079	16.9279	41.3372	44.4607	48.2782	50.9933
29	13.1211	14.2565	16.0471	17.7083	42.5569	45.7222	49.5879	52.3356
30	13.7867	14.9535	16.7908	18.4926	43.7729	46.9792	50.8922	53.6720
40	20.7065	22.1643	24.4331	26.5093	55.7585	59.3417	63.6907	66.7659
50	27.9907	29.7067	32.3574	34.7642	67.5048	71.4202	76.1539	79.4900
60	35.5346	37.4848	40.4817	43.1879	79.0819	83.2976	88.3794	91.9517
70	43.2752	45.4418	48.7576	51.7393	90.5312	95.0231	100.425	104.215
80	51.1720	53.5400	57.1532	60.3915	101.879	106.629	112.329	116.321
90	59.1963	61.7541	65.6466	69.1260	113.145	118.136	124.116	128.299
100	67.3276	70.0648	74.2219	77.9295	124.342	129.561	135.807	140.169

Table B.IV   d-Factors for Wilcoxon Signed Rank Test and Confidence Intervals for the Median[a],[b]

n	d	γ	α''	α'	n	d	γ	α''	α'
3	1	.750	.250	.125	12	8	.991	.009	.005
4	1	.875	.125	.062		9	.988	.012	.006
5	1	.938	.062	.031		14	.958	.042	.021
	2	.875	.125	.063		15	.948	.052	.026
6	1	.969	.031	.016		18	.908	.092	.046
	2	.937	.063	.031		19	.890	.110	.055
	3	.906	.094	.047	13	10	.992	.008	.004
	4	.844	.156	.078		11	.990	.010	.005
7	1	.984	.016	.008		18	.952	.048	.024
	3	.953	.047	.016		19	.943	.057	.029
	4	.922	.078	.039		22	.906	.094	.047
	5	.891	.109	.055		23	.890	.110	.055
8	1	.992	.008	.004	14	13	.991	.009	.004
	2	.984	.016	.008		14	.989	.011	.005
	4	.961	.039	.020		22	.951	.049	.025
	5	.945	.055	.027		23	.942	.058	.029
	6	.922	.078	.039		26	.909	.091	.045
	7	.891	.109	.055		27	.896	.104	.052
9	2	.992	.008	.004	15	16	.992	.008	.004
	3	.988	.012	.006		17	.990	.010	.005
	6	.961	.039	.020		26	.952	.048	.024
	7	.945	.055	.027		27	.945	.055	.028
	9	.902	.098	.049		31	.905	.095	.047
	10	.871	.129	.065		32	.893	.107	.054
10	4	.990	.010	.005	16	20	.991	.009	.005
	5	.986	.014	.007		21	.989	.011	.006
	9	.951	.049	.024		30	.956	.044	.022
	10	.936	.064	.032		31	.949	.051	.025
	11	.916	.084	.042		36	.907	.093	.047
	12	.895	.105	.053		37	.895	.105	.052
11	6	.990	.010	.005	17	24	.991	.009	.005
	7	.986	.014	.007		25	.989	.011	.006
	11	.958	.042	.021		35	.955	.045	.022
	12	.946	.054	.027		36	.949	.051	.025
	14	.917	.083	.042		42	.902	.098	.049
	15	.898	.102	.051		43	.891	.109	.054

Table B.IV (Continued)

n	d	γ	α''	α'	n	d	γ	α''	α'
18	28	.991	.009	.005	22	49	.991	.009	.005
	29	.990	.010	.005		50	.990	.010	.005
	41	.952	.048	.024		66	.954	.046	.023
	42	.946	.054	.027		67	.950	.050	.025
	48	.901	.099	.049		76	.902	.098	.049
	49	.892	.108	.054		77	.895	.105	.053
19	33	.991	.009	.005	23	55	.991	.009	.005
	34	.989	.011	.005		56	.990	.010	.005
	47	.951	.049	.025		74	.952	.048	.024
	48	.945	.055	.027		75	.948	.052	.026
	54	.904	.096	.048		84	.902	.098	.049
	55	.896	.104	.052		85	.895	.105	.052
20	38	.991	.009	.005	24	62	.990	.010	.005
	39	.989	.011	.005		63	.989	.011	.005
	53	.952	.048	.024		82	.951	.049	.025
	54	.947	.053	.027		83	.947	.053	.026
	61	.903	.097	.049		92	.905	.095	.048
	62	.895	.105	.053		93	.899	.101	.051
21	43	.991	.009	.005	25	69	.990	.010	.005
	44	.990	.010	.005		70	.989	.011	.005
	59	.954	.046	.023		90	.952	.048	.024
	60	.950	.050	.025		91	.948	.052	.026
	68	.904	.096	.048		101	.904	.096	.048
	69	.897	.103	.052		102	.899	.101	.051

[a]From *Introduction to Statistics: A Fresh Approach.*   Copyright 1971 by Houghton Mifflin Company.   Used by permission of the publisher.

[b]γ = confidence coefficient; α' = $(1/2)(1 - \gamma)$ = one-sided significance level; α'' = $2\alpha'$ = $1 - \gamma$ = two-sided significance level.

Table B.Va   Values of F 0.05[a]

Vertical degrees of freedom

	1	2	3	4	5	6	7	8	9	10	12	15	20	24	30	40	60	120	∞
1	161	200	216	225	230	234	237	239	241	242	244	246	248	249	250	251	252	253	254
2	18.5	19.0	19.2	19.2	19.3	19.3	19.4	19.4	19.4	19.4	19.4	19.4	19.4	19.5	19.5	19.5	19.5	19.5	19.5
3	10.1	9.55	9.28	9.12	9.01	8.94	8.89	8.85	8.81	8.79	8.74	8.70	8.66	8.64	8.62	8.59	8.57	8.55	8.53
4	7.71	6.94	6.59	6.39	6.26	6.16	6.09	6.04	6.00	5.96	5.91	5.86	5.80	5.77	5.75	5.72	5.69	5.66	5.63
5	6.61	5.79	5.41	5.19	5.05	4.95	4.88	4.82	4.77	4.74	4.68	4.62	4.56	4.53	4.50	4.46	4.43	4.40	4.37
6	5.99	5.14	4.76	4.53	4.39	4.28	4.21	4.15	4.10	4.06	4.00	3.94	3.87	3.84	3.81	3.77	3.74	3.70	3.67
7	5.59	4.74	4.35	4.12	3.97	3.87	3.79	3.73	3.68	3.64	3.57	3.51	3.44	3.41	3.38	3.34	3.30	3.27	3.23
8	5.32	4.46	4.07	3.84	3.69	3.58	3.50	3.44	3.39	3.35	3.28	3.22	3.15	3.12	3.08	3.04	3.01	2.97	2.93
9	5.12	4.26	3.86	3.63	3.48	3.37	3.29	3.23	3.18	3.14	3.07	3.01	2.94	2.90	2.86	2.83	2.79	2.75	2.71
10	4.96	4.10	3.71	3.48	3.33	3.22	3.14	3.07	3.02	2.98	2.91	2.85	2.77	2.74	2.70	2.66	2.62	2.58	2.54
11	4.84	3.98	3.59	3.36	3.20	3.09	3.01	2.95	2.90	2.85	2.79	2.72	2.65	2.61	2.57	2.53	2.49	2.45	2.40
12	4.75	3.89	3.49	3.26	3.11	3.00	2.91	2.85	2.80	2.75	2.69	2.62	2.54	2.51	2.47	2.43	2.38	2.34	2.30
13	4.67	3.81	3.41	3.18	3.03	2.92	2.83	2.77	2.71	2.67	2.60	2.53	2.46	2.42	2.38	2.34	2.30	2.25	2.21
14	4.60	3.74	3.34	3.11	2.96	2.85	2.76	2.70	2.65	2.60	2.53	2.46	2.39	2.35	2.31	2.27	2.22	2.18	2.13
15	4.54	3.68	3.29	3.06	2.90	2.79	2.71	2.64	2.59	2.54	2.48	2.40	2.33	2.29	2.25	2.20	2.16	2.11	2.07
16	4.49	3.63	3.24	3.01	2.85	2.74	2.66	2.59	2.54	2.49	2.42	2.35	2.28	2.24	2.19	2.15	2.11	2.06	2.01
17	4.45	3.59	3.20	2.96	2.81	2.70	2.61	2.55	2.49	2.45	2.38	2.31	2.23	2.19	2.15	2.10	2.06	2.01	1.96
18	4.41	3.55	3.16	2.93	2.77	2.66	2.58	2.51	2.46	2.41	2.34	2.27	2.19	2.15	2.11	2.06	2.02	1.97	1.92
19	4.38	3.52	3.13	2.90	2.74	2.63	2.54	2.48	2.42	2.38	2.31	2.23	2.16	2.11	2.07	2.03	1.98	1.93	1.88
20	4.35	3.49	3.10	2.87	2.71	2.60	2.51	2.45	2.39	2.35	2.28	2.20	2.12	2.08	2.04	1.99	1.95	1.90	1.84
21	4.32	3.47	3.07	2.84	2.68	2.57	2.49	2.42	2.37	2.32	2.25	2.18	2.10	2.05	2.01	1.96	1.92	1.87	1.81
22	4.30	3.44	3.05	2.82	2.66	2.55	2.46	2.40	2.34	2.30	2.23	2.15	2.07	2.03	1.98	1.94	1.89	1.84	1.78
23	4.28	3.42	3.03	2.80	2.64	2.53	2.44	2.37	2.32	2.27	2.20	2.13	2.05	2.01	1.96	1.91	1.86	1.81	1.76
24	4.26	3.40	3.01	2.78	2.62	2.51	2.42	2.36	2.30	2.25	2.18	2.11	2.03	1.98	1.94	1.89	1.84	1.79	1.73
25	4.24	3.39	2.99	2.76	2.60	2.49	2.40	2.34	2.28	2.24	2.16	2.09	2.01	1.96	1.92	1.87	1.82	1.77	1.71
30	4.17	3.32	2.92	2.69	2.53	2.42	2.33	2.27	2.21	2.16	2.09	2.01	1.93	1.89	1.84	1.79	1.74	1.68	1.62
40	4.08	3.23	2.84	2.61	2.45	2.34	2.25	2.18	2.12	2.08	2.00	1.92	1.84	1.79	1.74	1.69	1.64	1.58	1.51
60	4.00	3.15	2.76	2.53	2.37	2.25	2.17	2.10	2.04	1.99	1.92	1.84	1.75	1.70	1.65	1.59	1.53	1.47	1.39
120	3.92	3.07	2.68	2.45	2.29	2.18	2.09	2.02	1.96	1.91	1.83	1.75	1.66	1.61	1.55	1.50	1.43	1.35	1.25
∞	3.84	3.00	2.60	2.37	2.21	2.10	2.01	1.94	1.88	1.83	1.75	1.67	1.57	1.52	1.46	1.39	1.32	1.22	1.00

Horizontal degrees of freedom

[a]Adapted from The Biometrika Tables for Statisticians, Vol. 1, 3rd Ed. (E. S. Pearson and H. O. Hartley, eds.), with permission of the Biometrika Trustees.

Table B.Vb   Values of F 0.01[a]

Vertical degrees of freedom

	1	2	3	4	5	6	7	8	9	10	12	15	20	24	30	40	60	120	∞
1	4052	5000	5403	5625	5764	5859	5928	5982	6023	6056	6106	6157	6209	6235	6261	6287	6313	6339	6366
2	98.5	99.0	99.2	99.2	99.3	99.3	99.4	99.4	99.4	99.4	99.4	99.4	99.4	99.5	99.5	99.5	99.5	99.5	99.5
3	34.1	30.8	29.5	28.7	28.2	27.9	27.7	27.5	27.3	27.2	27.1	26.9	26.7	26.6	26.5	26.4	26.3	26.1	26.1
4	21.2	18.0	16.7	16.0	15.5	15.2	15.0	14.8	14.7	14.5	14.4	14.2	14.0	13.9	13.8	13.7	13.7	13.6	13.5
5	16.3	13.3	12.1	11.4	11.0	10.7	10.5	10.3	10.2	10.1	9.89	9.72	9.55	9.47	9.38	9.29	9.20	9.11	9.02
6	13.7	10.9	9.78	9.15	8.75	8.47	8.26	8.10	7.98	7.87	7.72	7.56	7.40	7.31	7.23	7.14	7.06	6.97	6.88
7	12.2	9.55	8.45	7.85	7.46	7.19	6.99	6.84	6.72	6.62	6.47	6.31	6.16	6.07	5.99	5.91	5.82	5.74	5.65
8	11.3	8.65	7.59	7.01	6.63	6.37	6.18	6.03	5.91	5.81	5.67	5.52	5.36	5.28	5.20	5.12	5.03	4.95	4.86
9	10.6	8.02	6.99	6.42	6.06	5.80	5.61	5.47	5.35	5.26	5.11	4.96	4.81	4.73	4.65	4.57	4.48	4.40	4.31
10	10.0	7.56	6.55	5.99	5.64	5.39	5.20	5.06	4.94	4.85	4.71	4.56	4.41	4.33	4.25	4.17	4.08	4.00	3.91
11	9.65	7.21	6.22	5.67	5.32	5.07	4.89	4.74	4.63	4.54	4.40	4.25	4.10	4.02	3.94	3.86	3.78	3.69	3.60
12	9.33	6.93	5.95	5.41	5.06	4.82	4.64	4.50	4.39	4.30	4.16	4.01	3.86	3.78	3.70	3.62	3.54	3.45	3.36
13	9.07	6.70	5.74	5.21	4.86	4.62	4.44	4.30	4.19	4.10	3.96	3.82	3.66	3.59	3.51	3.43	3.34	3.25	3.17
14	8.86	6.51	5.56	5.04	4.70	4.46	4.28	4.14	4.03	3.94	3.80	3.66	3.51	3.43	3.35	3.27	3.18	3.09	3.00
15	8.68	6.36	5.42	4.89	4.56	4.32	4.14	4.00	3.89	3.80	3.67	3.52	3.37	3.29	3.21	3.13	3.05	2.96	2.87
16	8.53	6.23	5.29	4.77	4.44	4.20	4.03	3.89	3.78	3.69	3.55	3.41	3.26	3.18	3.10	3.02	2.93	2.84	2.75
17	8.40	6.11	5.18	4.67	4.34	4.10	3.93	3.79	3.68	3.59	3.46	3.31	3.16	3.08	3.00	2.92	2.83	2.75	2.65
18	8.29	6.01	5.09	4.58	4.25	4.01	3.84	3.71	3.60	3.51	3.37	3.23	3.08	3.00	2.92	2.84	2.75	2.66	2.57
19	8.19	5.93	5.01	4.50	4.17	3.94	3.77	3.63	3.52	3.43	3.30	3.15	3.00	2.92	2.84	2.76	2.67	2.58	2.49
20	8.10	5.85	4.94	4.43	4.10	3.87	3.70	3.56	3.46	3.37	3.23	3.09	2.94	2.86	2.78	2.69	2.61	2.52	2.42
21	8.02	5.78	4.87	4.37	4.04	3.81	3.64	3.51	3.40	3.31	3.17	3.03	2.88	2.80	2.72	2.64	2.55	2.46	2.36
22	7.95	5.72	4.82	4.31	3.99	3.76	3.59	3.45	3.35	3.26	3.12	2.98	2.83	2.75	2.67	2.58	2.50	2.40	2.31
23	7.88	5.66	4.76	4.26	3.94	3.71	3.54	3.41	3.30	3.21	3.07	2.93	2.78	2.70	2.62	2.54	2.45	2.35	2.26
24	7.82	5.61	4.72	4.22	3.90	3.67	3.50	3.36	3.26	3.17	3.03	2.89	2.74	2.66	2.58	2.49	2.40	2.31	2.21
25	7.77	5.57	4.68	4.18	3.86	3.63	3.46	3.32	3.22	3.13	2.99	2.85	2.70	2.62	2.53	2.45	2.36	2.27	2.17
30	7.56	5.39	4.51	4.02	3.70	3.47	3.30	3.17	3.07	2.98	2.84	2.70	2.55	2.47	2.39	2.30	2.21	2.11	2.01
40	7.31	5.18	4.31	3.83	3.51	3.29	3.12	2.99	2.89	2.80	2.66	2.52	2.37	2.29	2.20	2.11	2.02	1.92	1.80
60	7.08	4.98	4.13	3.65	3.34	3.12	2.95	2.82	2.72	2.63	2.50	2.35	2.20	2.12	2.03	1.94	1.84	1.73	1.60
120	6.85	4.79	3.95	3.48	3.17	2.96	2.79	2.66	2.56	2.47	2.34	2.19	2.03	1.95	1.86	1.76	1.66	1.53	1.38
∞	6.63	4.61	3.78	3.32	3.02	2.80	2.64	2.51	2.41	2.32	2.18	2.04	1.88	1.79	1.70	1.59	1.47	1.32	1.00

Horizontal degrees of freedom

[a]Adapted from *The Biometrika Tables for Statisticians*, Vol. 1, 3rd Ed. (E. S. Pearson and H. O. Hartley, eds.), with permission of the Biometrika Trustees.

Table B.VI d-Factors for Wilcoxon–Mann–Whitney Test and Confidence Intervals for the Shift Parameter Δ

γ = confidence coefficient
$\alpha' = \frac{1}{2}(1 - \gamma)$ = one-sided significance level
$\alpha'' = 2\alpha' = 1 - \gamma$ = two-sided significance level

		$m = 3$			$m = 4$		
	d	γ	α''	α'	d	γ	α''	α'
$n = 3$	1	.900	.100	.050				
$n = 4$	1	.943	.057	.029	1	.971	.029	.014
	2	.886	.114	.057	2	.943	.057	.029
					3	.886	.114	.057
$n = 5$	1	.964	.036	.018	1	.984	.016	.008
	2	.929	.071	.036	2	.968	.032	.016
	3	.857	.143	.071	3	.937	.063	.032
					4	.889	.111	.056
$n = 6$	1	.976	.024	.012	1	.990	.010	.005
	2	.952	.048	.024	2	.981	.019	.010
	3	.905	.095	.048	3	.962	.038	.019
	4	.833	.167	.083	4	.933	.067	.033
					5	.886	.114	.057
$n = 7$	1	.983	.017	.008	1	.994	.006	.003
	2	.967	.033	.017	2	.988	.012	.006
	3	.933	.067	.033	4	.958	.042	.021
	4	.883	.117	.058	5	.927	.073	.036
					6	.891	.109	.055
$n = 8$	1	.988	.012	.006	2	.992	.008	.004
	3	.952	.048	.024	3	.984	.016	.008
	4	.915	.085	.042	5	.952	.048	.024
	5	.867	.133	.067	6	.927	.073	.036
					7	.891	.109	.055
$n = 9$	1	.991	.009	.005	2	.994	.006	.003
	2	.982	.018	.009	3	.989	.011	.006
	3	.964	.036	.018	5	.966	.034	.017
	4	.936	.064	.032	6	.950	.050	.025
	5	.900	.100	.050	7	.924	.076	.038
					8	.894	.106	.053
$n = 10$	1	.993	.007	.004	3	.992	.008	.004
	2	.986	.014	.007	4	.986	.014	.007
	4	.951	.049	.025	6	.964	.036	.018
	5	.923	.077	.039	7	.946	.054	.027
	6	.888	.112	.056	8	.924	.076	.038
					9	.894	.106	.053
$n = 11$	1	.995	.005	.003	3	.994	.006	.003
	2	.989	.011	.006	4	.990	.010	.005
	4	.962	.038	.019	7	.960	.040	.020
	5	.940	.060	.030	8	.944	.056	.028
	6	.912	.088	.044	9	.922	.078	.039
	7	.874	.126	.063	10	.896	.104	.052
$n = 12$	2	.991	.009	.004	4	.992	.008	.004
	3	.982	.018	.009	5	.987	.013	.007
	5	.952	.048	.024	8	.958	.042	.021
	6	.930	.070	.035	9	.942	.058	.029
	7	.899	.101	.051	10	.922	.078	.039
					11	.897	.103	.052

For sample sizes m and n beyond the range of this table use $d \doteq \frac{1}{2} \left[mn + 1 - z\sqrt{mn(m + n + 1)/3} \right]$, where z is read from Table C.

[a]From *Introduction to Statistics: A Fresh Approach by* Gottfried Noether. Copyright 1971 by Houghton Mifflin Company. Used by permission of the publisher.

Table B.VI (Continued)

		m = 5				m = 6				m = 7				m = 8		
	d	γ	α''	α'	d	γ	α''	α'	d	γ	α''	α'	d	γ	α''	α'
n = 5	1	.992	.008	.004												
	2	.984	.016	.008												
	3	.968	.032	.016												
	4	.944	.056	.028												
	5	.905	.095	.048												
	6	.849	.151	.075												
n = 6	2	.991	.009	.004	3	.991	.009	.004								
	3	.983	.017	.009	4	.985	.015	.008								
	4	.970	.030	.015	6	.959	.041	.021								
	5	.948	.052	.026	7	.935	.065	.033								
	6	.918	.082	.041	8	.907	.093	.047								
	7	.874	.126	.063	9	.868	.132	.066								
n = 7	2	.995	.005	.003	4	.992	.008	.004	5	.993	.007	.004				
	3	.990	.010	.005	5	.986	.014	.007	6	.989	.011	.006				
	6	.952	.048	.024	7	.965	.035	.018	9	.962	.038	.019				
	7	.927	.073	.037	8	.949	.051	.026	10	.947	.053	.027				
	8	.894	.106	.053	9	.927	.073	.037	12	.903	.097	.049				
					10	.899	.101	.051	13	.872	.128	.064				
n = 8	3	.994	.006	.003	5	.992	.008	.004	7	.991	.009	.005	8	.993	.007	.004
	4	.989	.011	.005	6	.987	.013	.006	8	.986	.014	.007	9	.990	.010	.005
	7	.955	.045	.023	9	.957	.043	.021	11	.960	.040	.020	14	.950	.050	.025
	8	.935	.065	.033	10	.941	.059	.030	12	.946	.054	.027	15	.935	.065	.033
	9	.907	.093	.047	11	.919	.081	.041	14	.906	.094	.047	16	.917	.083	.042
	10	.873	.127	.064	12	.892	.108	.054	15	.879	.121	.060	17	.895	.105	.052
n = 9	4	.993	.007	.004	6	.992	.008	.004	8	.992	.008	.004	10	.992	.008	.004
	5	.988	.012	.006	7	.988	.012	.006	9	.988	.012	.006	11	.989	.011	.006
	8	.958	.042	.021	11	.950	.050	.025	13	.958	.042	.021	16	.954	.046	.023
	9	.940	.060	.030	12	.934	.066	.033	14	.945	.055	.027	17	.941	.059	.030
	10	.917	.083	.042	13	.912	.088	.044	16	.909	.091	.045	19	.907	.093	.046
	11	.888	.112	.056	14	.887	.113	.057	17	.886	.114	.057	20	.886	.114	.057
n = 10	5	.992	.008	.004	7	.993	.007	.004	10	.990	.010	.005	12	.991	.009	.004
	6	.987	.013	.006	8	.989	.011	.006	11	.986	.014	.007	13	.988	.012	.006
	9	.960	.040	.020	12	.958	.042	.021	15	.957	.043	.022	18	.957	.043	.022
	10	.945	.055	.028	13	.944	.056	.028	16	.945	.055	.028	19	.945	.055	.027
	12	.901	.099	.050	15	.907	.093	.047	18	.912	.088	.044	22	.899	.101	.051
	13	.871	.129	.065	16	.882	.118	.059	19	.891	.109	.054				
n = 11	6	.991	.009	.004	8	.993	.007	.004	11	.992	.008	.004	14	.991	.009	.005
	7	.987	.013	.007	9	.990	.010	.005	12	.989	.011	.006	15	.988	.012	.006
	10	.962	.038	.019	14	.952	.048	.024	17	.956	.044	.022	20	.959	.041	.020
	11	.948	.052	.026	15	.938	.062	.031	18	.944	.056	.028	21	.949	.051	.025
	13	.910	.090	.045	17	.902	.098	.049	20	.915	.085	.043	24	.909	.091	.045
	14	.885	.115	.058	18	.878	.122	.061	21	.896	.104	.052	25	.891	.109	.054
n = 12	7	.991	.009	.005	10	.990	.010	.005	13	.990	.010	.005	16	.990	.010	.005
	8	.986	.014	.007	11	.987	.013	.007	14	.987	.013	.007	17	.988	.012	.006
	12	.952	.048	.024	16	.959	.041	.021	19	.955	.045	.023	23	.953	.047	.024
	13	.936	.064	.032	16	.947	.053	.026	20	.944	.056	.028	24	.943	.057	.029
	14	.918	.082	.041	18	.917	.083	.042	22	.917	.083	.042	27	.902	.098	.049
	15	.896	.104	.052	19	.898	.102	.051	23	.900	.100	.050	28	.885	.115	.058

Table B.VI (Continued)

		m = 9				m = 10				m = 11				m = 12		
	d	γ	α''	α'	d	γ	α''	α'	d	γ	α''	α'	d	γ	α''	α'
n = 9	12	.992	.008	.004												
	13	.989	.011	.005												
	18	.960	.040	.020												
	19	.950	.050	.025												
	22	.906	.094	.047												
	23	.887	.113	.057												
n = 10	14	.992	.008	.004	17	.991	.009	.005								
	15	.990	.010	.005	18	.989	.011	.006								
	21	.957	.043	.022	24	.957	.043	.022								
	22	.947	.053	.027	25	.948	.052	.026								
	25	.905	.095	.047	28	.911	.089	.045								
	26	.887	.113	.056	29	.895	.105	.053								
n = 11	17	.990	.010	.005	19	.992	.008	.004	22	.992	.008	.004				
	18	.988	.012	.006	20	.990	.010	.005	23	.989	.011	.005				
	24	.954	.046	.023	27	.957	.043	.022	31	.953	.047	.024				
	25	.944	.056	.028	28	.949	.051	.026	32	.944	.056	.028				
	28	.905	.095	.048	32	.901	.099	.049	35	.912	.088	.044				
	29	.888	.112	.056	33	.886	.114	.057	36	.899	.101	.051				
n = 12	19	.991	.009	.005	22	.991	.009	.005	25	.991	.009	.004	28	.992	.008	.004
	20	.988	.012	.006	23	.989	.011	.006	26	.989	.011	.005	29	.990	.010	.005
	27	.951	.049	.025	30	.957	.043	.021	34	.956	.044	.022	38	.955	.045	.023
	28	.942	.058	.029	31	.950	.050	.025	35	.949	.051	.026	39	.948	.052	.026
	31	.905	.095	.048	35	.907	.093	.047	39	.909	.091	.045	43	.911	.089	.044
	32	.889	.111	.056	36	.893	.107	.054	40	.896	.104	.052	44	.899	.101	.050

SELECTED BIBLIOGRAPHY AND REFERENCES

Burington, R. S., and D. C. May (1970). *Handbook of Probability and Statistics with Tables*, 2nd Ed., McGraw-Hill, New York, N. Y.

Cochran, W. G. (1963). *Sampling Techniques*, 2nd Ed., John Wiley and Sons, New York, N. Y.

Committee on the Undergraduate Program in Mathematics (1965). *A General Curriculum in Mathematics for Colleges.*

Dixon, W., and F. Massey, Jr. (1969). *Introduction to Statistical Analysis*, 3rd Ed., McGraw-Hill, New York, N. Y.

Downton, F. (1966). Linear Estimates with polynomial coefficients. Biometrika 53: 129.

Draper, N. R., and H. Smith (1966). *Applied Regression Analysis*, John Wiley and Sons, New York, N. Y.

Dwass, M. (1970). *Probability and Statistics: An Undergraduate Course*, W. A. Benjamin, New York, N. Y.

Feller, W. (1957), (1966). *An Introduction to Probability Theory and Its Applications*, Vols. 1 and 2, John Wiley and Sons, New York, N. Y.

Fisz, M. (1963). *Probability Theory and Mathematical Statistics*, 3rd Ed., John Wiley and Sons, New York, N. Y.

Gibbons, J. D. (1971). *Nonparametric Statistical Inference*, McGraw-Hill, New York, N. Y.

Gibbons, J. D. (1976). *Nonparametric Methods for Quantitative Analysis*, Holt, Rinehart, and Winston, New York, N. Y.

Goldberg, S. (1960). *Probability: An Introduction*, Prentice-Hall, Englewood Cliffs, N. J.

Graybill, Franklin A. (1961). *An Introduction to Linear Statistical Models*, Vol. I., McGraw-Hill, New York, N. Y.

Handbook of Tables for Probability and Statistics (1966). The Chemical Rubber Company, Cleveland, Ohio.

Hodges, J., and E. Lehman (1964). *Basic Concepts of Probability and Statistics*, Holden-Day, San Francisco, Calif.

Kemeny, J. G., and T. E. Kurtz (1972). *BASIC Programming*, 2nd Ed., John Wiley and Sons, New York, N. Y.

Kemeny, J. G., L. Snell, and G. L. Thompson (1974). *Introduction to Finite Mathematics*, 3rd Ed., Prentice-Hall, Englewood Cliffs, N. J.

Kendall, M. G. (1958), (1951). *The Advanced Theory of Statistics*, Vol. I, and Vol. II, 3rd Ed., Charles Griffin, London, England.

Kendall, M. G. (1955). *Rank Correlation Methods*, 2d Ed., Charles Griffin, London, England.

Larson, H. J. (1974). *Introduction to Probability and Statistical Inference*, 2d Ed., John Wiley and Sons, New York, N. Y.

Mather, D., and S. Waite (1974). *BASIC*, Trustees of Dartmouth College, Hanover, N. H.

Mood, A., and F. Graybill (1963). *Introduction to the Theory of Statistics*, McGraw-Hill, New York, N. Y.

Mosteller, F., R. Rourke, and G. Thomas, Jr. (1961). *Probability and Statistics*, Addison-Wesley, Reading, Mass.

Owen, D. (1962). *Handbook of Statistical Tables*, Addison-Wesley, Reading, Mass.

Parzen, E. (1960). *Modern Probability and Its Applications*, John Wiley and Sons, New York, N. Y.

Pearson, E. S., and H. O. Hartley (1962). *Biometrika Tables for Statisticians*, 2d Ed., Cambridge University Press, Cambridge, England.

Roberts, H. V., and W. A. Wallis (1956). *Statistics: A New Approach*, Macmillan, New York, N. Y.

Scheid, F. (1962). *Elements of Finite Mathematics*, Addison-Wesley, Reading, Mass.

Tanur, J. M., Frederick Mosteller, William H. Kruskal, Richard F. Link, Richard S. Pieters, Gerald R. Rising (1972). *Statistics: A Guide to the Unknown*, Holden-Day, San Francisco, Calif.

Wonnacott, T. H., and R. J. Wonnacott (1972). *Introductory Statistics for Business and Economics*, John Wiley and Sons, New York, N. Y.